Earth and Environmental Sciences Library

The *Earth and Environmental Sciences Library* (EESL) is a multidisciplinary book series that showcases innovative approaches and in-depth reviews, aimed at enhancing the role of the Earth and Environmental Sciences communities. The series also offers valuable guidance for stakeholders, decision-makers, policymakers, international organizations, and NGOs.

Topics of interest include oceanography, the marine environment, atmospheric sciences, hydrology, soil sciences, geophysics, geology, agriculture, environmental pollution, remote sensing, climate change, water resources, and natural resource management.

To further advance these critical areas, the *Earth and Environmental Sciences Library* invites contributions from the Earth and Environmental Sciences communities in the form of edited volumes, monographs, textbooks, conference proceedings, and more. The series fosters knowledge exchange and collaboration across disciplines, encouraging comprehensive discussions and solutions to pressing environmental challenges.

Alok Prasad Das · Sailaja Priyadarsini
Editors

Recovery and Recycling of Critical Metals from Electronic Waste for Environmental Sustainability

Springer

Editors
Alok Prasad Das
Department of Life Sciences
Rama Devi Women's University
Bhubaneswar, Odisha, India

Sailaja Priyadarsini
Department of Life Sciences
Rama Devi Women's University
Bhubaneswar, Odisha, India

ISSN 2730-6674 ISSN 2730-6682 (electronic)
Earth and Environmental Sciences Library
ISBN 978-3-032-29063-2 ISBN 978-3-032-29064-9 (eBook)
https://doi.org/10.1007/978-3-032-29064-9

This Springer imprint is published by the registered company Springer Nature Switzerland AG
The registered company address is: Gewerbestrasse 11, 6330 Cham, Switzerland

Preface

The world's reliance on sophisticated materials and electrical equipment has grown dramatically in recent decades due to the rapid pace of technological innovation. A wide range of important metals, including lithium, cobalt, nickel, copper, and rare earth elements, are essential to modern electronics, renewable energy systems, and electric transportation technologies. Although these materials are necessary to maintain technological advancement, concerns about resource depletion, environmental damage from primary mining, and the fragility of global supply chains have grown because of their rising demand. A crucial component of contemporary resource management techniques is the sustainable recovery and recycling of crucial metals from electronic waste. An interdisciplinary strategy that incorporates developments in materials science, biotechnology, environmental engineering, computational modelling, and sustainability assessment is needed to address these issues.

This edited volume brings together contributions from researchers and experts working at the forefront of critical metal recovery and recycling technologies. The scientific foundations, technological advancements, and sustainability issues related to extracting valuable metals from electronic waste and other secondary resources are all covered in the chapters. Additionally, it looks at microbial activity in bioleaching systems, closed-loop bioleaching procedures, and bio-circular strategies that help recover important metals from complicated waste streams effectively and sustainably. This book also discusses the effectiveness, sustainability, and environmental impact of various e-waste recycling technologies using fuzzy logic-based decision models, environmental footprint assessment, and life cycle assessment. AI-driven modelling and predictive tools present new possibilities for enhancing recovery paths, increasing process efficiency, and developing the idea of bio-circularity in sustainable resource recovery systems. The book seeks to add to the expanding body of knowledge that supports the shift toward circular resource systems and environmentally responsible technology growth by showcasing a variety of scientific viewpoints and cutting-edge methodologies.

Bhubaneswar, India

Alok Prasad Das
Sailaja Priyadarsini

Contents

Understanding Critical Metals: Definitions, Strategic Importance, and Applications in Modern Electronics

Parshant, Prachi Saxena, and Suman Sen

Abstract In cutting-edge technological environment, crucial metals are critical elements which are used as building blocks for the introduction of electronics, renewable strength resources, and defense applications. With a focus on their unique mixture of deliver chance and economic significance, this text examines the various and converting definitions of key metals. The standards used to identify a steel as "vital" vary from country to kingdom and organization to company, however they usually encompass elements like shortage, geopolitical risk, and the absence of appropriate alternatives. On this category, rare earth elements (REEs), lithium, cobalt, and tantalum are essential metals which can be utilized in semiconductors, high-overall performance magnets, mobile phones, and batteries, amongst other gadgets. Vital metals' strategic importance goes past their bodily traits to encompass country wide safety worries and international supply lines. However, the concentration of mining and refining capability in a small range of nations, mainly China, has sparked worries about monetary fragility and deliver stability. Countries and corporations are therefore making significant investments in recycling, fabric substitution, and domestic production packages on the way to lessen those dangers. Additionally, the observe explores the specific uses of crucial metals in present day electronics. As an example, the tiny, excessive-efficiency vehicles determined in wind turbines and electric powered vehicles rely upon neodymium and dysprosium. To create strong supply chains and sustainable innovation plans, it is critical to recognize the meanings, strategic significance, and technical makes use of critical metals. Coordinated international responses are required to this multidisciplinary hassle, which crosses the fields of materials technology, geopolitics, environmental policy, and industrial engineering. The advancement of generation and geopolitical stability will preserve to rely closely on important metals as the sector enters a generation of electrification and virtual connectedness.

Parshant · P. Saxena (✉)
Department of Chemistry, Lovely Professional University, Phagwara, Punjab, India

S. Sen
Department of Chemistry, Basirhat College, Basirhat, West Bengal, India

A. P. Das and S. Priyadarsini (eds.), *Recovery and Recycling of Critical Metals from Electronic Waste for Environmental Sustainability*, Earth and Environmental Sciences Library, https://doi.org/10.1007/978-3-032-29064-9_1

Keywords Critical metals · Strategic importance · Rare earth metals · Supply risks · Modern electronics

Abbreviations

REEs	Rare earth elements
DRC	Democratic Republic of the Congo
WEEE	Waste electronic and electrical equipment
EVs	Electric vehicles
CAGR	Compound annual growth rate
ITO	Indium tin oxide
LCD	Liquid crystal display
OLED	Organic light-emitting diode
GaN	Gallium nitride
IGZO	Indium gallium zinc oxide
TFT	Thin-film transistor
InP	Indium phosphide
YAG	Yttrium aluminum garnet
MoO	Molybdenum trioxide
PGMs	Platinum group metals
USGS	United States geological survey
ESG	Environmental, social, and governance

1 Introduction

Gaining an understanding of vital metals is critical to information the monetary and technological adjustments of the present-day period. important for economic increase, national protection, and the development of cutting-edge technology, essential metals also called important minerals or uncooked materials run the chance of causing deliver chain disruptions because of their scarcity, geopolitical occasions, or concentrated extraction assets. Critical metallic definitions would possibly differ relying at the location and situation. The U.S. department of energy and the U.S. Geological Survey define key materials as non-fuel minerals or additives which can be critical for strength manufacturing, transmission, garage, or conservation but have a high danger of disrupting the supply chain. Critical minerals are those which are crucial to countrywide security and the financial system, and their shortage or attention in specific areas can placed countries at threat in terms of supply chain vulnerabilities, consistent with India's Ministry of Mines. As a result, a mineral's criticality depends on its context and is frequently assessed using factors like monetary significance and supply disruption threat (Müller et al. 2025).

Category of minerals and factors with considerable strategic and monetary value, crucial metals play important roles in present day technology consisting of advanced electronics, energy structures, and protection. The delivery of these metals is prone to interruptions and geopolitical threats due to their restrained alternatives, focused geographic deliver, and relatively low natural abundance. The manufacturing of electronic additives, which enhance overall performance, durability, and miniaturization in gadgets like laptops, cell phones, and renewable strength technologies, relies upon closely on important metals like cobalt, lithium, tantalum, tungsten, and uncommon earth elements. Important procedures including effective power storage, heat dissipation, signal transmission, and mechanical balance in difficult electronic structures are made viable by their wonderful physical, chemical, and electric traits. To relaxed deliver chains, sell sustainable aid use, and enhance technological innovation in contemporary electronics, it's far critical to realize the definitions, strategic significance, and programs of critical metals, given the developing reliance of the global financial system on electronic gadgets and clean power technologies (Mahnoor et al. 2025). Demand in industries like wind strength, sun photovoltaic, and electric mobility is being pushed by using the world's swiftly growing decarbonization, digitization, and electrification, which highlights the strategic importance of essential metals via applications like the country wide crucial Minerals task to safeguard domestic deliver chains and inspire sustainable extraction, recycling, and innovation in processing technologies, nations inclusive of India have recognized this need (Ministry of Mines (mother), 2025; Coronary Interventions, Imaging, and Structural Programmed 2025 conference, IIT(ISM) Dhanbad) (Bielowicz 2025).

2 Classification of Critical Metals: Properties and Examples

A certain class of metals known as "critical metals" is necessary for national security, scientific progress, and economic growth. They are considered "critical" because of their high economic significance, crucial roles in new and green technologies, and substantial supply concerns, which are frequently brought on by restricted geographic availability, geopolitical unrest, or difficulties with extraction and processing (Xia et al. 2024). The economic significance of critical metals stems from their essential function in sustaining important contemporary industries, including advanced manufacturing, electronics, information technology, defense, and clean energy, because of their distinct physical and chemical characteristics that currently lack viable alternatives in high-performance applications. The foundation of both national economy and global technology growth, their supply disruption can cause significant losses for vital sectors (Liu et al. 2022). Critical metals are specifically needed for advanced electronics (such as gallium, indium, and tantalum for semiconductors and displays), defense systems (such as titanium for aerospace, rare earths for precision-guided munitions), renewable energy infrastructure (such as silicon and copper for solar panels, and rare earths for wind turbines and electric motors), and electric vehicle batteries (such as lithium, cobalt, and nickel). Their dire financial situation is highlighted even further because they make the shift to a low-carbon economy possible and are essential to reaching climate goals like the use of electric cars and renewable energy sources. Since these resources are crucial to economic competitiveness, many nations including the US and India have established national missions to ensure their secure supply. Economic vulnerability to global supply chain disruptions can result from a country's strong reliance on imports and lack of native resources, particularly for vital applications (Hu et al. 2024). Critical metals can be classified into some different types like supply risk, national and regional assessments, application domains, and economic importance are shown in Fig. 1.1

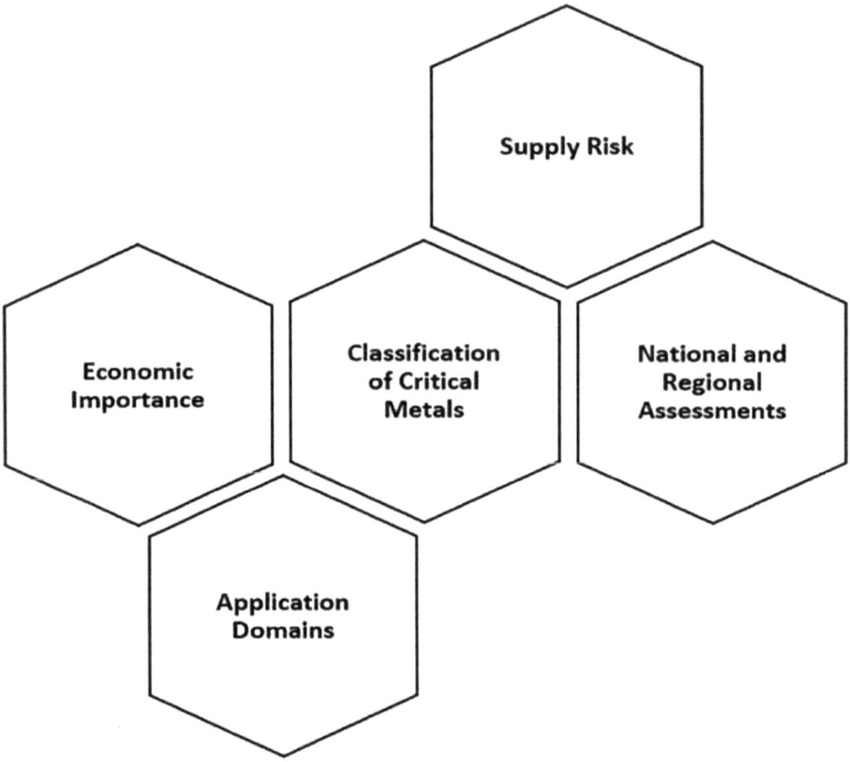

Fig. 1.1 Framework illustrating the multifactorial classification of critical metals (Liu et al. 2022; Hu et al. 2024; Jia et al. 2025; Dieaconescu et al. 2025; Born and Ciftci 2024; Hagelüken and Goldmann 2022; Sun 2022; Dou et al. 2023; Patil et al. 2023; Lamblin 2024; Mitrica et al. 2021; Qiu et al. 2025; Khalkho 2025; Kannan et al. 2025)

2.1 Supply Risk

The possibility and possible effects of interruptions in the supply of key metals are referred to as supply risk. These disruptions may have wide-ranging economic and technological repercussions. Numerous factors are taken into consideration while evaluating the risk, including: A limited handful of nations or perhaps a single mine provide the majority of key metals. More than half of the world's cobalt, for example, comes from the Democratic Republic of the Congo (DRC), whereas a small number of countries provide a significant amount of its nickel.

As a result, supply chains in these areas are susceptible to political unrest, conflict, and regulatory changes (Jia et al. 2025). Risk is significantly increased by excessive reliance on outside sources. For instance, eighty-six percent of the manganese, 91% of the nickel, and ninety percent of the cobalt in China are imported. Unexpected shortages and price volatility might result from disruptions brought on by trade restrictions, diplomatic relations, or economic policies (Dieaconescu et al. 2025). Many essential metals, including copper, nickel, lithium, and cobalt, are irreplaceable in digital, clean energy, and advanced technology applications due to their distinct physical and chemical characteristics. In the manufacturing of batteries, semiconductors, and permanent magnets, for instance, specific metals are essential, and substitutes are either nonexistent or lead to significant reductions in performance, efficiency, or dependability. Certain metals, such as gallium and indium, have no practical alternatives that meet the demands of modern technology. Most high-value or high-performance use cases still make switching to a different material either economically or technically impracticable, even in circumstances where substitution research is ongoing (Born and Ciftci 2024). Recycling essential metals is essential to creating a more robust, sustainable supply chain, however there are several barriers that restrict how effective and scalable recycling initiatives

The complicated multi-layered architecture, low amounts of key metals, and low collection rates of products such as Waste Electronic and Electrical Equipment (WEEE) frequently result in low recyclability. Numerous essential metals have structural integrations that make extraction and separation challenging and expensive. Advanced methods like hydrometallurgy and pyrometallurgy, which are chemically and energy-intensive and produce hazardous byproducts, are frequently needed for the recycling process. Furthermore, new, tailored solutions are required for every product design or generation because primary mineral extraction procedures are difficult to adapt to end-of-life products. Recycling rates are quite low for many important metals, especially rare earth elements (rarely above 1%).

The significant labor and energy costs involved in disassembling items and reusing complicated material streams usually prevent secondary (recycled) material from competing with main raw materials. There are now very few batteries and end-of-life devices that can be recycled for industries like electric automobiles. Years, if not decades, will pass before significant amounts of secondary feedstock are accessible for extensive recycling. The development of recycling infrastructure is further hampered by strict regulations, expensive investment costs, and operating expenses, particularly for metals embedded in layers of polymers or adhesives (Hagelüken and Goldmann 2022).

2.2 *Dynamics of Markets and Supply Chains*

The global energy transition, geopolitical threats, and supply chain concentration issues are driving rapidly increasing demand for important metals in 2025, which will impact market and supply chain dynamics. Key points consist of: Growing Need for Clean Technologies and Energy is the demand for vital metals such as lithium, cobalt, nickel, copper, and rare earth elements is greatly increased by the rise in electric vehicles (EVs), renewable energy sources (wind, solar, and battery storage), and sophisticated electronics. For instance, because copper is crucial to electrification, its demand is predicted to climb sharply, while the need for lithium is predicted to increase ninefold by 2040.

The global critical minerals market is expected to reach over US$586 billion by 2032 at a cumulative annual growth rate (CAGR) of roughly 7.5%, mainly due to energy transition technologies. The market is now estimated at about US$328 billion in 2024. Market volatility and economic factors include trade policies, inflation, interest rate changes, and macroeconomic uncertainty that affect investment choices and metal demand. The demand for metals may be cyclically impacted by the possibility of recessions or stimulus plans. The recycling of essential metals is increasing, but it is not enough to meet the rising demand, which increases reliance on primary extraction. This highlights the need to diversify supply sources and make investments in sustainability (Sun 2022). The world's rapid shift to sustainable energy, digitization, and advanced manufacturing are the main factors driving the growth in critical metal consumption. To achieve net-zero goals and support technologies like electric vehicles (EVs), battery storage, advanced electronics, and renewable energy infrastructure (wind turbines, solar panels), the demand for metals like lithium, cobalt, nickel, copper, and rare earth elements is predicted to increase dramatically over the next ten years. For example, the demand for copper is expected to increase by more than 40% by 2040, and between 2024 and 2025, the total amount of minerals needed for low-carbon solutions might reach billions of tones (Dou et al. 2023).

2.3 *Financial Limitations*

It is anticipated that the mining and processing industries will require $2.1 trillion in new investments by 2050 in order to scale up output and sustainably supply the increased demand. Economies are vulnerable to disruptions and price volatility when supply is concentrated in politically unstable or geopolitically tense areas (for example, the Democratic Republic of Congo for cobalt, or China's hegemony in rare earth processing). Interest rate swings, inflation, and uncertainty surrounding international commerce can all discourage investment and make supply-demand balance more difficult. Rich nations with abundant natural resources may prioritize their own needs over exports, which would increase prices and restrict access to vital metals on international markets (Patil et al. 2023).

Growing mining operations frequently encounter resistance because of their effects on local residents, water supplies, and biodiversity, particularly in areas that are vulnerable. Mining is associated with war, violations of human rights, and relocation in places like eastern Congo, raising ethical questions and supply chain hazards. Mining companies must adopt cleaner, more responsible methods in response to stricter environmental legislation and social governance standards. This can raise operating costs and delay project approvals. Although recycling and material efficiency innovation are promoted as ways to alleviate the strain on primary supplies, recycling rates are still low at the moment, and many technologies will continue to rely on primary extraction for many years to come (Lamblin 2024).

2.4 Application Domains

Many software sectors, particularly the ones bearing on sophisticated manufacturing, smooth power, modern era, and defense, rely heavily on important metals several crucial technologies on these metals for their improvement and operation, including lithium, cobalt, nickel, platinum organization metals, rare earth elements (REEs), and others. For the strength enterprise to supply renewable electricity technology like solar panels, wind generators, and strength storage devices, key metals are crucial. As an instance, lithium and cobalt are generally used in battery packs for electricity garage and electric vehicles because of their superb price density and efficiencies. On the way to create powerful everlasting magnets for electric powered cars and turbines crucial components of wind turbines and electric motors uncommon earth factors inclusive of neodymium, dysprosium, and terbium are wished.

Metals like copper, nickel, and aluminum are also important components of the infrastructure that supports the power shift (Mitrica et al. 2021). Essential metals are applied in country wide safety and protection to assist precision technology like missile propulsion structures, radar, and sonar. Superior defense programs, along with concentrated on sensors, optical devices, and communications structures, require uncommon earth elements. The particular traits of beryllium are essential for inertial navigation structures and satellite tv for pc optics, while titanium's wonderful electricity and thermal stability make it essential for missile components and plane airframes. In laser technologies important to army guns concentrated on and guiding, neodymium is utilized. The electronics area is some other crucial industry that relies upon on essential metals. Excessive-density statistics garage, show generation, microelectronics, and semiconductor production all rely upon these metals. Lithium-selenium batteries, solar panels, and semiconductors are some examples of merchandise that rent selenium (Qiu et al. 2025).

2.5 National and Regional Assessments

Countries and areas should behavior both country wide and regional opinions of essential metals to analyze the metals' availability, supply dangers, economic significance, and environmental outcomes. The goal of these critiques is to assure a safe and sustainable supply of important metals needed for excessive-tech, protection, and smooth energy structures. Countries like India have created thorough assessment frameworks on the country wide stage the use of multi-stage strategies. India's Ministry of Mines, for instance, completed a three-phase assessment manner that concerned researching international crucial minerals techniques, accomplishing inter-ministerial consultations to determine zone-particular essential minerals, and applying standards like supply threat, monetary significance, availability of home sources, and destiny technological necessities. This brought about the creation of a prioritized list of approximately 30 vital minerals for India, inclusive of uncommon earth factors, cobalt, lithium, and different metals which might be essential for the country's strategic industries (Khalkho 2025). Consistent with the United States Geological Survey (USGS), the USA employs a mineral structures approach for its indigenous important minerals. This entails appraising mineral structures of interest the use of updated fashions, reading geological procedures that listen those metals, and inventorying mineral abundance.

A comprehensive and information-rich technique to countrywide assessment is pondered in the USGS's worldwide collaborations, which includes its trilateral mapping paintings with Canada and Australia. Place critiques vary among continents, reflecting their wonderful deliver chain places and aid endowments. As an instance, the majority of lithium reserves are in South America, specifically in Chile and Argentina, while most people of cobalt reserves are within the Democratic Republic of the Congo (DRC). China is the arena's biggest producer and refiner of uncommon earth and graphite, and it is also a main center for processing critical metals. As a result of growing supply chains and global collaboration, Southeast Asian international locations like Thailand and Nigeria have become big producers of uncommon earth factors. To inform policy and guarantee robust and various sourcing within the place, the EU Union has a listing of crucial uncooked materials this is updated on an everyday basis. This listing is based totally on assessments of financial importance and supply risk. Advanced country wide criticality strategies, which contain deliver threat, environmental effect, and supply restrict threats to expand nuanced chance scores for metals, are examples of the way these checks regularly combine environmental and social threat concerns (Kannan et al. 2025).

These regional and country wide reviews paintings together to create a framework for comprehending the dynamics of the supply of important metals, directing useful resource exploration, recycling projects, strategic stockpiling, and worldwide collaboration to lessen dangers related to marketplace volatility, geopolitics, and environmental troubles.

3 Applications in Modern Electronics

The development and operation of modern electronics depend heavily on important metals, which allow for terrific overall performance, performance, and dependability in the whole thing from laptops and smartphones to facts servers and wearable technologies. Thanks to their notable electric conductivity, durability, and thermal management features, copper, gold, silver, tungsten, tantalum, molybdenum, beryllium, platinum, palladium, and other digital components along with motherboards, chips, charge garage gadgets, warmth sinks, connections, and radar make use of additional uncommon earth factors (Nazir et al. 2024). Applications in modern electronics are classified into different types like smart devices, semiconductor and displays, renewable energy and energy storage, and connectivity and communications are shown in Fig. 1.2

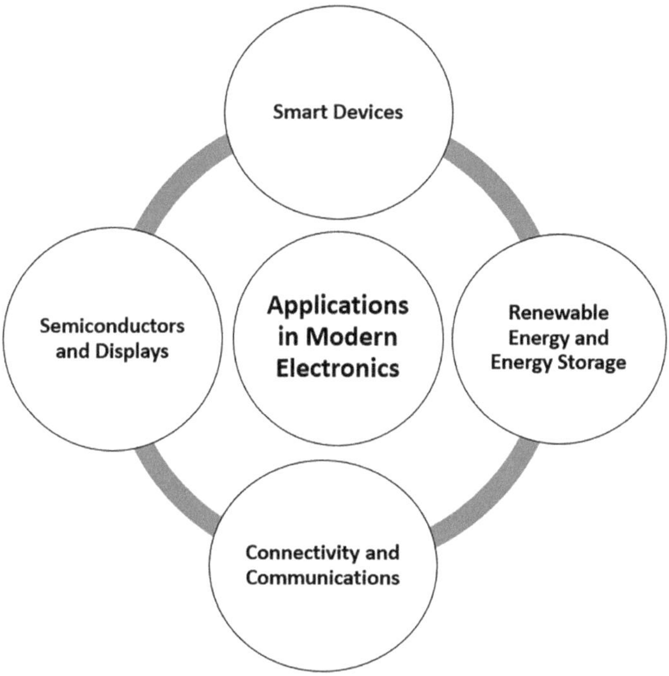

Fig. 1.2 Key application domains of critical materials in modern electronics, including smart devices, renewable energy storage, semiconductors and displays, and connectivity technologies (Xia et al. 2024; Gu et al. 2022; Teng et al. 2025; Parvini et al. 2024; Becci et al. 2024; Yandem and Jabłońska-Czapla 2024; Blanco et al. 2024; Ibrahim et al. 2025; Hossain et al. 2022; Shaikh et al. 2024; Awan et al. 2025; Mahnoor et al. 2025; D'Orazio 2024; Pathan et al. 2024; Tirmare et al. 2024)

3.1 Smart Devices

Numerous important metals are present in clever gadgets, together with wearables, capsules, and smartphones, dependability, performance, and functioning. The primary crucial metals, their features, and corroborating sources are broken down underneath.

3.1.1 Touchscreen

One of the maximum essential crucial metals is indium, that's important for the responsive touchscreens visible in clever devices like drugs and smartphones. Optical transparency and electrical conductivity are blended in indium tin oxide (ITO), a combination of indium, tin, and oxygen. Due to this special blend, ITO can create an obvious, conductive coating at the floor of monitors product of glass or plastic, making capacitive touch era viable. These screens create invisible electrodes inside the form of a grid by depositing a thin layer of ITO onto the glass. Tapping, swiping, and pinching are made possible by way of the tool registering the neighborhood electric powered area as an enter factor whilst your finger touches the screen. The subsequent evidence helps the efficacy and big use of ITO in monitors.

Despite the fact that studies on indium-free substitutes (such carbon nanotubes and silver nanowires) is in development because to viable deliver shortages and developing call for, ITO remains the industry preferred for the majority of modern-day touchscreen clever devices (Xia et al. 2024).

3.1.2 Display

The improvement of sophisticated displays in clever gadgets, which enable colorful photographs, precise coloration duplicate, and strength efficiency, relies upon heavily on important metals. Uncommon earth elements and unique semiconductor metals are necessary for the show modules, which can be regularly liquid crystal displays (liquid crystal display), natural LEDs (OLED), or the new micro-LED screens (Gu et al. 2022). Some of the critical essential elements involved are:

3.1.3 Gallium

The development of sophisticated displays for clever devices closely is predicated on gallium, a crucial metallic. The most significant use is in semiconductors made from gallium nitride (GaN) and indium gallium zinc oxide (IGZO), that are widely employed in show generation.

The production of shows changed into transformed by using gallium nitride (GaN), which made it feasible for green and blue mild-emitting diodes (LEDs) to operate efficiently. Thanks to this innovation, complete-coloration panels with vibrant shades and right strength efficiency may want to now be determined in TVs, monitors, smartphones, and tablets. Further, GaN serves as the premise for extremely-excessive brightness micro-LEDs, which promise reduced strength intake and brighter, better-decision photographs than conventional technology (Teng et al. 2025). Some other gallium-based substance, IGZO (Indium Gallium Zinc Oxide), is utilized as a thin-film transistor (TFT) fabric in flat-panel displays, consisting of LCDs and OLEDs. More electron mobility than amorphous silicon is furnished by means of the special structure of IGZO, enabling faster switching, better light transmission, and more resolutions. Each excessive-give up displays and transportable devices can advantage from its stability and coffee leakage cutting-edge, which also expand battery existence and improve photograph quality. The readability, refresh rate, and patience of present-day shows are all immediately impacted by means of gallium compounds, which also help combine sophisticated coloration-tuning and switching functions and suppress undesired defects (together with oxygen vacancies) in display transistors (Parvini et al. 2024).

3.1.4 Indium

Maximum generally applied as indium tin oxide (ITO), an obvious, electrically conductive skinny film implemented to glass or plastic surfaces in presentations, indium is an important metal required for the production of presentations in contemporary clever devices. ITO's notable optical transparency and conductivity permit liquid crystal displays (LCDs), organic LEDs (OLEDs), plasma panels, and touchscreens to perform their number one capability. This lets in the displays to sign up tactile enter and convey clean images without obstructing vision (Becci et al. 2024). Indium's salient traits in display innovation: The seen carrying out Layer, or ITO, serves as a transparent electrode that is vital for capacitive touch and pixel switching, permitting interactive consumer interfaces in televisions, drugs, and handsets. For flat-panel displays, indium's special combo of conductivity, transparency, excessive adherence to glass, and sturdiness makes it irreplaceable because there is no commercially available opportunity. The problem of sustainability is a on account that indium is used in greater than half of the arena's shows, attempts to recycle and create alternative transparent conductors have intensified because of its strategic importance and limited deliver. The excellent of color and quantum dots is a much broader shade gamut's are provided by indium compounds like indium phosphide (InP), which can be being applied increasingly in quantum dots for stylish presentations in high priced TVs and monitors (Yandem and Jabłońska-Czapla 2024).

3.1.5 Germanium

Due to its capacity to enable excessive-performance skinny-movie transistors, pixel control, and optical sensing in clever gadgets, germanium is a crucial steel within the manufacturing of contemporary shows. The display backplanes of OLED and AMOLED panels, where germanium improves pixel-switching speed and helps excessive refresh quotes, advantage substantially from its high carrier mobility and silicon compatibility. This permits extra sophisticated visual consequences, crisper visuals, and enhanced contact sensitivity (Blanco et al. 2024). Additionally, germanium performs a key role inside the improvement of photodetectors and quantum dot displays, increasing spectral performance and light absorption for brighter, sharper panels. Similarly, to OLED and LCD panels, germanium substrates and compounds facilitate using vertical cavity surface emitting lasers (VCSELs) and excessive-brightness LEDs, which can be vital improvements for sensor incorporation and display backlighting. Contemporary clever tool displays gain substantially from better decision, advanced coloration accuracy, and quicker response instances because of germanium's unique semiconductor and optical traits. Its inclusion in sensors makes facial recognition, biometric authentication, and complex picture programs even greater possible (Ibrahim et al. 2025).

3.1.6 Praseodymium, Europium, Gadolinium, Terbium, Dysprosium, Yttrium, and Lanthanum

The sophisticated show technology featured in modern-day clever devices rely heavily on uncommon earth metals along with praseodymium, europium, gadolinium, terbium, dysprosium, yttrium, and lanthanum. On the subject of growing first rate pink phosphors, which might be critical for displaying vivid, excessive-definition snap shots, Europium is the enterprise trendy. Terbium is also important for producing accurate inexperienced emissions, which, when paired with crimson and blue phosphors, form the center of full-color presentations. In yttrium aluminum garnet (YAG), which supports both contemporary LCDs and LEDs as well as older CRTs, yttrium acts as a number matrix for phosphors, particularly changing blue LED mild into white for backlighting. Lanthanum improves optical output and shade purity by means of increasing display brightness and power performance, especially whilst used as a matrix in phosphor materials doped with terbium or europium use is vital for producing wider spectra and extra effective white mild emission. In an effort to contribute to each blue and yellow phosphor transitions for warmer white mild, dysprosium stabilizes coloration emission below a diffusion of working situations. This is in particular beneficial in modes together with night show settings. Additionally, it promotes sturdiness and heat balance, which are essential for the rigorous and prolonged utilization of cell devices (Hossain et al. 2022).

3.1.7 Molybdenum

Molybdenum's awesome electric, optical, and chemical traits make it a vital metal this is extensively incorporated into the generation behind present day smart device monitors. Especially, electrochromic display panels and smart home windows use molybdenum oxides, particularly molybdenum trioxide (MoO_3), to permit for dynamic mild modulation, color modifications in response to carried out voltage, and strength-green functioning. Sputtering is regularly used to create molybdenum-primarily based skinny movies, which are critical additives of contact monitors and flat panel shows as conductive layers, capping, or barrier materials. The strength efficiency and multi-functionality of show devices are similarly progressed by smart windows and display prototypes made possible by way of advancements in nanocomposite molybdenum oxide substances that could concurrently feature as batteries and reversible optical switching. Due to their capacity to improve durability, electrical conductivity, and optical evaluation, molybdenum compounds are critical for next-technology clever tool shows and related optoelectronic additives (Shaikh et al. 2024).

3.1.8 Silver, Platinum, Palladium

Its miles acknowledged that silver, platinum, and palladium are essential metals which can be essential to the improvement and operation of clever tool displays. Due to the fact that powerful sign transmission and reflecting qualities are vital inside the development of touchscreens and display circuits, silver is particularly prized for its great electrical conductivity. Participants of the platinum group metal (PGMs), platinum and palladium are prized for their chemically stable nature, resisting corrosion, and catalysts traits. These characteristics guarantee the lifespan and reliable operation of digital display additives, especially in skinny-film layers and microelectronic circuits discovered in smartphones, capsules, and sophisticated panels. Similarly, to presenting dependable tool overall performance, the incorporation of those metals makes advancements possible, together with stepped forward touch sensitivity and more display resolutions. As smart devices turning into more advanced, the call for silver, platinum, and palladium is developing, underscoring their important function as factors in digital generation and current consumer electronics (Awan et al. 2025).

3.2 Renewable Energy and Energy Storage

The improvement of battery storage and power-efficient technologies depends significantly on important metals like lithium, cobalt, nickel, copper, platinum, palladium, and rare earth elements. Batteries for electric vehicles and grid-scale energy storage, as well as parts for sun panels, wind mills, and gas cells, depend on these metals for production. Cobalt, nickel, and lithium are essential for lithium-ion batteries that energy renewable energy solutions, while platinum and palladium are utilized in gasoline cell innovation. The require for such metals is rising alongside the need for renewable energy, which makes supply, extraction, and recycling greater hard. In a few components of the arena, mining is very focused, and growing production is hampered by using geopolitical and environmental troubles. Critical metallic recycling is turning into an increasing number of vast because it lessens dependency on primary extraction and promotes sustainability and resilience within the supply chain during the strength transition. But a few metals nevertheless have low recycling costs these days, especially those observed in batteries and magnets. To reduce supply demands, technological improvements in recycling and fabric substitution are required (Mahnoor et al. 2025).

Powerful resource control and elevated recycling centers are vital for addressing destiny energy desires responsibly, because the developing use of renewable strength and electricity storage technology is expected to bring about severe competition for critical metals (D'Orazio 2024).

3.3 Semiconductors and Displays

Due to the fact their special traits permit excessive-overall performance digital functions, essential metals and minerals are important additives of semiconductors and smart device shows. Critical metals like arsenic, gallium, palladium, platinum, indium, and cobalt are critical to the manufacturing of semiconductors. For the production of gallium arsenide (GaAs) semiconductors, which might be essential for excessive-frequency wi-fi communication gadgets, arsenic is important. Palladium and platinum are used inside the manufacturing of semiconductor gadgets because of their catalytic homes and chemical stability. For the creation of skinny-movie transistor screens and compound semiconductors, gallium and indium are important. Cobalt improves semiconductor device electrical performance and conductor patience (Pathan et al. 2024).

Due to its semiconducting traits, silicon is still the mainstay of semiconductor substrates. It is also seemed as an essential mineral that wishes to be extremely pure so one can be utilized in chip manufacture. Etching, doping, insulation, and circuit patterning all important steps within the advent of semiconductor chips also name for added important minerals and noble gases. Huge factors like palladium, platinum, and silver are regularly utilized in displays. Silver is crucial for touchscreen and circuit components because of its extremely good electric conductivity, while platinum and palladium assure corrosion resistance and dependable operation in the microelectronic layers of sophisticated displays. As a result, the critical metals that aid semiconductors and displays are closely associated, bearing in mind dependable, compact, and effective electronic components that allow modern-day high-resolution, high-sensitivity smart gadgets. Due to their concentrated geographic manufacturing and the requirement for sustainable sourcing, these metals' availability and purity pose a full-size difficulty (Tirmare et al. 2024).

3.4 Connectivity and Communications

The manufacturing of excessive-performance electronic components essential for stressed out and wi-fi verbal exchange networks is made viable through essential metals, that are vital to connectivity and communications. Indium tin oxide (ITO), a transparent conductive substance used in touchscreens and display components critical to communique gadgets, relies upon on metals like gallium, at the same time as semiconductors used in excessive-frequency packages, Indium is important for systems like satellite TV for pc and cell phones. In circuit forums, specialized digital components, and excessive-frequency connectors, silver, which has the most electrical conductivity, guarantees powerful signal transmission. Palladium and platinum beautify the general overall performance of communication devices by manner of supplying dependable conductivity and corrosion resistance in multilayer capacitors and digital additives. Furthermore, metals like copper, gold, tin, and bismuth are crucial for soldering, production of circuit forums that aid facts transmission networks, wiring, and connectors (Rahmani et al. 2023). Those metals offer the signal integrity, robustness, and strength efficiency needed for satellites, fiber optics, broadband, and superior digital verbal exchange infrastructure. Neodymium and dysprosium, uncommon earth factors, are also important for sturdy magnets located in data garage devices and different communique devices.

The planned use of those vital metals helps meet the developing want for quicker, extra reliable connectivity in a selection of settings, which include hard ones like underground mines, in which robust wireless networks improve running performance and safety. Those metals' concentrated production and important roles in growing verbal exchange technology make their sustainability and supply principal problems. Therefore, essential metals permit excessive-frequency semiconductor gadgets, strong circuit components, and lengthy-lasting community infrastructure required for easy records switch and wi-fi conversation globally, forming the foundation of present-day connection and communications technology (Khandekar et al. 2025).

4 Challenges and Future Prospectus

Various types of challenges and future prospectus are arranged in flowchart and shown in Fig. 1.3

4.1 Recycling and Urban Mining

The management of the planet's fast growing useful resource call for, particularly in light of supply chain weaknesses and environmental issues, has made recycling and

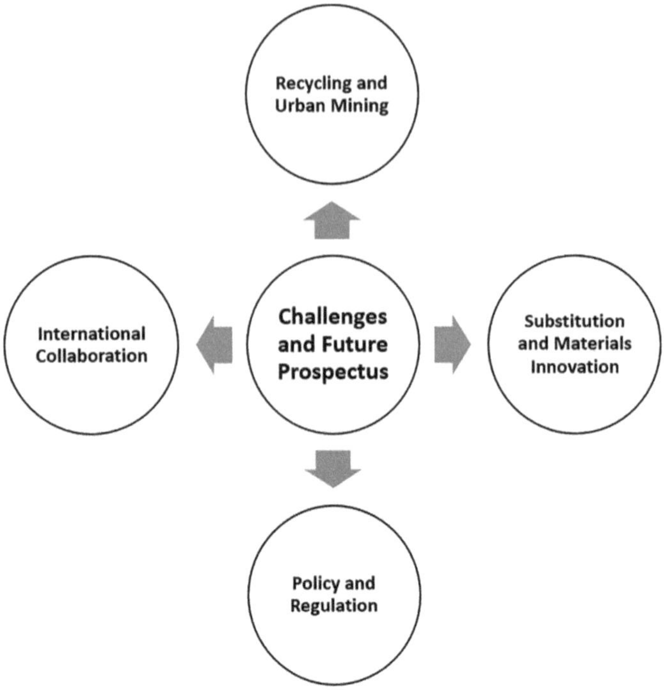

Fig. 1.3 Challenges and future prospects of critical materials management highlighting recycling, substitution, international collaboration, and policy frameworks (Yıldız 2024; Rousseau et al. 2025; Kursunoglu 2025; Mishra and Mancheri 2024; Su and Hu 2022; Zhang et al. 2022)

urban mining for vital metals critical strategies. Essential metals consisting of lithium, cobalt, nickel, copper, gold, and uncommon earth factors are recovered systematically from e-waste and other metropolis belongings thru a manner known as urban mining. Sorting, disassembling, mechanical and chemical extraction, and sophisticated approaches like bioleaching are all a part of this manner. For example, businesses like as Mint Innovation and Elemental protecting effectively recover metals from used car and digital additives, developing round fee chains and decreasing dependency on traditional mining.

Recycling crucial metals has numerous advantages, together with lowering greenhouse gas emissions, water and energy use, and the environmental effect of mining. It also reduces reliance on landfills and guarantees resources for crucial industries like electronics, electric automobiles, and renewable electricity projects. Recycling lithium-ion batteries reduces greenhouse fuel emissions by 58–81%, water intake by using 72–88%, and energy intake by means of 77–89% compared to mining virgin metals, in keeping with lifecycle research from Stanford college. It's far predicted that urban mining could get better up to 30% of the world's treasured metals from e-waste by 2025, highlighting its contribution to sustainability and the shift to a circular economy (Yıldız 2024).

4.2 Substitution and Materials Innovation

Innovation in substances and substitution are key to solving the increasing hassle of securing crucial metals, which are wished for everything from modern electronics to electric powered vehicles and renewable energy. The advent of new materials with equal or higher capability, the restructuring of industrial methods to reduce reliance, and the primary strategies of modern innovation contain substituting extra considerable factors for rare or geopolitically constrained metals. There have been exciting advancements in the sector. In electrolysis and gasoline cells, for example, nano technology has advanced nano-coating answers that reduce costs and growth longevity by substituting platinum and iridium.

The creation of synthetic graphite from methane, which could update natural graphite in lithium-ion batteries and in addition decarbonization targets, is being innovated by Molten Industries. By generating synthetic graphite for EV battery anodes using CO_2 from flue emissions, Up Catalyst in addition diversifies the supply and lessens reliance on conventional mining. Some other modern field is alloying, wherein scientists are growing substances that get rid of or lessen the need for high priced or deliver-hazard metals, like cobalt in batteries and uncommon earth elements in everlasting magnets, at the same time as retaining important overall performance characteristics (Rousseau et al. 2025). Further to looking beyond simple substitution, complete processes also do not forget recycling, healing from secondary sources including business wastes, and the ideas of the circular economy. By means of reducing waste and final useful resource cycles, those techniques no longer most effective boom cloth safety but additionally help environmental and sustainability goals. However, attaining great substitution is a hard system that frequently calls for global cooperation, regulatory assistance, and technological improvements to conquer supply chain structural and technical limitations (Kursunoglu 2025).

4.3 International Collaboration

Because sources like lithium, nickel, and uncommon earth elements critical components for green technologies and modern-day industrial sectors are in greater call for global, international cooperation on vital metals has grown more and more vital. Nations are developing worldwide coalitions, bilateral agreements, and strategic alliances to help environmental and social governance (ESG) norms, inspire innovation, and create sturdy deliver chains. Examples include Canada's numerous collaborations with international locations including Germany, France, Chile, Japan, Australia, South Korea, Indonesia, the United Kingdom, and Italy which can be centered on cooperative research and development, investment, and supply chain transparency. Due to its restricted home reserves, India is likewise collaborating with other international locations, especially through alliances with the us, Australia, and

Japan, that allows you to promote its transition to green energy and ensure lengthy-time period get right of entry to important minerals (Mishra and Mancheri 2024). Multilateral agencies just like the G7, G20, global electricity corporation, and the Minerals security Partnership offer boards for group movement, records exchange, and trendy-putting that cross past country-unique alliances. These agencies emphasize the not unusual need to assure reliable, sustainable flows of these critical assets inside the face of supply risks and geopolitical opposition (Su and Hu 2022).

4.4 Policy and Regulation

Law and coverage for critical metals are changing quickly to deal with their important role in current industries, from national defense to renewable electricity, and to control dangers related to supply disruptions, environmental consequences, and geopolitical concentration. Governments all across the world are growing strategic policies and legislative frameworks to guide recycling, technological innovation, and home manufacturing. For instance, India's country wide critical Mineral venture, which was mounted in 2025, offers the vital authorities the authority to put mining rentals for essential minerals up for auction, creates exploration projects specially for those minerals, and suggests the establishment of a middle of Excellence on crucial Minerals to regularly check and direct supply chain security features (Zhang et al. 2022). By means of 2030, the European Union's critical raw materials Act establishes strategic benchmarks for domestic extraction, processing, and recycling; restricts immoderate dependence on unmarried-source countries; expedites the allowing system; fortifies supply chain monitoring; and promotes sustainability and circularity via regulations along with environmental footprint pointers and recyclability standards. To be able to protect economic resilience and technical sovereignty in an international that is changing quick, these regulations combine sustainable, transparent practices, international alliances, home useful resource improvement, and deliver chain due diligence (Ruan et al. 2023).

5 Conclusion

To understand the development, upkeep, and development of present-day generation, one must have a stable knowledge of crucial metals. Aside from their natural scarcity, certain metals like lithium, cobalt, rare earth elements, and tantalum—are also characterized with the aid of the risks involved of their deliver and their necessary function in high-tech programs. They are critical to both national security and economic resilience due to the fact they strategically support crucial corporations, starting from protection technologies to customer electronics and renewable electricity structures. Essential metals provide for performance attributes in current electronics that would otherwise be impossible, which includes improved processing and verbal exchange abilities, strength efficiency, and downsizing. However, developing call for, geopolitical difficulties, and environmental concerns underscore the pressing need for recycling breakthroughs, opportunity material improvement, and sustainable sourcing. Ultimately, research on essential metals emphasizes two imperatives: promoting technological advancement and guaranteeing prudent use of those constrained sources. To relaxed the future of electronics and, extra usually, the global technological surroundings, it will likely be important to strike a stability between innovation and sustainability.

Acknowledgments The Lovely Professional University (LPU), Phagwara, Punjab, India, is greatly appreciated by the authors for providing the facilities and assistance required for this study. We would especially like to thank Basirhat College's Department of Chemistry for their help to this chapter on crucial metals.

Competing interests No conflicting interests are disclosed by the writers. There were no financial or non-financial conflicts pertaining to financing, connections, or commercial goals in key metals research when this work was carried out.

References

Awan TI, Afsheen S, Mushtaq A (2025) Role of noble metals in fuel cells and batteries. In: Advanced materials and methods for catalysis and electrocatalysis by transition metals editorial, pp 197–221. https://doi.org/10.1007/978-3-031-80983-5_9

Becci A, Amato A, Merli G, Beolchini F (2024) The green indium patented technology SCRIPT, for indium recovery from liquid crystal displays: bench scale validation driven by sustainability assessment. Sustainability 16(20):8917. https://doi.org/10.3390/su16208917

Bielowicz B (2025) Waste as a source of critical raw materials—a new approach in the context of energy transition. Energies 18(8):2101. https://doi.org/10.3390/en18082101

Blanco E, Martín P, Domínguez M, Fernández-Palacios P, Lombardero I, Sanchez-Perez C, García I, Algora C, Gabás M (2024) Refractive indices and extinction coefficients of p-type doped Germanium wafers for photovoltaic and thermophotovoltaic devices. Solar Energy Mater Solar Cells 264:112612. https://doi.org/10.1016/j.solmat.2023.112612

Born K, Ciftci MM (2024) The limitations of end-of-life copper recycling and its implications for the circular economy of metals. Resour Conserv Recycl 200:107318. https://doi.org/10.1016/j.resconrec.2023.107318

Can Sener SE, Thomas VM, Hogan DE, Maier RM, Carbajales-Dale M, Barton MD, Karanfil T, Crittenden JC, Amy GL (2021) Recovery of critical metals from aqueous sources. ACS Sustain Chem Eng 9(35):11616–11634. https://doi.org/10.1021/acssuschemeng.1c03005

D'Orazio P (2024) Assessing the fiscal implications of changes in critical minerals' demand in the low-carbon energy transition. Appl Energy 376:124197. https://doi.org/10.1016/j.apenergy.2024.124197

Dieaconescu RI, Gheorghe M, Petrariu I-R, Hurduzeu G (2025) The role of critical raw material imports in shaping economic competitiveness within the European Union. Roman Econ J 89:2. https://doi.org/10.24818/REJ/2024/89/02

Dou S, Xu D, Zhu Y, Keenan R (2023) Critical mineral sustainable supply: challenges and governance. Futures 146:103101. https://doi.org/10.1016/j.futures.2023.103101

Gu C, Jia A-B, Zhang Y-M, Zhang SX-A (2022) Emerging electrochromic materials and devices for future displays. Chem Rev 122(18):14679–14721. https://doi.org/10.1021/acs.chemrev.1c01055

Hagelüken C, Goldmann D (2022) Recycling and circular economy—towards a closed loop for metals in emerging clean technologies. Min Econ 35(3–4):539–562. https://doi.org/10.1007/s13563-022-00319-1

Hossain MK, Raihan GA, Akbar MA, Kabir Rubel MH, Ahmed MH, Khan MI, Hossain S, Sen SK, Jalal MIE, El-Denglawey A (2022) Current applications and future potential of rare earth oxides in sustainable nuclear, radiation, and energy devices: a review. ACS Appl Electr Mater 4(7):3327–3353. https://doi.org/10.1021/acsaelm.2c00069

Hu Z, Yu B, Daigo I, Tan J, Sun F, Zhang S (2024) Circular economy strategies for mitigating metals shortages in electric vehicle batteries under China's carbon-neutral target. J Environ Manag 352:120079. https://doi.org/10.1016/j.jenvman.2024.120079

Ibrahim M, Aouassa M, Hassan NK, Aladim AK, Saron KMA, Algarni SA, Bouabdellaoui M, Berbezier I (2025) High photosensitive amorphous germanium-based MIS photodetector. Phys B Cond Matter 710:417229. https://doi.org/10.1016/j.physb.2025.417229

Jia S, Meng W, Li S (2025) Risks of mineral resources in the supply of renewable energy batteries. Sci Rep 15(1):10142. https://doi.org/10.1038/s41598-025-94848-8

Kannan SG, Fleming M, Hirwa J, Castillo E, Eggert R (2025) Critical materials and manufacturing: comparing China, the European Union, Japan, and the United States. Resour Conser Recycl Adv 25:200250. https://doi.org/10.1016/j.rcradv.2025.200250

Khalkho R (2025) Critical minerals and national security: the role of strategic resource reserves in India's security strategy. World J Adv Eng Technol Sci 14(2):103–109. https://doi.org/10.30574/wjaets.2025.14.2.0054

Khandekar PP, Dash B, Yadav MD (2025) Recent advances in the recovery of neodymium and dysprosium from industrial waste. In: Metal value recovery from industrial waste using advanced physicochemical treatment technologies. Elsevier, pp 295–316. https://doi.org/10.1016/B978-0-443-21884-2.00011-3

Kursunoglu S (2025) A review on the recovery of critical metals from mine and mineral processing tailings: recent advances. J Sustain Metal 2:1126. https://doi.org/10.1007/s40831-025-01126-y

Lamblin V (2024) La demande de métaux critiques liés à la mobilité électrique. Futuribles 460(3):84–92. https://doi.org/10.3917/futur.460.0084

Liu W, Li X, Wang M, Liu L (2022) Research trend and dynamical development of focusing on the global critical metals: a bibliometric analysis during 1991–2020. Environ Sci Pollut Res 29(18):26688–26705. https://doi.org/10.1007/s11356-021-17816-5

Mahnoor M, Chandio R, Inam A, Ahad IU (2025) Critical and strategic raw materials for energy storage devices. Batteries 11(4):163. https://doi.org/10.3390/batteries11040163

Mishra N, Mancheri N (2024) Critical rare metal and collaboration of India and Japan. In: India, Japan and Beyond. Springer, Singapore, pp 65–87. https://doi.org/10.1007/978-981-97-3282-1_5

Mitrica D, Badea IC, Serban BA, Olaru MT, Vonica D, Burada M, Piticescu R-R, Popov V, v. (2021) Complex concentrated alloys for substitution of critical raw materials in applications for extreme conditions. Materials 14(5):1197. https://doi.org/10.3390/ma14051197

Moita Neto JM, Leal RC, Araújo NL, da Silva EA (2025) Challenges and opportunities for the development of urban mining in Brazil. Minerals 15(6):593. https://doi.org/10.3390/min15060593

Müller D, Groves DI, Santosh M, Yang C-X (2025) Critical metals: their applications with emphasis on the clean energy transition. Geosyst Geoenviron 4(1):100310. https://doi.org/10.1016/j.geogeo.2024.100310

Nazir S, Zhang J-M, Junaid M, Saleem S, Ali A, Ullah A, Khan S (2024) Metal-based nanoparticles: basics, types, fabrications and their electronic applications. Zeitschrift Für Physikalische Chemie 238(6):965–995. https://doi.org/10.1515/zpch-2023-0375

Parvini E, Hajalilou A, Gonçalves-Vilarinho JP, Alhais-Lopes P, Maranha M, Tavakoli M (2024) Gallium–carbon: a universal composite for sustainable 3D printing of integrated sensor–heater–battery systems in wearable and recyclable electronics. ACS Appl Mater Interf 16(25):32812–32823. https://doi.org/10.1021/acsami.4c02706

Pathan MKW, Weston NR, Mavris DN (2024) Critical materials in aerospace: a comprehensive analytic study on the strategic importance of semiconductors. In: AIAA SCITECH 2024 Forum. https://doi.org/10.2514/6.2024-1536

Patil AB, Struis RPWJ, Ludwig C (2023) Opportunities in critical rare earth metal recycling value chains for economic growth with sustainable technological innovations. Circ Econ Sustain 3(2):1127–1140. https://doi.org/10.1007/s43615-022-00204-7

Qiu Z-J, Xing Y, Brugger J, Etschmann B, Pintér Z, Fougerouse D, Xu J, Yu Z, Guo G-J, Fan H-R (2025) Vacancies in sulfides facilitate fluid-induced solid-state diffusion and critical metals accumulation. Nat Commun 16(1):1835. https://doi.org/10.1038/s41467-025-57171-4

Rahmani H, Shetty D, Wagih M, Ghasempour Y, Palazzi V, Carvalho NB, Correia R, Costanzo A, Vital D, Alimenti F, Kettle J, Masotti D, Mezzanotte P, Roselli L, Grosinger J (2023) Next-generation IoT devices: sustainable eco-friendly manufacturing, energy harvesting, and wireless connectivity. IEEE J Microwaves 3(1):237–255. https://doi.org/10.1109/JMW.2022.3228683

Rousseau F, Nominé A, Zavašnik J, Cvelbar U (2025) Is alloying a promising path to substitute critical raw materials? Mater Today 83:1–8. https://doi.org/10.1016/j.mattod.2025.01.015

Ruan S, Song Y, Cheng J, Zhan C (2023) Green eco-innovation and supply of critical metals: evidence from China. Sustainability 15(17):12730. https://doi.org/10.3390/su151712730

Shaikh AA, Bhattacharjee J, Datta P, Roy S (2024) A comprehensive review of the oxidation states of molybdenum oxides and their diverse applications. Sustain Chem Environ 7:100125. https://doi.org/10.1016/j.scenv.2024.100125

Su Y, Hu D (2022) Global dynamics and reflections on critical minerals. E3S Web Confer 352:03045. https://doi.org/10.1051/e3sconf/202235203045

Sun X (2022) Supply chain risks of critical metals: sources, propagation, and responses. Front Energy Res 10:957884. https://doi.org/10.3389/fenrg.2022.957884

Teng D, Wu J, Ma Q, Wang W, Zhou G, Fan G, Cao Y, Li P (2025) Advances in the recovery of critical rare dispersed metals (gallium, germanium, indium) from urban mineral resources. ACS Omega 10(1):76–92. https://doi.org/10.1021/acsomega.4c08689

Tirmare AH, Gowda VD, Dhabarde RJ, Tirmare HA, Kale SB, Suryawanshi VA, Kumar NA (2024) Modulated advancements in semiconductor-based nanomaterials for environmental solutions. Nanotechnol Environ Eng 9(4):525–537. https://doi.org/10.1007/s41204-024-00371-y

Xia D, Lee C, Charpentier NM, Deng Y, Yan Q, Gabriel JP (2024) Drivers and pathways for the recovery of critical metals from waste-printed circuit boards. Adv Sci 11(30):635. https://doi.org/10.1002/advs.202309635

Yandem G, Jabłońska-Czapla M (2024) Review of indium, gallium, and germanium as emerging contaminants: occurrence, speciation and evaluation of the potential environmental impact. Arch Environ Protect 4:84–99. https://doi.org/10.24425/aep.2024.151688

Yıldız TD (2024) Opportunities for the recovery of rare earth elements from mining tailings and urban mining. In: Trash or treasure entrepreneurial opportunities in waste management, pp 183–205. https://doi.org/10.1007/978-3-031-55131-4_7

Zhang L, Chen Z, Yang C, Xu Z (2022) Global supply risk assessment of the metals used in clean energy technologies. J Clean Product 331:129602. https://doi.org/10.1016/j.jclepro.2021.129602

Critical Metals: Importance, Scenario and Strategic Roadmap Towards Sustainability

Gaurav Sharma, Sailaja Priyadarsini, Alok Prasad Das, and V. P. Sharma

Abstract Critical metals are essential to the continuing energy transition and research activities aimed at developing sustainable, low-carbon systems. They facilitate the transition to greener economies by allowing the replacement and upgrade of traditional technology. Neodymium, titanium, cobalt, nickel, lithium, and other rare earth elements are essential in a wide range of applications, including magnets, cryocoolers, circuit boards, and lasers, making them critical to electronics and sustainable energy solutions. Lithium, in particular, is critical for rechargeable batteries that power electric cars and enable large-scale energy storage. These metals are working together to improve renewable energy, futuristic mobility, and digitalization. The demand for platinum-based group elements (PGEs) and lanthanides is continually rising, driven by their significance in addressing both technological and environmental difficulties. These component recovery and recycling are frequently emphasized as attractive prospects in the intermediate raw materials chain. Their scarcity of replacements, unpredictable supply networks, and great economic value make them strategically essential. Countries may become global leaders in the essential minerals market through the implementation of an inclusive approach that balances sustainability, innovation, and resource safety. To do this, it is critical to predict risks, detect future possibilities, and establish effective regulations using scenario analysis and proactive approaches to mineral governance. Future-oriented mineral governance must address a variety of key dimensions, including increasing demand for critical resources, the implementation of sustainable extraction practices,

G. Sharma
Mineral Economist (I), Office of CCOM (MES), Indian Bureau of Mines, Ministry of Mines, Nagpur, India
e-mail: gsharma@ibm.gov.in

S. Priyadarsini · A. P. Das
Department of Life Sciences, Rama Devi Women's University, Bhubaneswar, India
e-mail: alokprasaddas@rdwu.ac.in

V. P. Sharma (✉)
CSIR-Indian Institute of Toxicology Research, Lucknow, India
e-mail: vpsitrc1@rediffmail.com

A. P. Das and S. Priyadarsini (eds.), *Recovery and Recycling of Critical Metals from Electronic Waste for Environmental Sustainability*, Earth and Environmental Sciences Library, https://doi.org/10.1007/978-3-032-29064-9_2

the readiness of political and regulatory frameworks, the use of advanced innovations, and the strengthening of global cooperation. While these minerals are critical for allowing renewable energy systems and promoting resilient economies, they also pose threats to the environment and public health. For example, neodymium, which has no biological use, can cause skin and eye irritation, and extended exposure to its dust or compounds may impair lung and liver function. Environmentally, it tends to accumulate in soil and water, affecting ecosystems. Innovative applications in aerospace, medical implants, and advanced engineering highlight their value for future technologies. To mitigate associated risks, it is essential to implement comprehensive safety standards, many of which are still in the early stages of formulation.

Keywords Critical chemicals · Development · Sustainability · Transition · Transformative

1 Introduction

Critical minerals are essential to a nation's financial growth and security. Their scarcity, along with the concentration of extraction and processing in a few places, renders supply chains very fragile and poses hazards to critical businesses (Castillo et al. 2024). While vital for contemporary technology and security, its uneven distribution poses strategic problems to long-term stability. Figure 1 depicts glimpses of vital substances that fuel clean energy, modern electronics, and advanced safeguarding technologies.

For several of these critical metals, worldwide information on reserves, potential resources, and production are either incomplete or ambiguous (Aiyar et al. 2023a; b). Current estimates rely heavily on information supplied by the mining sector,

Fig. 1 Glimpse of critical minerals

implying that better openness and consistent reporting procedures are critical for improving understanding of the internationally distribution and life cycles of these resources (Müller et al. 2025).

Acquisition, extraction, and exploitation of important minerals provide enormous economic, social, and environmental issues. These complications are exacerbated by a lack of strong international collaboration and efficient multilateral governance structures. To manage such complex supply chains, authorities must carefully weigh the benefits and disadvantages between trade, climate action, energy security, and sustainable (Zhang et al. 2026a; b). Trade constraints, whether motivated by geopolitical rivalry, energy security concerns, or protectionist objectives, have the potential to divide markets, increase costs, inhibit investment, and hinder clean energy progress (Vikrant et al. 2026).

There are still insufficient supplies of rare earth elements and metals needed for solar and battery technologies, underscoring the urgency of implementing ethical extraction and recycling methods. Integrating knowledge from materials science, environmental science, and energy science is essential to developing environmentally friendly technologies. The focus of current research is on creating materials for enhanced energy storage (batteries, fuel cells), renewable energy generation (wind turbines, solar cells), and minimising the environmental effect of production (Prabhakar et al. 2019; Priyadarsini and Das 2024). Reducing waste and emissions requires a focus on sustainability, efficiency, and durability. Innovations that promote a greener energy future are also being encouraged by rapid progress in nanomaterials and nanomaterials (Priyadarsini et al. 2025).

The combination of energy, environmental, and materials research remains the core of green technology growth. Designing materials that are more durable, efficient, and sustainably sourced decreases the ecological impact and resource reliance. Emerging advancements in nanotechnology and biotechnology have the potential to transform renewable energy and energy storage systems, opening up new avenues for sustainable innovation (Mishra et al. 2023).

Critical Energy Transition Minerals (CETMs) are essential for sustainable technologies including photovoltaics, vehicular batteries, and smart grids. They constitute the foundation of India's clean energy transition, in which supply chain resilience and sustainability are becoming increasingly important as foreign economies strive to accomplish climate neutrality ambitions (Mallick et al. 2024).

Nonetheless, the high concentration of supply chains and underinvestment in CETM exploration and processing have resulted in substantial hazards. Geopolitical developments have contributed to supply interruptions and pricing volatility (Ali et al. 2021). These shocks emphasize the vulnerability of global mineral supply chains. India has announced aims for boosting renewable energy and increasing electric car deployment, as well as the ambition of being self-sufficient in clean energy production. Our country's mineral reserves are low, therefore rendering it challenging to satisfy anticipated local demand for vital materials (Tian et al. 2025).

The success of the low-carbon transition is dependent on the development and localization of low-carbon technologies (LCTs) such as solar PV, wind power, electrolyzer systems, electric vehicle motors, and grid-scale battery packs. The adoption rate and market share of diverse technical versions within each category will considerably impact the volume and composition of mineral necessities in the decades ahead (Zhang et al. 2024).

2 OECD Countries' Status and Scenario for Critical Minerals

Despite efforts by advanced countries to encourage domestic investigation and mining, the significant gap between resource discovery and output remains a hurdle. As a result, the countries of the OECD, who are already heavily reliant on imports, are expected to keep depending on innovative and emerging economies for crucial mineral requirements for many years (Bian et al. 2026). The current supply of key minerals is insufficient to meet the rapidly increasing demand. Despite continuous expenditures in mining exploration and production, industrialized countries are anticipated to continue to rely on imports due to rising consumption patterns and the long lead time frames needed to bring new mining operations fully operational.

The global spread of clean energy and technological advances, from OECD members to rapidly developing nations, emphasizes the importance of reducing reliance on virgin material extraction. Establishing effective recycling systems and boosting the use of secondary raw materials are viable methods for relieving the strain on finite natural resources (Zhang et al. 2026a; b). The majority of the critical metals are classified as transition metals (d-block), rare earth elements (f-block), and just a few are designated as s-block, such as lithium (Fig. 2). Their exact positions in the periodic table define their unique chemical and physical characteristics, which explains their critical relevance in renewable energy, electrification, and digital infrastructure (Zeng et al. 2026).

The United States Geological Survey (USGS), which is part of the Department of the Interior, has issued an updated list of fifty formally recognized essential minerals. This comprehensive list includes elements such as aluminium, antimony, and beryllium, as well as cobalt, nickel, lithium, rare earths, platinum group metals, and many more, highlighting the wide range of resources required for industrial and technological developments.

3 Importance in the Current Scenario

Future economies worldwide depend on essential minerals including lithium, graphite, cobalt, titanium, and the rare earth group. They are essential in fields including defense, transportation, electronics, and telecommunications. Additionally, they are essential to the worldwide movement for low-carbon growth and the use of renewable energy technology to reach Net Zero climate goals. Using a three-stage

Fig. 2 Placements in periodic table—critical metals

evaluation system, the Indian government has designated thirty salient minerals as strategically essential. Lithium, cobalt, nickel, rare earth elements, titanium, molybdenum, and vanadium were amongst the elements selected. The selection criteria were based on availability, dependency on imports, and relevance to developing technologies, renewable energy, and agriculture. Global concerns regarding energy are intertwined, and the future of energy is shaped by interconnected features of sustainability, technology, geopolitics, and climate change (Fig. 3).

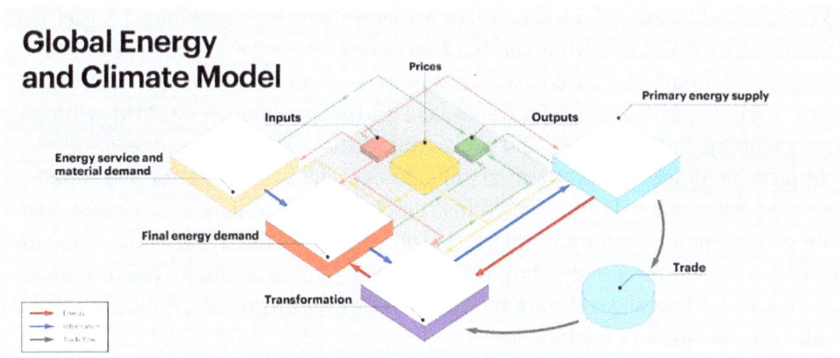

Fig. 3 Global energy and interrelated aspects: an overview

4 Market Scenario

Critical minerals are the fundamental building blocks of renewable energy systems, and their value can be observed across a variety of sectors.

- Solar energy requires minerals that consist of silicon, tellurium, indium, and gallium for making photovoltaic (PV) cells. India's currently operational solar power of 64 GW is primarily reliant on these natural resources.
- Wind turbines rely on elements like dysprosium and neodymium for their permanent magnets. India's plan to increase wind capacity from 42 to 140 GW by 2030 would necessitate a sustainable supply of these natural assets. Critical minerals are the backbone of clean energy systems, and their value can be visualised in multiple industries.
- Lithium-ion batteries are used in electric vehicles (EVs) and consist of lithium, cobalt, and nickel. India's National Electric Mobility Mission Plan (NEMMP) intends to have 6–7 million EVs on the road by 2024, creating an enormous market for these minerals.
- Large-scale energy storage systems rely largely on lithium, cobalt, and nickel, highlighting their strategic utility.

5 Future Trends

To build long-term resilience, the National Critical Mineral Mission (NCMM) was launched in 2025 by India to establish an extensive framework for self-sufficiency in this sector and coordinate with the well-known Geological Survey of India and other stakeholders with national relevance. The mission also proposed establishing a Centre of Excellence on Critical Minerals (CECM) to maintain the list current and advise long-term planning. Because essential minerals are the backbone of technologies such as solar panels, EVs, wind turbines, and grid-scale batteries, NCMM is intended to assure uninterrupted supply and processing skills.

Improving domestic exploration and processing is critical for the safety of resources. Adoption of sophisticated instruments like as remote sensing, automated drilling, and digital exploration platforms will boost productivity, while expanding the scope of licenses and streamlining clearances will promote private investment. The National Mineral Exploration Trust (NMET) will provide financial support, accelerating projects to the next level.

Another top aim is to boost domestic processing capacity. India may reduce its reliance on foreign refineries by giving targeted incentives and removing regulatory barriers. Establishing authorized mineral processing laboratories at the state level would increase technological capabilities, environmental standards, and infrastructure investment, integrating India into the international supply network (Roy et al. 2025).

International collaboration is also essential in developing safe and resilient supply chains for Critical Energy Transition Minerals (CETMs). Investing in resource-rich nations, particularly those located in Africa, has the potential to diversify supplies. Establishing strategic reserves will protect India from supply interruptions and price volatility. Furthermore, simplifying contracting rules and promoting long-term partnerships would boost investor trust. Participation in multilateral platforms such as the Mineral Security Partnership (MSP) and bilateral agreements with significant nations would help India secure access to crucial resources and strengthen its geopolitical stature (Xin et al. 2009).

One of the most important aspects of sustainable CETM management is encouraging recycling and circularity. Recycling performance will be enhanced by establishing Resource Recovery Parks (RRPs), formalizing the work of informal recyclers, and introducing graded incentives for recycling efficiency. These projects will increase material efficiency, create secondary resource markets, and further India's circular economy goals (Kordloo et al. 2025).

Building robust and dependable supply systems requires integrating strong Environmental, Social, and Governance (ESG) principles throughout the mineral value chain. Employing international ESG standards in mining operations would contribute to transparency and sustainability while easing exports. Social acceptability and trust will increase when local communities actively participate in planning and benefit-sharing.

Adoption of global ESG standards in mining operations would improve sustainability and transparency while facilitating exports. Involving neighborhood populations in decision-making and benefit-sharing will boost trust and social acceptance. At the same time, technologies like as digital battery passports may assure accountability from manufacture to recycling, and restoring mining-affected land for profitable use will help to secure long-term growth.

Finally, encouraging innovation through academic-industry partnership will be important to the advancement of CETM technology. Dedicated information exchange platforms, as well as incentives for cooperative R&D initiatives, can help to speed advances in exploration, processing, and sustainable mining. Expanding government programs like SATYABHAMA (Science and knowledge Yojana for Aatmanirbhar Bharat in Mining Advancement) would assist to promote knowledge transfer, encourage innovation, and promote India's competitiveness and self-reliance in the important mining industry.

Driving innovation through academic-industry partnership will be important to the advancement of CETM technology. Dedicated information exchange platforms, as well as incentives for cooperative R&D initiatives, can help to speed advances in exploration, processing, and sustainable mining. Expanding government schemes, such as the Science and knowledge Yojana for Aatmanirbhar Bharat in Mining Advancement, would allow knowledge transfer, stimulate innovation, and boost India's competitiveness and independent status in the important minerals manufacturing sector.

6 Environmental Issues

6.1 Regulations

The Mines and Minerals (Development and Regulation) Act 2023 is significant for all of us in terms of potential advancement and vision accomplishments. To boost domestic production of important and strategic minerals, the Geological Survey of India (GSI) has launched 195 exploration projects in fiscal year 2024–2025, focusing on possible mining locations around the nation. In addition, the Ministry of Mines has increased its funding for innovative exploration initiatives through the National Mineral Exploration Trust (NMET). To far, the NMET has sponsored 139 key mineral research projects through collaboration with several entities. Furthermore, to promote increased private sector engagement, the Ministry has licensed 28 private exploration agencies (NPEAs), who are now carrying out interdisciplinary projects with NMET's support.

Beyond internal changes, the Ministry of Mines is actively involved in a number of bilateral and international projects aimed at strengthening the global critical minerals value chain. These engagements address topics such as mineral processing, recycling, and supply chain resilience. Key platforms include the Minerals Security Partnership (MSP), the Indo-Pacific Economic Framework (IPEF), and the Initiative on Critical and Emerging Technologies (iCET), among others, demonstrating India's desire to strengthen international collaboration in the sector.

6.2 Geopolitical Scenarios

India is aggressively preparing for long-term resilience and sustainability in the crucial minerals industry, employing a dual strategy that prioritizes self-reliance through national efforts and global integration through strategic foreign partnership. This strategy focuses on creating varied, reliable, and responsible supply chains, which are essential for the shift to greener energy and industrial growth. India's engagement with Australia under the crucial Minerals Investment Partnership to procure lithium and cobalt from outside demonstrates the crucial necessity of global collaborations in supply chain security.

7 Key Elements of India's Critical Minerals Strategy

7.1 Domestic Mineral Exploration and Production

Domestic mineral exploration and extraction have attracted significant investment and incentives, including the use of contemporary technologies such as remote sensing and automated drilling. The National Critical Mineral Mission (NCMM) intends to carry out 1200 exploration projects by 2030–2031 to identify and utilize mineral resources such as lithium, rare earths, cobalt, and nickel. Reforms to the Mines and Minerals Act are going to accelerate auctions and standardize private sector participation, allowing for speedier mineral block allocations and supporting junior miners. In order to decrease reliance on foreign processors, domestic processing and refining infrastructure will be enhanced, including the creation of Mineral Processing Parks and Centers of Excellence. Recycling and secondary mineral recovery are examples of circular economy operations that are supported by incentives and resource recovery parks.

7.2 Strategic International Partnerships and Acquisitions

Active acquisition of critical mineral assets overseas through the PSU joint venture Khanij Bidesh India Ltd. (KABIL), with a focus on lithium and cobalt-rich countries like as Australia, Argentina, and Chile. Membership along with involvement in multinational partnerships such as the US-led Mineral Security Partnership (MSP), the Quad Critical Minerals Initiative, and contacts with the EU's Critical Raw Materials Club, which promote supply diversification, sustainability, and technology transfer. Bilateral and minilateral agreements with key resource-rich and geopolitical partners to improve supply chains while mitigating geopolitical concerns, specifically by reducing reliance on China, which dominates global production and processing.

We need to emphasize coordinated diplomacy to take advantage on trade, investment, and technology transfer possibilities while also ensuring multi-source supply resilience in the event of global disruptions (pandemics, conflicts, and export bans).

Critical minerals help to reduce India's overreliance on imports, particularly those from China, which controls the global rare earth supply chain. India now gets 60% of its rare earth imports from China. By expanding domestic mining, India can boost its production capacity in critical industries like as electronics, aerospace, and defence.

7.3 Innovation, Capacity Building, and Policy Frameworks

Support for mineral extraction and processing, and sustainable mining technologies, as well as improved metallurgical education and training. A comprehensive regulatory framework will be developed to promote responsible mining and outreach to communities, with a focus on environmental, social, and governance (ESG) standards.

7.4 Long-Term Vision and Targets

Critical minerals such as lithium, cobalt, and nickel are required for India's shift to renewable energy and electric car use. They are the basis of technologies including lithium-ion batteries, solar panels, and turbines for wind power. The 2023 changes to the Mines and Minerals Act, as well as the discovery of 5.9 million tonnes of lithium resources in Jammu and Kashmir, show India's commitment to increasing domestic capacity. According to World Bank study, output of minerals such as graphite, lithium, and cobalt may expand by 500% by 2050 to satisfy rising demand for environmentally friendly energy sources, with India intending to become a net-zero emitter of greenhouse gases (GHG) by 2070.

Mineral policy should be in line with India's climate goals, which include using electric vehicles extensively, achieving net-zero emissions by 2070, and obtaining 50% of power from non-fossil sources by 2030. By combining both domestic and foreign efforts, we aim to decrease our reliance on imports of vital minerals by 30–40% by 2035. assisting India in achieving its goal of dominating the global green energy production market by creating sustainable, safe supply chains.

8 Reasoning and Evidence Synthesis

8.1 Supply Chain Vulnerability

The extent to which social sustainability standards are satisfied has a substantial impact on mining's long-term growth prospects. Local, regional, and national communities are becoming aware of this dynamic, and they are demanding greater accountability from mining companies. National-level limits, particularly in developing countries, have repeatedly proven insufficient and ineffective in preventing overexploitation of resources of natural origin, displacement, and detrimental effects on the environment. Limited indigenous reserves, along with global concentrations of critical minerals, notably China's dominance, necessitate a balanced strategy. Geopolitical volatility (conflicts, export prohibitions, pandemics) highlights the importance of supply chain variety and resilience.

8.2 Policy and Institutional Action

Policy and Institutional Action: India created the NCMM and changed mining rules, allowing the private sector to develop and function more effectively in exploration and extraction. Growing expenditures in R&D and supporting infrastructure demonstrate a strong and continuous commitment to achieving a higher level of autonomy.

8.3 International Cooperation

India's active engagement in global initiatives and bilateral agreements enables technological collaboration, cash, and access to worldwide natural resources. These alliances assist in preventing geopolitical risks and ensure uniform global sourcing.

Sustainability Focus: The emphasis on ESG adoption, recycling incentives, and community participation underlines India's commitment to building a responsible and sustainable mining business that is consistent with global environmental goals.

Capacity Building: Recognizing metallurgical and processing capacity gaps, India is encouraging specialist education, professional development, and research collaborations to build indigenous skills and competencies.

Dominant control of essential minerals by individual nations, which are crucial for high-tech sectors, is increasing global economic and geopolitical disputes. In India, despite Mines and Minerals Amendment Act revisions, just 48% of identified blocks have been auctioned, with relatively low levels of international investment. Minerals like lithium and rare earths are increasingly tightly tied to national security concerns. However, ongoing hurdles in exploration and resource categorization continue to impede development. To fully realize its mineral potential, India must implement stronger monetary incentives and upfront financial support, akin to the measures utilized in semiconductor industry investments.

9 Chemistry of Critical Metals

9.1 Patents in Critical Mineral Value Chain

Promoting innovation by encouraging patent development throughout the essential mineral value chain.

9.2 Enhancement of Abilities

Training and upgrading the personnel to support mining, processing, and R&D operations. There are chances for developing countries to capitalize on their resource endowments, which include strategic advantages such as significant mineral reserves, geographic location, and technological expertise.

9.3 Mineral Processing Park

Dedicated mineral processing zones with cutting-edge infrastructure and facilities. Promoting innovation by fostering the development of patents throughout the entire critical mineral value chain.

9.4 Centre of Excellence

Institutions established for advanced research and technological development in the sector. Mining technology advancements enhance extraction and production efficiency, while reducing environmental effect. They can enhance value and create cross-functional relationships. Over the last two decades, there has been an important development in technologies.

9.5 Mineral Stockpile

Maintaining important reserves of vital minerals is critical to ensuring a steady supply for key sectors. Local content restrictions are critical tools for mineral-exporting countries to maximize the developmental impact of natural resources. Their success is dependent on meticulously evaluating economic efficiency, political feasibility, and social inclusion. Countries like Australia, Canada, and Finland show how flexible and deliberate policies may lead to long-term prosperity and technology ties. Meanwhile, developing and least developed countries continue to experiment with different approaches to maximising local benefits from resource availability.

Critical mineral supply is failing to keep pace with the rapid growth in demand. Even with efforts to step up domestic exploration and production, the considerable lead time between discovery and production for new mining operations means that OECD countries, who are net importers of critical minerals, will continue to rely on exports from developing and emerging economies in the coming decades.

The growing demand for green and digital technologies in countries around the world, from OECD members to emerging economies such as China, India, and

Indonesia, means that countries at all stages of development can help reduce pressure on primary extraction by recycling and increasing the supply of secondary raw materials. Adopting circular economy techniques can reduce material demand by promoting the use of alternative materials, supporting new designs, and fostering business models that extend product lifespans while improving utilization of resources efficiency.

10 Inference

India's goal of self-sufficiency in low-carbon technology is inextricably linked to the establishment of a safe and sustainable supply chain for Critical Energy Transition Minerals (CETM). As global competition for these resources increases up, preemptive initiatives may reduce risks and boost India's position as a leader in clean energy innovation. The nation can secure its mineral security, speed its energy transition, and maintain commitment to the environment and economic progress by increasing domestic exploration, fostering international cooperation, and implementing more sustainable mining practices.

India is purposefully integrating local innovation and exploration with international collaboration to create self-sufficient and globally varied essential mineral supply chains. This comprehensive plan prepares India to achieve a clean energy transition, reduce import reliance, and increase geopolitical and economic resilience while promoting sustainability. The changing rules and regulations, mission-driven investments, and international linkages all demonstrate India's readiness for long-term mining sector sustainability and technological supremacy.

Acknowledgements The authors are thankful to the parent organisations for support and encouragement. The books, journals and databases of the Knowledge Resource Centres of the CSIR, Universities and other scientific institutions were extremely beneficial.

Conflict of Interest No conflict of interest.

References

Aiyar S et al (2023a) Geoeconomic fragmentation and the future of multilateralism. Staff discussion notes, 15 January. International Monetary Fund, Washington

Aiyar S, Ilyina A, Chen J, Kangur A, Trevino J, Ebeke C, Gudmundsson T, Soderberg G, Schulze T, Kunaratskul T, Ruta M, Garcia-Saltos R, Rodriguez S (2023b) Geo-economic fragmentation and the future of multilateralism. Staff Disc Notes 2023(001):1. https://doi.org/10.5089/9798400229046.006

Ali I, Cheng Q, Ding T, Yiguang Q, Yuechao Z, Sun H, Peng C, Naz I, Li J, Liu J (2021) Micro-and nanoplastics in the environment: occurrence, detection, characterization and toxicity: a critical review. J Clean Prod 313:127863. https://doi.org/10.1016/j.jclepro.2021.127863

Bian K, Feng L, Yan Y, Geng J, Zhang Y, Sun M (2026) Temperature-dependent enrichment, occurrence modes and speciation evolution of critical metals (In, Li, Ga, REY) during Ningxia coal combustion. Fuel 422:139115. https://doi.org/10.1016/J.FUEL.2026.139115
Castillo E, del Real I, Roa C (2024) Critical minerals versus major minerals: a comparative study of exploration budgets. Miner Econ 37:433–444
Kordloo M, Boroumand Z, Abdollahi H, Rezaei A, Ghorbani Y, Solgi A (2025) Lithium recovery from waste aluminum electrolyte via bioleaching: comparative insights into acidic and alkaline pathways using *Aspergillus niger* and *Glutamicibacter nicotianae*. Waste Manage 206:115043. https://doi.org/10.1016/J.WASMAN.2025.115043
Mallick K, Sahu S, Sahu A, Dey S, Das AP (2024) A critical review of marine microfiber pollution routes, toxicity, and its sustainable remediation. Environ Sci Eng F2333:189–211. https://doi.org/10.1007/978-3-031-51792-1_10
Mishra S, Ghosh S, van Hullebusch ED, Singh S, Das AP (2023) A critical review on the recovery of base and critical elements from electronic waste-contaminated streams using microbial biotechnology. Appl Biochem Biotechnol 2023:1–30. https://doi.org/10.1007/S12010-023-04440-X
Müller D, Groves DI, Santosh M, Yang CX (2025) Critical metals: their applications with emphasis on the clean energy transition. Geosyst Geoenviron 4(1):100310. https://doi.org/10.1016/J.GEOGEO.2024.100310
Prabhakar A, Mishra S, Das AP (2019) Isolation and identification of lead (Pb) solubilizing bacteria from automobile waste and its potential for recovery of lead from end of life waste batteries. Geomicrobiol J 36(10):894–903. https://doi.org/10.1080/01490451.2019.1654044
Priyadarsini S, Das AP (2024) Analytical and structural characterization of waste lithium-ion batteries for their effective recycling strategy. Environ Chem Ecotoxicol 2:4. https://doi.org/10.1016/j.enceco.2024.12.004
Priyadarsini S, Alok T, Das P, Journal G, Das AP (2025) Isolation and identification of native lithium-resistant bacteria from electronic waste contaminated sites: a focused investigation on tolerance assessment with potential lithium bio-recovery application. Geomicrobiol J 4:509
Roy S, Vishwakarma A, Hait S (2025) Enhanced bioleaching of spent lithium-ion battery cathode powder by *Acidithiobacillus ferroxidans* for selective metal recovery: influence of ferrous iron and elemental sulfur as energy sources. Sep Purif Technol 376:134165. https://doi.org/10.1016/J.SEPPUR.2025.134165
Tian X, Guo Z, Zhu D, Pan J, Yang C, Li S (2025) Recovery of valuable elements from coal fly ash: a review. Environ Res 282:121928. https://doi.org/10.1016/j.envres.2025.121928
Vikrant K, Giannakoudakis DA, Nguyen-Dang T, Nguyen DT, Phan MH (2026) Navigating the catalytic landscape for methyl ethyl ketone abatement: a critical review of noble metal versus metal oxide strategies. Environ Res 302:124659. https://doi.org/10.1016/J.ENVRES.2026.124659
Xin B, Zhang D, Zhang X, Xia Y, Wu F, Chen S, Li L (2009) Bioleaching mechanism of Co and Li from spent lithium-ion battery by the mixed culture of acidophilic sulfur-oxidizing and iron-oxidizing bacteria. Biores Technol 100(24):6163–6169. https://doi.org/10.1016/J.BIORTECH.2009.06.086
Zeng L, Shen H, Huang H, Zhang W (2026) Green leaching strategies for recovering critical metals from spent lithium-ion batteries using fruit peels and coffee waste. Waste Manage 219:115571. https://doi.org/10.1016/J.WASMAN.2026.115571
Zhang W, Zhao L, Wang W, Nechaev VP, French D, Graham I, Lang Y, Li Z, Dai S (2024) Enrichment of critical metals (Li, Ga, and rare earth elements) in the early Permian coal seam from the Jincheng Coalfield, southeastern Qinshui Basin, northern China: with an emphasis on cookeite as the Li host. Ore Geol Rev 167:105939. https://doi.org/10.1016/j.oregeorev.2024.105939
Zhang M, Hu W, Li Z, Zhang Y, Zhang H, Zeng Z, Wang S, Ma Z, Liu J (2026a) Efficient regeneration of $LiFePO_4$ through low-temperature roasting with ammonium bisulfate for spent $LiFePO_4$ cathode scraps. Mater Chem Phys 348:131610. https://doi.org/10.1016/j.matchemphys.2025.131610

Zhang X, Qi Z, Chen B, Pang Y, Mao J, Wen J, Cong H, Li Z (2026b) Review of novel hydrogel resins for the deep separation of critical metals: focus on mechanism-driven material design. Sep Purif Technol 401:138571. https://doi.org/10.1016/J.SEPPUR.2026.138571

Role of Microorganisms in Bioleaching: Implications for Critical Metal Recovery and Recycling

N. Aarthi

Abstract The recovery of critical metals from electronic waste—including spent batteries, mobile phones, laptops, tablets, and other devices—has emerged as a major challenge in response to growing industrial demand and the environmental consequences of conventional mining. Bioleaching, a promising alternative rooted in microbial metabolism, offers a sustainable and low-impact solution by employing specific bacteria, archaea, fungi, and microalgae to extract metals from complex waste matrices. This chapter explores how these microorganisms interact with metal-bearing substrates to mobilize elements such as lithium, lead, cadmium, mercury, chromium, arsenic, and nickel. Particular emphasis is placed on the underlying biochemical mechanisms, the ecological and operational factors that enhance microbial activity, and the practical application of bioleaching in managing electronic waste and industrial residues. By examining advances in microbial consortia, bioreactor design, and integrated recovery systems, this chapter highlights the growing significance of microorganisms in enabling a circular economy and advancing environmentally responsible metal recycling.

Keywords Bioleaching · Microbial metal recovery · Electronic waste (e-waste) · Microbial consortia · Sustainable recycling · Circular economy

1 Introduction

Metal values for economic uses symbolize the state's global mineral resource expansion. The aspiration for metals has escalated substantially and is anticipated to proceed to do so in the upcoming years to come due to the upliftment in industrialization brought upon with increase in population. As a result, solid wastes containing metals are simultaneously produced together with effluents, which had detrimental

---

N. Aarthi (✉)
Department of Zoology, Sri Sarada College for Women (Autonomous), Salem, India
e-mail: aarthissc.ac.in@gmail.com

A. P. Das and S. Priyadarsini (eds.), *Recovery and Recycling of Critical Metals from Electronic Waste for Environmental Sustainability*, Earth and Environmental Sciences Library, https://doi.org/10.1007/978-3-032-29064-9_3

consequences on the high-grade ores' depleting reserves. Thus, it is substantial to attack the issue of cost-effective contamination abatement and metal recovery.

Electronic industry is the world's largest and innovative industry for its kind. Electronic gadgets and technology are meant to make our lives simple and happy and to save our time, but they contain hazardous substances, their disposal and recycling becomes a health turmoil. It has invaded into every aspect of our lives and most of us do not think what happens to these gadgets when we dump the old electronic gadgets as scrap. The use of electronic devices has escalated in recent. The metals and hazardous substances accumulate in the ecosystem. Ecosystem is both biotic and abiotic factors and hence the chemicals and heavy metals found in the electronic gadgets starts accumulating both in biotic and abiotic factors. The continuous accumulation leads to biomagnification or bioamplification or biological magnification happens across the food chain. Heavy metals like mercury, methylmercury, lead, Nickel, Chromium from the e-waste accumulates the terrestrial and aquatic ecosystem.

Bioleaching is a biologically mediated process that mobilizes metals from ores or waste materials through microbial oxidation and solubilization reactions. It provides a low-energy, environmentally sustainable alternative to traditional pyrometallurgical and hydrometallurgical extraction methods (Vera et al. 2022a, b). The growing demand for critical and strategic metals has revived interest in microbial metal recovery from low-grade ores, tailings, and e-waste (Castro et al. 2024) (Fig. 1).

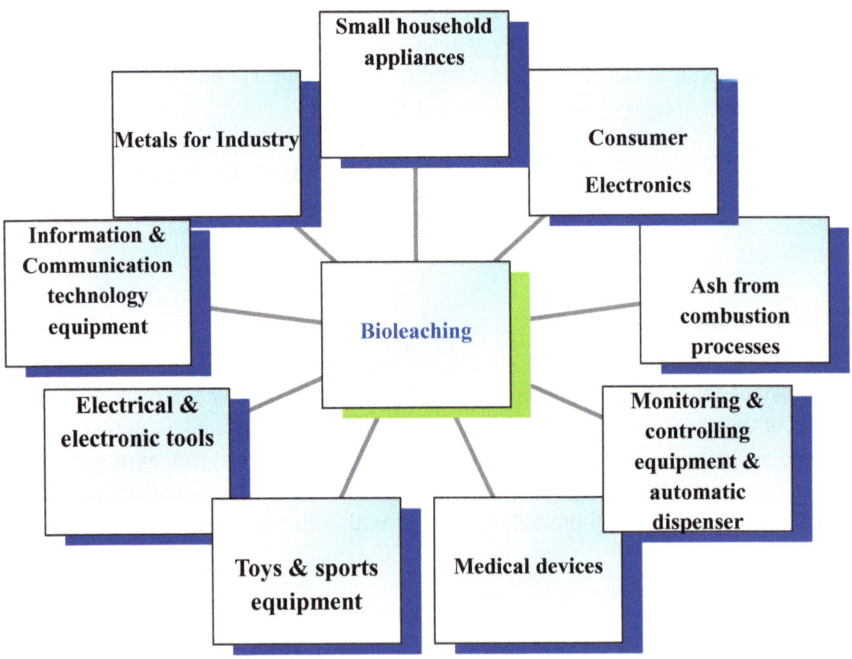

Fig. 1 Bioleaching applicable in depicted waste metals

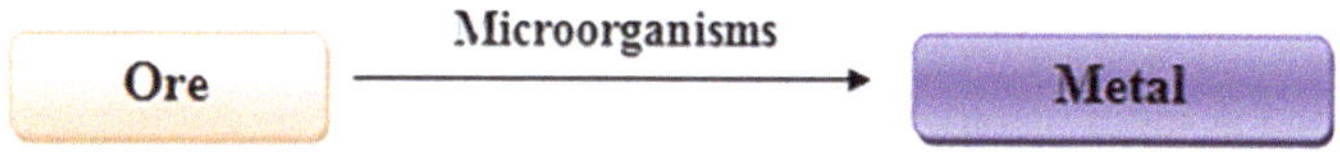

Fig. 2 Hydrometallurgy

1.1 *Hydrometallurgy*

Hydrometallurgy paves way for the recycling of materials, like higher efficiency, excellent metal selectivity, lower energy consumption, and low emissions of hazardous gases. Hydrometallurgy looks forward many challenges viz., increased production of waste water, high cost. In this regards it emphasizes the researchers globally to seek alternate mechanism. Synthetic leaching process which implies the disintegrates the metals in the waste using the relevant solvent (Behnamfard et al. 2013) or concentrated nitric acids (HNO_3) (Maguyon et al. 2012) hydrochloric acid (HCl), sulfuric acid (H_2SO_4) on industrial scale, aqua regia and few alkalis followed by pretreatment of the waste (Ghosh et al. 2015). The process of hydrometallurgy implicates the electrolytic treatment of metals which is achieved through various techniques and combinations of chemical leaching using halide, cyanide, Thiourea and Thiosulphate. Ethylenediaminetetraacetic acid (EDTA) is also considered as major chelating agent to adsorb the heavy metals like lead, cadmium, copper, chromium and zinc. Though it helps to leach the metals, it consumes high amount of energy and concurrently it generates more quantity of acidic waste water which is highly hazardous to the ecosystem and the living organisms. Ethylenediaminetetraacetic acid (EDTA) acts as a base transformer to increase the adsorption capacity of the medium with its chelating structure or acts as a ligand to form multiple bonds and synchronize with a wide range of metal ions (Kumar et al. 2017). This Hydrometallurgy leaves the residues of chemical used in the environment. Based on this, the researchers and other scientists started concentrating on some other methods as an alternative to the chemical. They thought of applying living organisms for this leaching process and termed as Bioleaching (Priyadarsini and Das 2024) (Priyadarsini et al. 2025).

Unlike conventional methods, biohydrometallurgy offers additional advantages since it enables the extraction of metals through aqueous media and metabolites generated by specific microorganisms (Fig. 2).

1.2 *Pyro-metallurgy*

Investigating an alternative to hydrometallurgy has marched towards the implementation of thermal (high temperature) to extract the metals. This method pertains 400–700 °C to melt the metals which liberates oil and other analogous volatile substances.

This regenerated oil fraction and volatile constituents find application both as alternative fuels and as input materials for allied chemical industries. Although pyrometallurgy has proven effective in extracting metals, including precious ones, from various waste materials, the process releases highly toxic by-products such as dioxins and furans. These hazardous emissions contaminate the environment, enter the food chain, and result in bioaccumulation and biomagnification across trophic levels. From microorganisms to higher mammals, such pollutants pose severe health risks, induce genetic mutations, and ultimately threaten overall biodiversity. These limitations have redirected research efforts toward sustainable metal recovery through bioleaching.

1.3 Bioleaching

Bioleaching, for instance, is recognized as an environmental-friendly approach to metal recovery, characterized by minimal installation and operational expenses, reduced energy demand, absence of hazardous by-products, and relatively low capital requirements (Priyadarsini and Das 2025, 2026). During bioleaching, metals are mobilized from otherwise insoluble solid materials either through the direct action of microbial metabolism or indirectly via metabolic by-products. These by-products include compounds such as organic acids, chelators, amino acids, and other complexing agents that are commonly secreted by heterotrophic bacteria and fungi.

2 Microorganisms in Bioleaching: An Overview

Microbial Diversity in Bioleaching Systems

Microorganisms that assist bioleaching belong primarily to acidophilic bacteria and archaea capable of oxidizing ferrous iron and reduced sulfur compounds.

2.1 Iron- and Sulfur-Oxidizing Bacteria

Members of the genera *Acidithiobacillus*, *Leptospirillum*, and *Sulfobacillus* overshadow in both natural and manipulate leaching environments (Dopson and Johnson 2022). *Acidithiobacillus ferrooxidans* oxidizes ferrous to ferric iron, discharge or generate the oxidants which directs the mineral dissolution, while *Leptospirillum ferrooxidans* fetches hands specially to iron cycling (Vera et al. 2022a, b).

2.2 Thermophilic Archaea

At elevated temperatures (>65 °C), thermophilic archaea such as *Sulfolobus metallicus* and *Acidianus brierleyi* become pivotal bioleaching agents, particularly for refractory sulfide ores (Colmer et al. 2023). These archaea withstand extreme acidity and temperature; restore biooxidation activity in industrial reactors and heaps.

2.3 Heterotrophic Microorganisms

Fungi (*Aspergillus niger*, *Penicillium simplicissimum*) and some bacteria (*Pseudomonas aeruginosa*, *Bacillus subtilis*) play a role indirectly through the manufacture of organic acids, chelating compounds, and biosurfactants (Tezyapar Kara et al. 2023). Such organisms are essential in bioleaching of non-sulfidic minerals, oxide ores, and complex waste materials including printed circuit boards (Awasthi et al. 2024).

3 Mechanisms of Microbial Action

3.1 Indirect (Ferric-Iron Mediated) Mechanism

Indirect oxidation functions primarily through oxidants generated by microbial metabolism rather than physical cell–mineral contact. During this process, microorganisms oxidize ferrous iron (Fe^{2+}) to ferric iron (Fe^{3+}) and reduced sulfur compounds to sulfuric acid. The ferric ions and protons then chemically attack the sulfide mineral, liberating metal ions into the leachate. The microbes subsequently regenerate Fe^{3+}, maintaining a self-sustaining oxidation cycle (Shen et al. 2020a; Vera et al. 2022a, b). This mechanism is especially important in large-scale heaps and reactors, where direct microbial contact with every mineral particle is not feasible (Tripathy et al. 2026). Together, these two mechanisms create a dynamic bioleaching system in which microbial catalysis and chemical oxidation work synergistically to enhance metal recovery efficiency. This reaction releases soluble metal ions and regenerates ferrous iron, which is again converted to ferric iron by microbial activity, maintaining a continuous leaching cycle (Rawlings et al. 2023).

3.2 Direct and Sulfur-Oxidation Mechanisms

In **direct oxidation**, microorganisms adhere to the mineral surface and catalyze the oxidation of metal sulfides through membrane-bound enzymatic reactions (Falagán

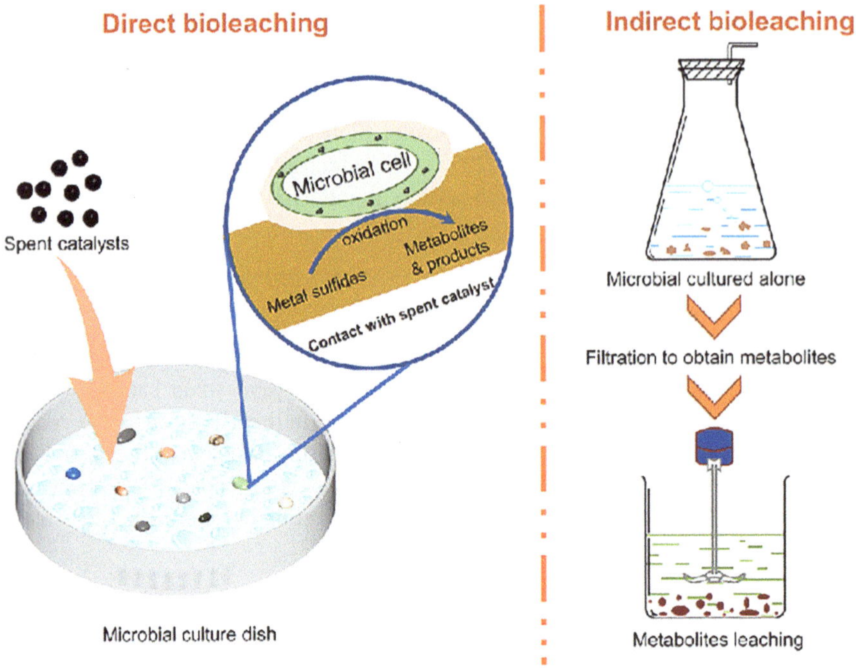

Fig. 3 The direct and indirect mechanism of bioleaching

et al. 2024). This intimate contact allows for direct electron transfer between the sulfide mineral and the microbial cell. Species such as *Acidithiobacillus ferrooxidans* and *Leptospirillum ferriphilum* are known to form structured biofilms on sulfide particles, enabling efficient electron transport and mineral dissolution (Rawlings and Johnson 2007). As a result, the metal ions are released into solution, while the microbes gain energy from the oxidation process to sustain their growth (Fig. 3).

3.3 Role of Biofilms and EPS

A biofilm is a structured assemblage of microbial cells that irreversibly attach to biotic or abiotic surfaces and to one another, becoming encased within a self-secreted matrix of extracellular polymeric substances (EPS). This matrix confers mechanical stability, mediates nutrient exchange, and enhances microbial resistance to environmental stressors.

In bioleaching systems, biofilms play a pivotal role by facilitating close contact between microbial cells and mineral surfaces. The EPS matrixes not only anchors the microorganisms but also mediates the localized concentration of oxidizing agents, such as ferric ions or sulfuric acid, enhancing mineral dissolution and modulate

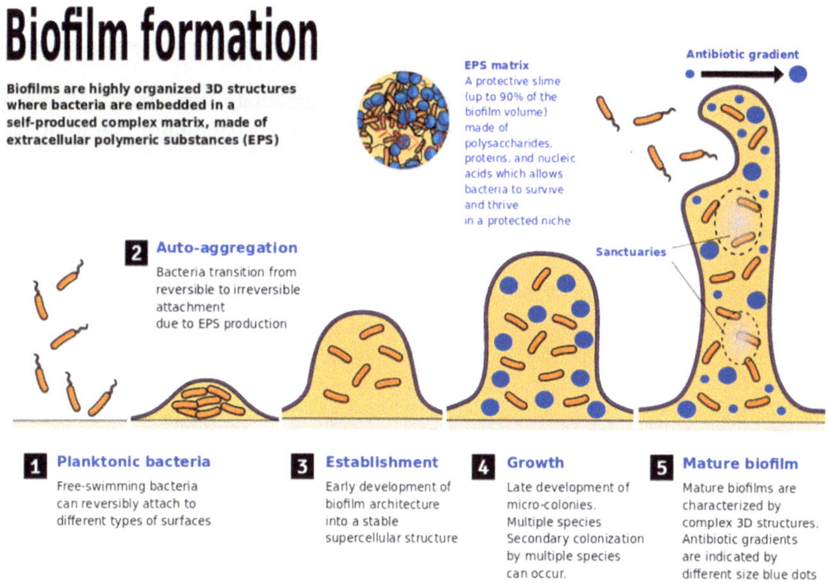

Fig. 4 Schematic representation of the different steps required for the formation of a mature biofilm (Meroni et al. 2021)

electron transfer. Furthermore, biofilms create microenvironments that can stabilize pH and redox conditions, improving the efficiency and consistency of the leaching process (Saldaña et al. 2023; Meroni et al. 2021) (Fig. 4).

Biofilms are structured microbial communities embedded in extracellular polymeric substances (EPS) that attach to mineral surfaces (Harshan et al. 2025). In bioleaching, biofilms augment metal extraction by increasing microbial adherence, maintaining a localized acidic environment, and facilitating oxidation of metal sulfides such as copper, nickel, and cobalt (Tonietti et al. 2024). Biofilm formation also protects microbial cells from toxic metal concentrations and fluctuating environmental conditions, sustaining activity over extended periods (Cozma et al. 2024).

From a sustainability perspective, bioleaching using biofilms supports SDG 12 by promoting responsible consumption and production. This approach reduces reliance on conventional mining methods, minimizes waste generation, and lowers environmental impact, making metal recovery processes more sustainable (Harshan et al. 2025; Cozma et al. 2024).

7 Industrial Applications

Bioleaching has evolved from a laboratory curiosity to a viable industrial technique for metal extraction. Its application spans traditional mining operations as well as innovative recycling of electronic waste. Industrial-scale bioleaching relies on the metabolic activity of specialized microorganisms to solubilize metals, providing a cost-effective and environmentally sustainable alternative to conventional chemical methods.

7.1 Mining Industry

In the mining sector, bioleaching is primarily applied to the extraction of copper, gold, cobalt, and nickel from low-grade ores and sulfide concentrates. For example, in Chile, copper extraction is routinely carried out using heap bioleaching, where crushed ore is stacked in large heaps and irrigated with microbial solutions. Acidophilic bacteria such as *Acidithiobacillus ferrooxidans* oxidize sulfide minerals, producing ferric ions and sulfuric acid that convert metal sulfides into soluble forms (Rawlings and Johnson 2007).

Similarly, in African mining operations, bioleaching is employed for cobalt and nickel recovery. These metals are often found in sulfide ores that are difficult to process using traditional smelting. By harnessing microbial oxidation, companies can recover metals from ores that would otherwise be economically unviable (Brierley and Brierley 2013).

7.2 Electronic Waste Recycling

The rapid accumulation of electronic waste has created a demand for alternative recycling methods. Bioleaching offers a method to recover valuable metals from e-waste without generating hazardous chemical residues. Microbial consortia, including acidophilic bacteria and heterotrophic fungi, are employed to extract metals such as gold, copper, and rare earth elements from printed circuit boards and other electronic components. Pilot-scale bioreactors in China and Europe have demonstrated the potential of this approach to operate continuously, achieving high recovery efficiencies while reducing environmental impact (Das et al. 2023b; Su et al. 2024).

7.3 *Industrial Bioreactors and Process Optimization*

Industrial bioleaching systems typically employ either heap leaching or stirred-tank reactors, depending on the ore type and scale. Heap leaching involves irrigation of large ore piles with microbial solutions over several months, suitable for low-grade ores. Stirred-tank reactors, in contrast, allow precise control of pH, temperature, and oxygen levels, enabling faster metal recovery from higher-grade concentrates (Rawlings et al. 2003). Process optimization often involves maintaining acidic conditions (pH 1–2) and temperatures below 40 °C, where microbial activity is maximal.

7.4 *Other Applications*

Beyond copper, cobalt, nickel, and e-waste metals, bioleaching has been explored for zinc, uranium, and rare earth element recovery. In some cases, microbial-assisted leaching is applied to reclaim metals from industrial sludge and spent catalysts, contributing to circular economy strategies and reducing the environmental footprint of industrial waste disposal (Brierley and Brierley 2013).

8 Bioleaching in Primary and Secondary Sources

Bioleaching from Primary Sources

Primary sources refer to naturally occurring ores containing metals, including sulfide and oxide ores. Bioleaching is particularly effective for sulfide ores, where chemolithotrophic bacteria, such as *Acidithiobacillus ferrooxidans* and *Leptospirillum ferrooxidans*, oxidize metal sulfides to soluble metal sulfates (Brierley and Brierley 2013).

Mechanism

1. Bacterial oxidation of Fe^{2+} to Fe^{3+}.
2. Fe^{3+} reacts with metal sulfide to release metal ions and sulfuric acid.
3. Solubilized metal is recovered from the leachate (Johnson 2014).

Example

Copper extraction from chalcopyrite ($CuFeS_2$) can be achieved via bioleaching, producing copper sulfate in solution, which is subsequently recovered by electrowinning (Rawlings 2002).

Bioleaching from Secondary Sources

Secondary sources include industrial wastes, slag, and electronic waste (e-waste) that contain residual metals. Bioleaching from secondary sources offers a sustainable method for recycling valuable metals, reducing environmental pollution from waste disposal (Kumar et al. 2017).

Mechanism

- Microorganisms attack metal-containing compounds in the waste, releasing metals into a bioavailable soluble form.
- Metal recovery is then performed using conventional methods such as precipitation or electrochemical processes.

Example

E-waste, such as printed circuit boards, contains copper, gold, and silver. Specific bacteria, such as *Acidithiobacillus thiooxidans*, facilitate metal solubilization, enabling recovery of metals without harsh chemicals (Li et al. 2007).

9 Bioprocess Optimization for Metal Recovery

The recovery of metals through bioprocessing techniques has emerged as a sustainable and eco-friendly alternative to conventional methods such as chemical leaching and pyrometallurgy. Microorganisms such as bacteria, fungi, and algae can mobilize, accumulate, or precipitate metals through processes like biosorption, bioleaching, and biomineralization (Gadd 2010). Optimization of these bioprocesses is crucial to maximize metal recovery efficiency while minimizing operational costs.

Factors Affecting Bioprocess Efficiency

Several factors influence the efficiency of microbial metal recovery:

1. **Microbial Strain Selection**
   The choice of microorganism plays a pivotal role in metal recovery. Metal-tolerant strains with high biosorption capacity or acidophilic bioleaching properties can significantly enhance recovery rates (Ahluwalia and Goyal 2007).
2. **PH and Temperature**
   Microbial activity is highly sensitive to pH and temperature. For instance, bioleaching bacteria such as *Acidithiobacillus ferrooxidans* perform optimally at acidic pH (~2–3) and moderate temperatures (~30 °C) (Rawlings and Johnson 2007).
3. **Nutrient Availability**
   The presence of carbon and nitrogen sources, trace elements, and energy substrates is essential for microbial growth and metabolism, directly affecting metal solubilization (Brandl 2001a).
4. **Metal Concentration and Toxicity**
   High concentrations of metals may inhibit microbial growth or metabolic activity. Controlled feeding strategies and gradual adaptation are often employed to mitigate toxicity (Brierley 2003).

5. **Agitation and Aeration**
   Oxygen availability and mixing can influence the efficiency of bioleaching or biosorption. Adequate aeration supports aerobic microbial metabolism, while agitation ensures uniform contact between microbes and metal substrates (Vieira et al. 2011).

10 Process Optimization Techniques

To enhance metal recovery, bioprocesses can be optimized using several strategies:

1. **Design of Experiments (DOE)**
   Statistical approaches like Response Surface Methodology (RSM) and factorial designs help in understanding the interaction between process variables and in identifying optimal conditions for metal recovery (Montgomery 2017).
2. **Immobilization of Microorganisms**
   Immobilizing cells on carriers can improve microbial stability, tolerance to metals, and reusability in continuous operations (Volesky 2003).
3. **Use of Bioreactors**
   Scaling up bioprocesses in bioreactors allows controlled parameters such as pH, temperature, aeration, and substrate feed rates, leading to consistent and enhanced metal recovery (Rawlings 2002).
4. **Genetic and Metabolic Engineering**
   Genetic modifications can enhance metal tolerance, biosorption capacity, or bioleaching efficiency, paving the way for next-generation microbial metal recovery systems (Nies 1999).

11 Applications

Optimized bioprocesses are employed in the recovery of precious metals (gold, silver), base metals (copper, nickel), and rare earth elements from ores, industrial waste, and electronic waste. The combination of process optimization and microbial biotechnology provides a cost-effective and environmentally sustainable approach to metal recovery (Gavrilescu 2004).

11.1 Environmental Implications

Bioleaching, the microbial extraction of metals from ores or waste, has distinct environmental advantages over traditional methods such as pyrometallurgy and chemical leaching. One major benefit is the reduction of toxic chemical usage. Conventional methods often employ strong acids or cyanide, which can generate hazardous effluents and contaminate soil and water. Bioleaching, in contrast, primarily relies on naturally occurring acidophilic microorganisms, producing less environmentally persistent residues (Rawlings and Johnson 2007).

Moreover, bioleaching processes generally operate at ambient temperatures and pressures, which lowers energy consumption and reduces greenhouse gas emissions associated with metal extraction (Brierley 2003). The process also facilitates the treatment of low-grade ores and mining residues, transforming otherwise waste materials into recoverable metals, thereby decreasing the need for extensive mining operations and mitigating landscape degradation (Gadd 2010).

However, potential environmental risks exist if bioleaching solutions containing high concentrations of dissolved metals or acidic leachates are released untreated into the environment. Careful containment, neutralization, and effluent treatment are essential to prevent contamination of soil and surface water (Ahluwalia and Goyal 2007) (Fig. 6).

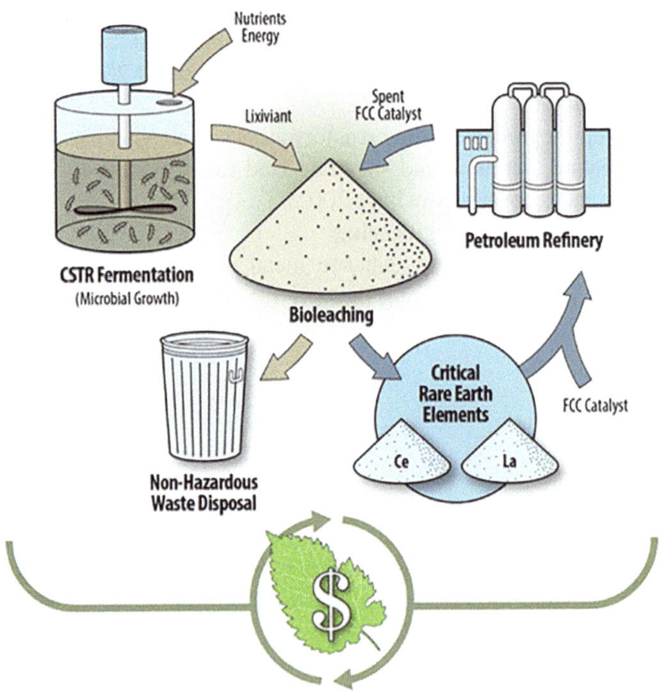

Fig. 6 The industrial process

11.2 Economic Implications

From an economic perspective, bioleaching presents several advantages. The technique allows the extraction of metals from low-grade ores or electronic waste, which would otherwise be uneconomical using conventional methods. This reduces the dependence on high-grade mining sites and diversifies the metal supply chain (Brandl 2001b).

Bioleaching also incurs lower operational costs because it uses ambient conditions, reducing energy expenditure associated with high-temperature smelting or chemical-intensive processing. Additionally, microbial consortia can be reused or scaled up in bioreactors, further improving cost-effectiveness in industrial applications (Rawlings 2002).

On the other hand, bioleaching may require longer processing times compared to conventional metallurgical methods, which can delay production cycles. Initial investment in bioreactor design, monitoring systems, and microbial cultivation may also represent a higher upfront cost. Nevertheless, the long-term environmental compliance savings and the ability to process otherwise non-viable ores can offset these initial expenditures (Vieira et al. 2011). Bioleaching represents a sustainable and economically viable approach to metal recovery. While environmental benefits include lower chemical usage, reduced energy consumption, and remediation of waste materials, economic advantages arise from the ability to process low-grade resources and reduce operational costs. Proper management of effluents and process optimization can further enhance the sustainability and profitability of bioleaching applications (Dey et al. 2024).

12 Challenges and Limitations

Bioleaching, the microbial extraction of metals from ores, has emerged as an eco-friendly alternative to conventional metallurgical methods. Despite its promise, several challenges limit its large-scale application and efficiency.

12.1 Slow Process Rate

The kinetics of bioleaching is generally slower compared to traditional pyrometallurgical and hydrometallurgical methods. Microbial growth and metabolic activity depend on environmental conditions such as pH, temperature, and nutrient availability, which can lead to prolonged extraction times (Rawlings and Johnson 2007).

Fig. 7 Indicates the SDG of government

promising pathway to address the growing demand for critical metals, contributing to resource conservation and circular economy initiatives. Continued research into microbial diversity, metabolic optimization, and process engineering is essential to unlock the full potential of bioleaching for large-scale metal recovery and sustainable material management.

E-waste management plays a vital role in fulfilling the Sustainable Development Goals (SDGs) outlined in the 2030 Agenda. During the global assembly, many nations emphasized the necessity of adopting sustainable practices in handling E-waste, with special attention on minimizing its generation. One of the most promising approaches is mining, which not only conserves valuable resources but also enhances economic value, supports urban infrastructure planning, promotes public health and safety, creates employment opportunities, and reduces greenhouse gas emissions. Collectively, these benefits contribute directly to achieving the SDGs and their associated targets.

The SDG targets connected to E-waste management include:

SDG Target 3: By 2030, with the help of bioleaching process will try to reduce the deaths and illnesses caused by the emission of hazardous chemicals and pollution of air, water, and soil.

SDG Target 8: Encourage policies that foster productive activities, decent employment, entrepreneurship, creativity, and innovation, while strengthening micro, small, and medium-sized enterprises through financial access.

SDG Target 11: In another 5–7 years, through the bioleaching process reduces the negative per capita environmental footprint of cities, focusing on air quality and sustainable waste management systems.

SDG Target 12: In near future, we try to achieve a significant reduction in waste generation through prevention, reuse, recycling, repair, and recovery methods. This can be achieved with the bioleaching process (Fig. 7).

References

Ahluwalia SS, Goyal D (2007) Microbial and plant-derived biomass for removal of heavy metals from wastewater. Biores Technol 98(12):2243–2257

Ahmad I, Rehman F, Khan M (2017) Advances in microbial bioleaching: applications and challenges. J Environ Manage 203:1–12

Awasthi AK, Li J, Kumar M (2024) Microbial valorization of electronic waste: green recovery of critical metals. Environ Sci Pollut Res 31(12):18845–18860. https://doi.org/10.1007/s11356-023-29344-1

Behnamfard A, Salarirad MM, Veglio F (2013) Process development for recovery of copper and precious metals from waste printed circuit boards with emphasize on palladium and gold leaching and precipitation. Waste Manage 33(11):2354–2363

Bosecker K (1997) Bioleaching: metal solubilization by microorganisms. FEMS Microbiol Rev 20(3–4):591–604

Brandl H (2001a) Microbial leaching of metals from solid materials (bioleaching). Hydrometallurgy 59(2–3):319–325. https://doi.org/10.1016/S0304-386X(00)00188-X

Brandl H (2001b) Metal-biotechnology in the environment: microbe-metal interactions. Curr Opin Biotechnol 12(3):248–253

Brierley CL (2003) Bioleaching: progress in metal recovery from ores. Hydrometallurgy 71(1–2):1–14

Brierley CL, Brierley JA (2013) Progress in bioleaching: fundamentals and mechanisms of bacterial metal sulfide oxidation. Hydrometallurgy 139:103–123. https://doi.org/10.1016/j.hydromet.2013.08.013

Castro L, Roldán MD, Vera M (2024) Bioleaching and biorecovery of critical raw materials: editorial overview. Front Microbiol 15:1359842. https://doi.org/10.3389/fmicb.2024.1359842

Colmer JA, Zhang T, Johnson DB (2023) Thermophilic archaea in bioleaching operations: roles and adaptive mechanisms. Extremophiles 27(4):245–259. https://doi.org/10.1007/s00792-023-01235-0

Cozma P et al (2024) Bio-recovery of metals through biomining within the framework of sustainable development. MDPI J Environ 12:785

Das S, Li Y, Zhang Y (2023a) Microbial bioleaching of e-waste: advances in metal recovery using microbial consortia. J Hazard Mater 456:131280. https://doi.org/10.1016/j.jhazmat.131280

Das S, Kumar P, Singh R, Mukherjee A (2023b) Bioleaching of electronic waste by microbial consortia: current status and future prospects. Chemosphere 332:138870

Dey S, Dash N, Ray SK, Priyadarsini S, Chhotaray C, Das AP (2024) Occurrence and quantification of tire wear plastics pollutants from highway city dust of Hhubaneswar city of Odisha. Sustain Microb Technol Synth Cellulosic Microfiber Bioremediat, 77–89. https://doi.org/10.1007/978-3-031-62600-5_5

Dopson M, Johnson DB (2022) Acidophilic microorganisms in biomining and bioleaching. FEMS Microbiol Rev 46(6):34. https://doi.org/10.1093/femsre/fuac034

Dresher K, Schippers A, Sand W (2013) Microbial interactions in bioleaching systems. Hydrometallurgy 133:76–83

Falagán C, Roffey R, Nancucheo I (2024) Nutrient optimization and metabolic regulation in large-scale bioleaching systems. Front Microbiol 15:1420571. https://doi.org/10.3389/fmicb.2024.1420571

Gadd GM (2010) Metals, minerals and microbes: geomicrobiology and bioremediation. Microbiology 156(3):609–643

Garcia L, Thompson R (2020) Emerging technologies and their transformative potential. Springer

Gavrilescu M (2004) Biotechnology: a sustainable approach to environmental protection. Biotechnol Adv 22(3):157–183

Ghosh B, Ghosh MK, Parhi P et al (2015) Waste printed circuit boards recycling: an extensive assessment. J Clean Prod 94:5–19

Guo H, Min Z, Hao Y, Wang X, Fan J, Shi P, Min Y, Xu Q (2021) Sustainable recycling of $LiCoO_2$ cathode scrap on the basis of successive peroxymonosulfate activation and recovery of valuable metals. Sci Total Environ 759:143478
Harshan K et al (2025) Unveiling the potential of microbial biominers in sustainable metal recovery. Environ Sci Technol 59(2):1234–1245. https://doi.org/10.1021/es5056789
Jain S, Mehta D (2023) Bio-extraction of lithium and cobalt from spent Li-ion batteries using mixed microbial cultures. J Environ Chem Eng 11(9):110994. https://doi.org/10.1016/j.jece.2023.110994
Johnson DB (2014) Biomining—biotechnologies for extracting and recovering metals from ores and waste materials. Curr Opin Biotechnol 30:24–31. https://doi.org/10.1016/j.copbio.2014.05.003
Johnson P, Nguyen T, Lee H (2020) Artificial intelligence in healthcare: trends and applications. J Med Syst 44(5):88–97
Kirk RD, Hagan B, Falagán C (2024) Microbial recovery of lithium and rare-earth elements from ores and secondary resources. Front Microbiol 15:1389621. https://doi.org/10.3389/fmicb.2024.1389621
Kumar A, Holuszko M, Espinosa DCR (2017) E-waste overview on generation, collection, legislation and recycling practices. Resour Conserv Recycl 122:32–42
Lee S, Kim J (2022) Sustainable energy innovations for a carbon-neutral future. Renew Energy Rev 158:112–125
Li J, Lu H, Guo JI, Xu Z, Zhou Y (2007) Recycle technology for recovering resources and products from waste printed circuit boards. Environ Sci Technol 41(6):1995–2000
Maguyon MCC, Alfafara CG, Migo VP et al (2012) Recover of copper from spent solid printed circuit board (PCB) wastes of a PCB chemical deposition. J Environ Sci Manage 15:17–27
Meroni G, Panelli S, Zuccotti G, Bandi C, Drago L, Pistone D (2021) Probiotics as therapeutic tools against pathogenic biofilms: have we found the perfect weapon? Microbiol Res 12(4):916–937. https://doi.org/10.3390/microbiolres12040068
Montgomery DC (2017) Design and analysis of experiments, 9th edn. Wiley
Müller A, Schmidt B, Weber C (2019) Circular economy and waste valorization: innovations in practice. Environ Sci Technol 53(14):8056–8068
Nies DH (1999) Microbial heavy-metal resistance. Appl Microbiol Biotechnol 51(6):730–750
Patel R, Sharma V (2021) Precision agriculture and climate-smart innovations. Agric Syst 189:103–116
Priyadarsini S, Das AP (2024) Analytical and structural characterization of waste lithium-ion batteries for their effective recycling strategy. Environ Chem Ecotoxicol. https://doi.org/10.1016/j.enceco.2024.12.004
Priyadarsini S, Das AP (2025) Lithium bioleaching: a review on microbial-assisted sustainable technology for lithium bio-circularity. J Water Process Eng, 69, 106744. https://doi.org/10.1016/j.jwpe.2024.106744
Priyadarsini S, Das AP (2026) Lithium in waste water: impacts on pollution and human health. Decarbonization of Wastewater Pollutants as a Sustainable Solution, 353–364. https://doi.org/10.1016/B978-0-443-33062-9.00010-2
Priyadarsini S, Alok, Das P, Journal G, Das AP (2025) Isolation and identification of native lithium-resistant bacteria from electronic waste contaminated sites: a focused investigation on tolerance assessment with potential lithium bio-recovery application. Geomicrobiol J. https://doi.org/10.1080/01490451.2025.2539509
Rawlings DE (2002) Heavy metal mining using microbes. Annu Rev Microbiol 56:65–91
Rawlings DE, Johnson DB (2007) The microbiology of biomining: development and optimization of mineral-oxidizing microbial consortia. Microbiology 153(2):315–324. https://doi.org/10.1099/mic.0.2006/002052-0
Rawlings DE, Dew D, du Plessis C (2003) Biomineralization of metal-containing ores and the industrial applications of bioleaching. Trends Biotechnol 21(1):38–44. https://doi.org/10.1016/S0167-7799(02)00002-2

Rawlings DE, Johnson DB, Falagán C (2023) Advances in the understanding of bioleaching chemistry and microbial ecology. Miner Eng 197:108236. https://doi.org/10.1016/j.mineng.2023.108236

Saldaña M, Benavides D, Pizarro R (2023) Biofilm dynamics and extracellular polymeric substances in biomining systems. Microorganisms 11(5):1207. https://doi.org/10.3390/microorganisms11051207

Shen Y, Zhang R, Liu X (2020a) Microbial oxidation mechanisms in bioleaching systems. Hydrometallurgy 195:105372. https://doi.org/10.1016/j.hydromet.2020.105372

Shen Y, Guo J, Liang Y, Liu H (2020b) Functional diversity of acidophilic microorganisms in metal bioleaching. Front Microbiol 11:605889

Smith J, Brown D (2021) Interdisciplinary approaches to global challenges. Routledge

Su Y, Wang H, Chen L (2024) Industrial-scale bioleaching for electronic waste recycling: pilot studies in China and Europe. Waste Manage 180:12–25. https://doi.org/10.1016/j.wasman.2024.01.005

Tezyapar Kara I, Özdemir O, Şenol S (2023) Bioleaching of metal-bearing wastes: Sustainable recovery for circular economy. Environ Sci Pollut Res 30(45):97654–97672. https://doi.org/10.1007/s11356-023-28937-x

Tonietti L et al (2024) Unveiling the bioleaching versatility of Acidithiobacillus ferrooxidans: mechanisms and applications. Microorganisms 12(12):2407. https://doi.org/10.3390/microorganisms12122407

Tripathy B, Singh S, Behera ID, Mishra S, Das AP (2026) Metagenomic profiling of diversified marine microbiome across microplastic-contaminated niches of Bay of Bengal, India. Mar Pollut Bull, 222, 118872. https://doi.org/10.1016/J.MARPOLBUL.2025.118872

Vera M, Sabater M, Roldán MD (2022a) Fundamentals and mechanisms of microbial metal recovery in bioleaching. Appl Microbiol Biotechnol 106(14):5671–5689. https://doi.org/10.1007/s00253-022-12187-5

Vera M, Schippers A, Sand W (2022b) Progress in bioleaching: fundamentals and applications of microbial mineral oxidation. Biotechnol Adv 60:107978. https://doi.org/10.1016/j.biotechadv.2022.107978

Vieira RHSF, Santelli RE, Oliveira LCA (2011) Recovery of metals using microorganisms: an overview. Environ Technol 32(12):1301–1310

Volesky B (2003) Biosorption of heavy metals. Biotechnol Prog 11(3):235–250

Yin S, Xu Y, Chen Y, Huang J (2022) Fungal bioleaching of rare earth elements from industrial waste: mechanisms and perspectives. J Clean Prod 350:131442

Analysis of the Activity of Microbes in the Bioleaching Approach that Executes Recycling and Metal Recovery

B. Jyotirmayee, Ipsita Priyadarsini Samal, Sameer Jena, and Gyanranjan Mahalik

Abstract Bioleaching refers to the effectiveness of microbes such as bacteria, fungi, and yeast, converting solid metals into soluble extractable forms using advanced bioprocesses. It has emerged as a green, sustainable technique for metal recovery that has a low impact on climate change. It serves as an environmentally conscious substitute for recovering metals from solid substances, which include agricultural, municipal, and industrial waste, thereby advancing cycle closures and, ultimately, environmentally friendly approaches. There is a probability to upgrade the metal bioleaching effectiveness through innovative and sustainable technological methods for e-waste administration. The biomass produced can be utilized as soil improvers, as well as in biosorption/desorption approaches for metal concentration and extraction. It can be a critical component in transitioning to an ecologically friendly, circular economic system by providing productive, inexpensive, and environmentally friendly techniques for recovering metals from solid materials. Cyanogenic bioleaching (biocyanidation) has significant capability, but challenges must be addressed before it can be scaled up from the lab to the commercial level. Restoration and recycling of base and essential materials from electronic waste not only lessens the withdrawal of these elements from natural sources, but also lower contamination caused by hazardous chemicals released when electronic waste is improperly disposed of. The finding of new leaching microbes, the development of biogenomics, and the integration of biocyanidation give new capabilities for producing heavy-metal industrial products from bioleaching tailings. The prime future development goal is to expand industrial production of heavy metal recoveries through bioleaching of mine tailings, continually strengthening the process, and attaining greater economic and ecological benefits.

Keywords Bioleaching · Sustainable · Metal recovery · Cyanogenic bioleaching · Biogenomics · Biocyanidation

B. Jyotirmayee · I. P. Samal · S. Jena · G. Mahalik (✉)
Department of Botany, School of Applied Sciences, Centurion University of Technology and Management, Bhubaneswar, Odisha, India
e-mail: gyanranjan.mahalik@cutm.ac.in

A. P. Das and S. Priyadarsini (eds.), *Recovery and Recycling of Critical Metals from Electronic Waste for Environmental Sustainability*, Earth and Environmental Sciences Library, https://doi.org/10.1007/978-3-032-29064-9_4

1 Introduction

Health emergencies across nations have recently led the way to temporary smelter shutdowns for maintenance and upgrades, lowering global mining sector output (Saldana et al. 2023). Microorganisms have recently gained popularity as a replacement for traditional leaching approaches in mineral treatment. This is because they improve the economic benefits of low-grade mining deposits, do not cause emissions or pollution, save energy, have low process costs, are environmentally friendly, and are congruent with products (Manesh et al. 2024). One of the newer techniques of mineral leaching, bioleaching, involves recovering inorganic metals from aqueous solutions using microorganisms. It involves using specific bacteria or fungi to recover valuable resources from a mine, ultimately generating soluble metal in an acid blend (Satapathy and Behera 2025). The procedure of removing metals from ores by employing naturally occurring components, such as air, water, and microbes, describes bioleaching (Manesh et al. 2024).

Microorganisms release vital metals from ores, offers a substitute for harmful mining techniques, particularly for difficult-to-extract sediment (Petersen 2023). It can aid the solubilization and removal of selected metals from complex mineral sources by catalyzing biochemical changes (Vera et al. 2022). Copper is broadly used compared to others. Directly and indirectly bioleaching is two ways to accomplish this process (Saldana et al. 2023). It is a business execution that uses naturally occurring microbes, such as bacteria and archaea, to oxidize minerals and treat harmful metals and metalloids in mining waste and remediate areas (Manesh et al. 2024). It is considered the gold standard for removing valuable metals from e-waste because it has a metal separation rate and has minimal environmental effect (Waghmode et al. 2021). It is also cost-effective for lowering raw material procurement requirements and preventing non-renewable resources degradation (Adetunji et al. 2023).

Organic compounds and heavy metal waste are on the rise due to the expansion of industries and emerging technologies (Naseri et al. 2023). The increase in the scrapping of sophisticated electronic devices leads to environmental pollution (Mokarian et al. 2022). In the recent era, there has been a rise in disquiet over the harmful impact of metals on polluted environments (Naseri et al. 2023). The proper functioning of electronic pollutants has appeared as a major challenge in both developed and developing countries (Waghmode et al. 2021). The recovery of metal-bearing wastes becomes difficult due to their complex and refractory nature and the unavailability of economical or environmentally friendly recovery methods (Yaashikaa et al. 2022; Tezyapar Kara et al. 2023). Consequently, environmentally friendly and economically viable techniques for recovering metals from wastes must be developed (Tezyapar Kara et al. 2023). Reductive bioleaching is worth recovering metals from industrial and mining waste (Roberto and Schippers 2022). There is a need to recover heavy metals from their various sources to secure and maintain a sustainable future (Fig. 1).

Bioleaching provides an intriguing innovation with economic benefits and environmental friendliness for recovering hazardous metals from contaminated soil. It

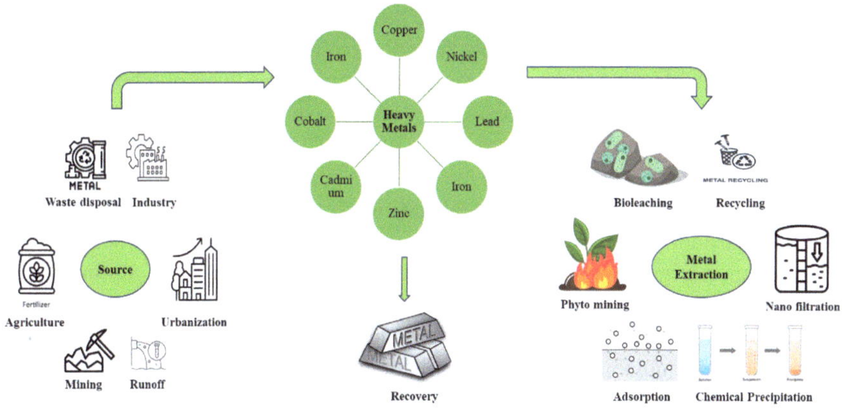

Fig. 1 Essential heavy metals recovery techniques from different sources

can increase the environmental remediation and produce new strains for large-scale bioleaching of polluted soil when combined with omics (Das and Ghosh 2023). This detailed study of microbial activity in recovering metals through bioleaching outlines the most up-to-date techniques, functions, and mechanisms implicated in recycling and metal recovery. Furthermore, the analysis of the different microorganisms involved, their operation, the variables influencing each bio-process, and aspects related to their scalability and industrial administration. Recent biotechnological approaches, with their new aspects, capabilities and research avenues in this emerging field of bioleaching will influence future studies.

2 Worldwide Current Scenario

Biomining, a new and promising technique for metal recovery, has been gaining popularity in recent decades (Panyushkina and Muravyov 2025). Resource recovery is a significant factor in both short-term cost savings and long-term resource efficiency. Bacteria and fungi played a significant role in modern biotechnological developments that made bioleaching a reality (Devi and Ganesh 2024). Limitations should be addressed, and the focus should be on developing this process for industrial usage in future studies. The dumping of solar panel waste has risen significantly as photovoltaic (PV) technology has become more widespread and the push towards renewable energy sources has increased. Improper disposal of solar panel waste poses a threat to human and ecological health (Johnston et al. 2026).

The combination of contemporary acidophilic microbiology with sustainable biotechnologies provides a novel, ecofriendly and low-carbon approach for treating sulfide-rich raw materials, such as industrial waste. Acidophilic bioleaching can increase the removal of nickel, a non-ferrous metal and has promising future administrations in mineral processing, which can decrease environmental issues through enhanced bioprocesses and secure mine-waste recycling (Panyushkina and Muravyov 2025). An environmentally friendly, circular economy might be built on bioleaching, which could make solar panel waste recycling a fundamental component (Johnston et al. 2026). It is a promising new alternative because it utilizes microbes' inherent capability to dissolve Rare Earth Elements across a variety of industrial waste streams (Zhao et al. 2025).

Ever-increasing waste from electrical and electronic equipment (WEEE) contains precious substances that can be studied to develop sustainable processes for their exploitation. An innovative bioleaching method for optimizing urban mining strategies is the recovery of Copper from Printed Circuit Boards (PCBs). It has the potential to reduce the carbon footprint of bioleaching processes by more than 50% by incorporating sustainability assessment, mathematical frameworks, and experiments (Becci et al. 2025). Factors such as nutrient availability and microbial activity can limit the efficiency of these mechanisms. Genetically modifying microbes will be required to optimize growth conditions, increase REE production and fix efficiency problems. Microbes can be genetically engineered to be more resistant to harmful metals, thereby making them even more productive in harsh environments. A detailed analysis of microbial consortia and the complementary roles of different microbial strains could yield insights to increase the efficiency (Joshi et al. 2025).

Bioleaching is a crucial part of sustainable resource application for metal recovery. It has grown in popularity as it is less environmentally harmful for extracting metals from low-grade ores. Sustainable mining applications can be greatly influenced by advancements in these methods (Naaz et al. 2024). As global sewage sludge production increases, demand for efficient administration increases rapidly. Sulfur-oxidizing bacteria are proposed as an essential part of this method for effectively removing heavy metals (HMs). Sustainable sludge administration was achieved by reducing metal toxicity in the resulting sludge (Pooja and Philip 2025). In India and other nations, agricultural biomass has excellent promise as a renewable energy source. Recovery of Potassium from biomass ash is an economically viable alternative that has a positive current value and an acceptable payback period. Biogenic soluble Potassium is used as a partial substitution for traditional K fertilizer in agricultural fields, since it is readily absorbed by plants (Gahane and Mandavgane 2024).

It is an excellent technique for waste control, as it manages waste products and decreases environmental contaminants. Intractable metals can be regenerated and effectively dissolved using this new eco-friendly technology. Metals can be recovered from numerous sources such as commercial trash, polluted soils, mineral-rich ores, mining tailings, and electronic waste (Mathew et al. 2024). The method of bioleaching phosphorus from waste biomass involves microorganisms that store polyphosphate in their cells. Microorganisms undergo biological phosphorus removal when they take up more amount of phosphorus. In the next step, magnesium salt was added to the water, forming a precipitate of struvite, which simultaneously removes ammonia and phosphates (Roy and Vurimindi 2024).

3 Bioleaching and the Role of Microbes

Low carbon emissions, cleaner protection, low cost, and high efficiency are among the benefits of bioleaching, which has gained significant attention. Table 1 lists various microbial strains which involved in the bioleaching technique for metal recovery. A more environmentally friendly alternative to traditional mining processes is the potential for beneficial relationships between plants and bacteria. Analyzing the interplay between plants and bacteria decreases environmental burden and, by extension, the worldwide decline in high-grade ore (Samson et al. 2025). A practical, environmentally friendly approach for extracting rare earth elements is a key direction in the field. The biogenic lixiviant, which consists of the produced and purified metabolites, allows for the efficient recovery of REEs (leaching efficiency > 90%) (Zhao et al. 2024). When acidophiles are used in bioleaching, sulfuric acid and ferric (Fe III) acid are produced. Cyanogenic microbes, in instance, may produce hydrogen cyanide, which they use to extract metals (Poornima et al. 2024).

In recent years, the Gram-negative bacterium *Acidithiobacillus ferrooxidans* has become an integral part of biomining, bioleaching, trace element bioleaching, and the biorecovery sector as a whole (Tonietti et al. 2024). In direct bioleaching, *Aspergillus niger* responded to environmental objections by regulating two processes, namely the one-step metabolism of tryptophan and the two-step production of secondary metabolites (Zhao et al. 2025). In current years, bio-hydrometallurgical leaching has gained popularity as a viable substitute for conventional chemical extraction techniques. The two most common fungal species used in microbial bioleaching are *Aspergillus niger* and *Penicillium simplicissimum*. This can disintegrate metals from ores or waste materials by converting glucose and sucrose into various organic acids (Nnaemeka et al. 2025).

Many types of microbes, such as autotrophic, chemolithotrophic, and mixotrophic bacteria, are used, which are well-adapted to different environments and metals. In mining bioleaching, microbial consortia (groups of interdependent bacteria) are widely used. The productivity of metal recovery can be increased by these consortia, which can adapt to new environments (Mora et al. 2025). The most effective strains for Rare Earth Elements leaching were *Aspergillus niger* and *Aspergillus fumigatus*, followed by *Bacillus subtilis*, *Candida lipolytica*, and *Penicillium pinophilum* (Zhou et al. 2025). Microbial populations and bioleaching efficacy can be improved by incorporating sterile medical waste. This was mediated by *Acidithiobacillus ferrooxidans* and *Leptospirillum ferriphilum* (Chen et al. 2025).

Bioleaching of anaerobic digestion and sewage sludge with native sulfur-oxidizing strains can be highly cost-effective. Autotrophic *A. thiooxidans* growth and activity can be moderated by maintaining pH (lowering sulfur supply); thus, maintain a balance between heavy metal removal and dewatering performance (Xu et al. 2025). The recovery process relies on various microbial strains, and the bacterial makeup of sulfide-associated habitats is considered a vital part in biomining and bio-oxidation procedures (Aimagambetov et al. 2025). In recent years, *Bacillus thuringiensis*, *Mycobacterium caseolyticus*, and *Candida funkei* strains have been utilized in the

Table 2 Various processes of bioleaching for effective metal recovery

Sl. no	Methods	Field	Model	Recovery rate	Recovery
1	Machine learning response surface methodology	Artificial intelligence	Regression-based ML algorithms	77%	Copper, Gold
2	Bacterial bioleaching	Biotechnology	*Acidithiobacillus ferrooxidans*, *Acidithiobacillus thiooxidans*	–	Iron, Copper, Cadmium, Lead
3	Fungal bioleaching	Biotechnology	*Aspergillus niger*, *Penicillium simplicissimum*	–	Aluminum, Lead, Vanadium, Iron, Zinc, Copper
4	Biohydrometallurgy	Microbial leaching	*Thiobacillus ferrooxidans*, *Thiobacillus thiooxidans*	95%	Copper, Manganese, Aluminum
5	Microbial biominers	Microbial mediated	Acidophilic ferrous iron-oxidizing bacteria, Extremophilic microalgae strains (*Chlamydomonas*)	100%	Copper, Cobalt, Aluminum, Zinc
6	Recycling of electronic waste	Microbial recycling	–	–	Copper, gold
7	Tailing	Tailing bioleaching	*Leptospirillum ferriphium*	5.46%	Copper
8	Lithium-ion battery (LIB)	Microbial mediated	*Acidithiobacillus ferrooxidans*, *Leptospirillum ferrooxidans*	75%	Cobalt, manganese
9	Redoxolysis, Acidolysis, Complexolysis	Microbial mediated	*Fusarium oxysporum*, *Acidithiobacillus ferrooxidans*, *Aspergillus niger*	–	Mickel, Copper, Zinc, Manganese
10	Cyanogenic bioleaching	Microbial mediated	*C. violaceum*, *Bacillus megaterium*, *P.* aeruginosa, *P. brassicacearum*	75.7%	Gold, Platinum, Palladium
11	Chalcopyrite leaching	Microbial activity	*Pseudomonas*, *Chromobacterium*	94–97%	Platinum, Palladium
12	Heap bioleaching	Acidophilic autotrophic iron-oxidizing microbes	*A. ferrooxidans*, *L. ferrooxidans*, *A. caldus*, *A. thiooxidans*	–	Copper, Gold, Uranium
13	Heterotrophic bioleaching	Microbial mediated	*Pseudomonas* spp., *Bacillus mageterium*, *Aspergillus* sp., *Penicillium* sp.	66–78%	Iron, Copper

(continued)

Table 2 (continued)

Sl. no	Methods	Field	Model	Recovery rate	Recovery
14	Chemolithotrophic bioleaching	Microbial mediated	Chemolithotrophs (*A. ferrooxidans*, *A. thiooxidans*)	93–94%	Copper, Iron, Nickel, Cobalt
15	Magnetic field-assisted bioleaching	Microbial mediated	*A. ferrooxidans*, *Acidithiobacillus thiooxidans*, *Aspergillus niger*	> 80%	Lithium, Nickel, Cobalt
16	Gold biosorption with brewer's yeast	Microbial mediated	*Acidithiobacillus ferrooxidans*, *Leptospirillum ferrooxidans*, *Leptospirillum ferriphilum*	98%	Gold
17	AI-Driven RGB-Based	Artificial intelligence	RGB sensor system	84%	Copper
18	Indirect bioleaching	Microbial mediated	*A. thiooxidans*,	54–60%	Cobalt, Nickel
19	Kaolin bioleaching	Microbial mediated	*Clostridium pasteurianum*, *Lysinibacillus*	–	Iron

ferrooxidans and *Acidithiobacillus thiooxidans* are the most common bacteria used in this process. Moderate thermophiles such as iron-oxidizing *Leptospirillum ferriphilum* and sulfur-oxidizing *Acidithiobacillus caldus* are also commonly used (Priyadarsini and Das 2024, 2026). Fungi such as *Penicillium chrysogenum*, *Penicillium simplicissimum,* and *Aspergillus niger* accelerate metal dissolution in either acidic or nearly neutral environments (Roy et al. 2023). It is an economical implementation of bio-hydrometallurgy by recovering valuable metals from low-grade ores with low metal concentrations and sulfide minerals. It can effectively recover excess metals from sediments and soil other than solid waste (Kour et al. 2021; Abdalla et al. 2023).

Microbial bioleaching using acidophilic ferrous iron-oxidizing bacteria and subsequent metal biouptake using extremophilic microalgae strains signify the viability of a biotechnological tool for e-scrap metal recovery. Extremotolerant bacteria can be used to fully valorize e-scraps, providing hope for a microbial sequential mechanism (García-Balboa et al. 2022). The interconnection between CirBioeco and e-waste recycling through microbial mechanisms enables a new eco-friendly biotechnological method for e-waste (Ilyas et al. 2021).

In the tailings bioleaching process, copper recovery and bacterial concentration can be improved, and excess harmful ions (SO_4^{2-} and Fe^{3+}) can be eliminated from the mixture by infusing tailings and bacteria at proper intervals (Chen et al. 2022). High metal dissolution was attained with the acidophilic mixed culture. Manganese and cobalt can be dissolved using both contact and noncontact bioleaching methods (Panda et al. 2024). Mesophilic fungus *Fusarium oxysporum* can bioleach stable, crystalline, fluorescent, water-soluble, and extracellularly distributed protein-capped silica nanoparticles from fly ash. Bioleaching of nanoparticles from electrical waste

Table 3 List of heavy metals derived from various sources causing serious health risks (Shokoohi et al. 2025; Zhang et al. 2025a, b; Halder et al. 2025; Acar et al. 2025; Pang et al. 2025; Roy et al. 2025; Kondaiah et al. 2024; World Health Organization 2025; Hassan et al. 2025; Munusamy et al. 2025; Kolandhasamy et al. 2024; Punia et al. 2025; Brown et al. 2024; Sandwal et al. 2025)

Sl. no	Heavy metals	Symbol	Sources	Health hazards
1	Aluminum	Al	Sewage sludge, Printed circuit board, Municipal solid waste incineration, Silicon photovoltaic cells	Alzheimer's disease, Cancer, Kidney dysfunction, Anemia
2	Lead	Pb	Sewage sludge, Coal tailings, Fly ash, Printed circuit board	Brain damage, Abdominal pain, Hypertension, and Kidney problems
3	Copper	Cu	Sewage sludge, Coal tailings, Fly ash, Mixed electroplating sludge, Printed circuit board, Municipal solid waste incineration	Abdominal pain, Diarrhea, Skin irritation, Respiratory issues
4	Zinc	Zn	Sewage sludge, Coal tailings, Fly ash, Mixed electroplating sludge, Printed circuit board, Municipal solid waste incineration	Nausea, Diarrhea, Headaches, Gastrointestinal discomfort
5	Chromium	Cr	Coal tailings, Fly ash, Mixed electroplating sludge, Low-grade ores, Municipal solid waste incineration	Cancer, Respiratory issues, Skin irritation, and Kidney and liver damage
6	Nickel	Ni	Coal tailings, Mixed electroplating sludge, Printed circuit board, Low-grade ores, Municipal solid waste incineration, Lithium-ion batteries	Cancer, Respiratory issues, Reproductive issues, and Kidney and liver damage
7	Cadmium	Cd	Fly ash, Municipal solid waste incineration, Industrial batches	Muscle pain, Lung damage, Cancer, and Bone health issues
8	Cobalt	Co	Printed circuit board, Mining	Cancer, Respiratory issues, Skin and eye irritation, Liver and kidney congestion
9	Iron	Fe	Printed circuit board, Mining	Liver damage, Cancer, Sickle cell disease, Cirrhosis
10	Silver	Ag	Printed circuit board, Silicon photovoltaic cells, Mining	Eye irritation, Kidney and liver damage, Respiratory problems, Renal problems
11	Manganese	Mn	Municipal solid waste incineration, Lithium-ion batteries	Respiratory issues, Organ damage, Neurological disorder, Manganism

(continued)

Table 3 (continued)

Sl. no	Heavy metals	Symbol	Sources	Health hazards
12	Gold	Au	Mining	Eye and skin irritation, Respiratory issues, Organ damage, Musculoskeletal Disorders
13	Lithium	Li	Lithium-ion batteries, Mining	Renal failure, Cardiac complications, Neurological disorder, Lethargy

and tailing ores has gained attention in sustainable development and the circular bioeconomy (Jaiswal et al. 2025).

Cyanogenic bioleaching of PGM mixture using *P. brassicacearum* results in 75.7% gold extraction. This systematic mechanism for UG2 (Upper Group 2 Reef) PGM aggregates begins with base metal bioleaching, followed by cyanogenic bioleaching of valuable metals (Kumar et al. 2025). Microbial usage is efficient for chalcopyrite leaching and bio-oxidation of gold-containing minerals. A remarkable reduction in carbon footprints resulted from sustainability in process metallurgy and utilization of green biotechnology in hydrometallurgy (Ilyas et al. 2024). The industrialization of uranium ores, gold, and primary copper (secondary ore) is among the few mechanisms currently involved in heap bioleaching. The evolution and commercialization of heap bioleaching on conceptual, computational, and experimental platforms depend on multidisciplinary cooperation (Jia et al. 2024).

Heterotrophic and chemolithotrophic microbes are the two main types that are involved in bioleaching. Cyanogenic bacteria *Bacillus megaterium* and heterotroph *Aspergillus niger* both produce secondary metabolites that immobilize metals by converting organic carbon sources (Erkmen et al. 2025). The leaching kinetics are much improved, and the used LIBs can be recycled in an environmentally acceptable manner using the suggested magnetic-field-assisted bioleaching process with *A. ferrooxidans*. An external magnetic field can positively influence *A. ferrooxidans* metabolism, opening up new possibilities for genetic engineering of acidophile bacteria with the express purpose of magnetic enhancement and for bioleaching engineering in general (Kim et al. 2024). The complete valorization of PCBs derived from heavily metal-containing laptops was achieved using an integrated technology that involved bioleaching base metals, followed by gold recovery via biosorption on brewer's yeast, and chemical leaching with aqua regia. Gold extraction was achieved selectively via biosorption with brewer's yeast, with an efficiency of 98% (Adam et al. 2024).

By incorporating data from RGB sensors and machine learning, AI-driven optimization enables continuous monitoring and predictive maintenance, with a focus on bioleaching to recover copper from old electronics (Mishra et al. 2022; Ghosh and Das 2018). It provides a web-based, open-source digital twin designed to track the bioleaching plant. Real-time data analysis and historical records enable predictive maintenance. It boosts recovery rates, reduces environmental problems, and promotes sustainable resource administration in e-waste recycling industries at various operational scales (Vives Pons et al. 2024). Nickel and cobalt can be removed from superalloy residual powders (extremely corrosion-resistant) using sustainable bioleaching, with minimal cost and energy requirement (Constantin et al. 2025). Microbial bioleaching provides environmental remediation and commercial mineral processing benefits, and is highly effective for kaolin whitening. It provides a substantial enhancement in kaolin quality, selectively recovering iron and conserving crystalline integrity, resulting in increased whiteness and decreased Iron concentration (Li et al. 2025).

5 Conclusion

Recent years of environmental depletion require sustainable mining and pollution cleanup for the future of the green economy, which depends on rising microbial activity. Environmental problems and restoration pose a significant challenge worldwide. Bioleaching is a low-cost and environmentally friendly approach that is currently an area of extensive research for metal recovery. Microorganisms such as *Aspergillus niger* and acidophilic bacteria enable efficient recovering without damaging mineral substances, which solubilize metals through direct enzymatic action and the production of biodegradable substances. There has been advancement in bioengineering and technical engineering, which has helped the here and now, but there is still room for enhancement in future opportunities. Scientific and technological advancements in biology and engineering may soon give answers to pressing questions about how to extract vital minerals from unconventional raw materials. Bioleaching effectiveness can be enhanced through microbial adaptation and the careful selection of process parameters.

6 Future Prospectus

Metals removed from inorganic waste can be recovered in the future through bioleaching, a prominent, environmentally friendly, and essential method. The field of spent Lithium-ion battery leaching could see new possibilities through additional research into genetically modified microbes that create strains more effectively at extracting metals, finding cheaper nutrient sources, designing better bioreactors, and identifying novel, metal-tolerant microbial strains. New approaches for visualizing inter- and intraparticle flow, as well as the development of suitable biological models with fully optimized parameters, are necessary. To efficiently and cleanly recover rare earth ore from ion-adsorption, bio-leaching is a promising future shift. The potential uses of *A. ferrooxidans* in space travel are rather intriguing. With its remarkable adaptability and capability to support metal extraction on Earth, it provides a revolutionary opportunity to use resources on other planets in the future. Soil remediation, low-grade ore processing, and metal scrap recovery are among bioleaching's current limitations. Future implementation requires robust industrial mechanisms and copper and gold recovery from substandard ores. Future studies should emphasis increasing bioleaching scalability and efficacy by upgrading microbes through the integration of genetic mutation with an advanced fermentation method.

Abbreviations

E-waste	Electronic waste
PV	Photovoltaic Technology
WEEE	Waste from Electrical and Electronic Equipment
Cu	Copper
PCBs	Printed Circuit Boards
REE	Rare Earth Element
HMs	Heavy Metals
K	Potassium
BPR	Biological Phosphorus Removal
EPS	Extracellular Polymeric Substances
Pb	Lead
Zn	Zinc
ML	Machine Learning
Fe	Iron
SO_4	Sulphuric acid
PGM	Platinum Group Metals
UG2	Upper Group 2 Reef
LIBs	Lithium-Ion Batteries
RGB	Red Green Black
AI	Artificial Intelligence

References

Abdalla AM, Abdullah MF, Dawood MK, Wei B, Subramanian Y, Azad AT, Azad AK (2023) Innovative lithium-ion battery recycling: sustainable process for recovery of critical materials from lithium-ion batteries. J Energy Storage 67:107551

Acar İ, Gülmez Ö, Barış Ö, Başman G (2025) Sustainable nickel extraction from low-grade ores via trametes versicolor: a bioleaching case study on olivine and serpentine. J Sustain Metall, 1–14

Adam F, Piret A, Canonne M, Decock R, Nicolay X (2024) Complete recycling of printed circuit boards: from base metal bioleaching in a semi-continuous reactor with dual regulation to gold biosorption with brewer's yeast. Hydrometallurgy 223:106190

Panyushkina A, Muravyov M (2025) Biomining nickel for a greener low-carbon future: progress in bioleaching technologies and acidophilic microbiology. Reviews Environ Sci Bio/Technol, 1–51

Petersen J (2023) From understanding the rate limitations of bioleaching mechanisms to improved bioleach process design. Hydrometallurgy 221:106148

Pooja P, Philip L (2025) Bioleaching and chemical fractionation of heavy metals from municipal and industrial dewatered sewage sludge (No. EGU25–9260). Copernicus Meetings

Poornima R, Poornachandhra C, Karthikeyan G, Kannan TG, Karthika S, Keerthana S, Dhevagi P (2024) Role of microbes and microbiomes in bioleaching and bioremediation for polluted ecosystem restoration. In: Microbes as agents of change for sustainable development, pp 29–64. Bentham Science Publishers

Priyadarsini S, Das AP (2024) Analytical and structural characterization of waste lithium-ion batteries for their effective recycling strategy. Environ Chem Ecotoxicol. https://doi.org/10.1016/j.enceco.2024.12.004

Priyadarsini S, Das AP (2025) Lithium bioleaching: a review on microbial-assisted sustainable technology for lithium bio-circularity. J Water Process Eng 69:106744. https://doi.org/10.1016/j.jwpe.2024.106744

Priyadarsini S, Das AP (2026) Lithium in waste water: impacts on pollution and human health. Decarbonization of Wastewater Pollutants as a Sustainable Solution, 353–364. https://doi.org/10.1016/B978-0-443-33062-9.00010-2

Punia A, Bharti R, Kumar P, Kumar PV (2025) Evaluation of iron, zinc, and manganese sources and associated health risks in groundwater of the mining-impacted semi-arid Udaipurwati Granite Region, Rajasthan, India. Sustain Water Res Manag 11(3), 1–13

Rendón-Castrillón L, Ramírez-Carmona M, Ocampo-López C, Gómez-Arroyave L (2023) Bioleaching techniques for sustainable recovery of metals from solid matrices. Sustainability 15(13):10222

Roberto FF, Schippers A (2022) Progress in bioleaching: part B, applications of microbial processes by the minerals industries. Appl Microbiol Biotechnol 106(18):5913–5928

Roy JJ, Zaiden N, Do MP, Cao B, Srinivasan M (2023) Microbial recycling of lithium-ion batteries: challenges and outlook. Joule 7(3):450–456

Roy C, Vurimindi H (2024) Bioleaching of municipal wastes: recovery of phosphorus. In: Advanced bioseparation of industrial wastes, pp 220–230. CRC Press

Roy S, Vishwakarma A, Hait S (2025) Enhanced bioleaching of spent lithium-ion battery cathode powder by *Acidithiobacillus ferroxidans* for selective metal recovery: influence of ferrous iron and elemental sulfur as energy sources. Separat Purificat Techn, 134165

Saldaña M, Jeldres M, Galleguillos Madrid FM, Gallegos S, Salazar I, Robles P, Toro N (2023) Bioleaching modeling—A review. Materials 16(10):3812

Samson OJ, Popoola BM, Olanbiwoninu AA (2025) Synergistic plant-microbe interactions in biomining and bioleaching: implications for sustainable resource recovery. Microbial Allies, 83–107

Sandwal SK, Jakhar R, Styszko K (2025) E-Waste challenges in India: environmental and human health impacts. Appl Sci 15(8):4350

Satapathy SS, Behera R (2025) Concept and application of bioleaching and biomining. Adv Environ Biotechn Solut Global Chall 143

Sethi S, Jonwal H, Parihar R (2025) Metal solubilization through bioleaching: microbial action. In: Extraction of precious metals from industrial waste using microbial technology, pp. 21–41. Cham: Springer Nature Switzerland

Shokoohi R, Najafi-Vosough R, Shabanloo A, Torkshavand Z, Mahmoudi MM, Abasi MA (2025) Removing heavy metals and improving the dewaterability of sewage sludge with the bioleaching process by *Thiobacillus ferrooxidans* bacteria. App Water Sci 15(2):34

Tezyapar Kara I, Kremser K, Wagland ST, Coulon F (2023) Bioleaching metal-bearing wastes and by-products for resource recovery: a review. Environ Chem Lett 21(6):3329–3350

Tonietti L, Esposito M, Cascone M, Barosa B, Fiscale S, Muscari Tomajoli MT, Giovannelli D (2024) Unveiling the bioleaching versatility of *Acidithiobacillus ferrooxidans*. Microorganisms 12(12):2407

Veeramalini JB, Muthukumar PV, Indhuja S (2025) Recycling e-waste by co-culture bioleaching process. Cuestiones de Fisioterapia 54(3)

Vera M, Schippers A, Hedrich S, Sand W (2022) Progress in bioleaching: fundamentals and mechanisms of microbial metal sulfide oxidation–part A. Appl Microbiol Biotechnol 106(21):6933–6952

Vives Pons J, Comerma A, Escobet T, Dorado AD, Tarrés-Puertas MI (2024) Optimizing bioleaching for printed circuit board copper recovery: an AI-driven RGB-based approach. Appl Sci 15(1):129

Waghmode MS, Gunjal AB, Patil NN (2021) Bioleaching of electronic waste. Pollution 7(1):141–152

Wang J, Cui Y, Qian C, Zhang S, Guan M, Wen L, Xin B (n.d.) Bioleaching of heavy metals from the municipal solid waste incineration fly ash in membrane bioreactor and microbial community analysis. Available at SSRN 5358796

World Health Organization (2025) Exposure to lead: a major public health concern. World Health Organization, Preventing disease through healthy environments

Xu S, Jiang Y, Zou R, Zhu X, Liu H (2025) Bioleaching process of sewage sludge and anaerobically digested sludge via indigenous sulfur-oxidizing bacteria to improve dewaterability and reduce heavy metal content. Fermentation 11(6):321

Yaashikaa PR, Priyanka B, Kumar PS, Karishma S, Jeevanantham S, Indraganti S (2022) A review on recent advancements in recovery of valuable and toxic metals from e-waste using bioleaching approach. Chemosphere 287:132230

Zhang L, Li Z, Fan Y, Ye Z, Zeng F, Ren L, Zhang J (2025a) Leaching behaviour and chemical speciation changes of heavy metals in mixed electroplating sludge during ferrous oxidation-assisted two-stage bioleaching. Miner Eng 229:109371

Zhang Q, Yan X, Peng Y (2025b) Heavy metal toxicity in coal tailings: implications for the bioleaching of effective silicon by *Bacillus mucilaginosus*. J Environ Chem Eng 13(1):115098

Zhang Q, Ma L, Peng Y, Yan X (2024) Sustainable bioleaching of heavy metals from coal tailings using Bacillus inaquosorum B. 4: mechanistic insights and environmental implications. J Environ Chem Eng 12(5):113400

Zhao Y, Zhao H, Shen L, Qiu G, Wang Y (2024) Study on the role of microbial metabolites in in-situ noncontact bioleaching of ion-adsorption rare earth ore. J Environ Manage 368:122184

Zhao Y, Zou K, Meng X, Shen L, Qiu G, Wang Y, Zhao H (2025) Study on bioleaching methods and microbial-mineral interaction of ion-adsorption type rare earth ore. J Environ Manage 382:125422

Zhou H, Shao S, Yu X, Huang M, Qiu G, Chen Z, Shen L (2025) Evaluation of microbial metabolites in the bioleaching of rare earth elements from ion-adsorption type rare earth ore. J Environ Chem Eng, 117851

Recovery and Recycling of E-Waste: Closed-Loop Bioleaching Processing for Enhanced Critical Metal Recovery

Kamalesh Sen

Abstract The exponential growth of electronic waste represents two problems. Firstly, electronic waste can cause environmental contamination. Secondly, it can result in the loss of many precious metals such as gold, copper, cobalt and rare earth elements. Traditional pyrometallurgical and hydrometallurgical methods for recovering metals are efficient but typically require substantial energy and contribute to environmental problems. Bioleaching is an eco-friendly and low-impact biotechnological process that extracts valuable metals from the complex matrices of electronic waste. This review discusses recent advances made in closed-loop bioleaching systems as a means to recover metals in a more sustainable manner with a focus on the use of acidophilic bacteria and fungi that solubilise metals through biooxidation and complexation. In addition, closed-loop bioleaching systems provide a means to enhance resource recovery, minimise waste and improve process efficiency through combining methods of metal recovery from the bioleachate, reusing the microbes involved and recycling the substrates used to produce the bioleachate. Finally, the review also offers comparisons of the microbial strains, the operational parameters of each strain, and their suitability for large-scale application. To facilitate the optimum take-up of bioleaching in a circular economy system, a number of techno-economic and regulatory considerations must be addressed, including the inclusion of closed-loop bioleaching as an emerging form of electronic waste recycling and recovery of important metals, as well as the introduction of green technologies within the closed-loop bioleaching process. In addition, closed-loop bioleaching supports and strengthens waste-to-value and circular economy principles.

Keywords E-waste · Bioleaching · Critical metals · Closed-loop recycling · Circular economy

---

K. Sen (✉)
Department of Environmental Science, Syamsundar College, Shyamsundar, Burdwan, India
e-mail: ksen.envs@gmail.com

A. P. Das and S. Priyadarsini (eds.), *Recovery and Recycling of Critical Metals from Electronic Waste for Environmental Sustainability*, Earth and Environmental Sciences Library, https://doi.org/10.1007/978-3-032-29064-9_5

1 Introduction

As technological advancements in the form of information and communication technologies (ICTs), digitalisation and rapid technological growth have helped create a new social landscape, they have also presented us with a significant environmental challenge with respect to electronic waste (E-waste) (Islam et al. 2025; Quinto et al. 2025Rajesh et al. 2022a). There has been an increase in discarded electrical (E&E) and electronic equipment around the world as a result of shorter individual product lifetime, increasing user upgrades and implementing of planned obsolescence as well as an increase in consumer demand for new products (Grandhi et al. 2024). In fact, global assessments in recent years estimate that over tens of millions of tonnes of e-waste has been generated annually from developing countries alone where informal recycling operations are the norm (Rajpoot & Thakur, 2026). Inadequate management of the dismantling, handling, and disposal of electronic waste presents serious environmental and human health concerns and is considered an urgent waste management issue facing the world in the 21st Century (Goel et al. 2024; Rajpoot & Thakur, 2026).

E-waste consists of multiple materials that are mixed together to create a ‘mixture’ (complex), heterogenous and toxic (Dash et al. 2025a). E-waste contains a variety of toxic materials, including heavy metals (such as lead), and hazardous chemicals like persistent organic pollutants (Brominated Flame Retardants (BRFs), and PCB’s) (Islam et al. 2025). There is a great deal of concern over the hazardous nature of e-waste, and there exists a very real problem when it comes to keeping contaminants associated with e-waste from entering the environment (Beula & Sureshkumar, 2021; Rajesh et al. 2022a). Contaminants associated with e-waste may end up in soil and ground water, or they may otherwise be released into the atmosphere via inappropriate methods of disposal (i.e. uncontrolled burning and acid leaching) (Frazzoli et al. 2022; Rajesh et al. 2022b). Chronic exposure to toxic substances leading to long-term effects may include neurological disorders cancer-endocrine disruption and developmental disorders. Therefore, the need for environmentally-sound management strategies for e-waste is imperative (Castro et al. 2024; Frazzoli et al. 2022).

E-waste is paradoxically a source of valuable secondary materials (“urban mines”) because it is so rich in precious, base, and critical metals (Cayumil et al. 2016). Printed circuit boards (PCBs) in particular have an exceptionally high concentration of metals such as copper, gold, silver, palladium, and many other strategic materials (> 50% higher than typical metal grades found in naturally occurring ores) (Abbadi et al. 2024; Gonzalez Baez et al. 2022). Increasing demand for and limited global supplies of certain primary minerals, coupled with volatile prices for these metals, have increased the need to recover metals from e-waste (Gonzalez Baez et al. 2022). Urban mining through the efficient recycling of e-waste preserves our finite supply of natural resources while at the same time conserving energy, reducing land degradation, and minimizing the negative environmental impact of traditional mining activities (Grandhi et al. 2024).

The prevailing methods used to recycle electronic waste (E-waste) are pyrometallurgy and hydrometallurgy (Fatema et al. 2025; Magoda et al. 2022). Pyrometallurgy recovers metals through smelting and refining at high temperatures, making it a very efficient way to recover bulk metals (Magoda et al. 2022); however, as it is energy intensive and results in high carbon emissions in addition to producing toxic fumes, it is not considered to be environmentally sustainable (Quinto et al. 2025). Hydrometallurgy uses acids or bases to leach out metals from e-waste and is a more selective way of recovering metal(s) from e-waste than pyrometallurgy (Magoda et al. 2022); on the other hand, as hydrometallurgy produces large quantities of secondary liquid waste and has a number of different chemicals used as solvents, it does present some environmental problems (Goel et al. 2024). Because of the aforementioned disadvantages, the standard e-waste recycling techniques are not generally considered to be environmentally sustainable, economically viable for low-grade materials, or compliant with the increasingly stringent environmental regulatory requirements (Gonzalez Baez et al. 2022).

Due to the limitations of previous methods, biohydrometallurgical technology is now being researched as an efficient method of reclaiming metals from electronic waste stream with very little effort or cost (Desmarais et al. 2020). One biohydrometallurgical technique that has shown much promise in terms of reclaimed metal is bioleaching. Bioleaching uses microorganisms to leach out metals from electronic waste (Minimol et al. 2021). As these microorganisms metabolize organic matter and oxidize chemical compounds found in electronic waste substrates, they generate organic acids (e.g., fulvic/ humic acids), inorganic acids (e.g., sulfuric acid), chelating agents (e.g., EDTA), or redox reactions that help to extract metals (Song et al. 2023). Several species of acidophilic bacteria such as *Acidithiobacillus ferrooxidans* and *Acidithiobacillus thiooxidans* and several species of heterotrophic fungi such as *Aspergillus* and *Penicillium* have demonstrated potential to extract metals from electronic waste substrates (Olaniran et al. 2013). Compared with other conventional techniques, bioleaching is performed in milder temperature ranges, uses significantly lower levels of energy, and produces far less hazardous byproducts than the other technologies. As a result, bioleaching represents an environmentally friendly and economically beneficial alternative to existing techniques (Minimol et al. 2021).

While bioleaching presents a unique way to reclaim materials from e-waste, there are limitations from both a technical and operational standpoint that impede the successful adoption of bioleaching for large-scale e-waste recycling (Quinto et al. 2025). These impediments include slow leaching kinetics; metal toxicity toward bacteria; a broad range of material composition; and the formation of inhibitory chemicals during leaching (Olaniran et al. 2013). Further complicating this issue is the fact that proper management of leaching agents and processing water is essential to achieving the environmental benefits of bioleaching (Li et al. 2022); the mismanagement of these resources can negate the sustainability advantages of a bioleaching system. Innovative methods must be developed to solve these challenges through the design and optimization of integrated processes (Dash et al. 2025b). In recent years, there has been an increasing emphasis on implementing closed-loop processing as part of the sustainable resource recovery and circular economy paradigms (Magoda et al. 2022). In general, a closed-loop bioleaching system continually recycles and reuses the critical components of the bioleaching process (microorganisms, nutrients, lixiviants, and process water) thereby limiting material loss and waste production in a system (Fathima et al. 2022). Reduced operational costs, improved stability, and decreased environmental impact through low toxicity are some of the major benefits of closed-loop processing. The concepts behind closed-loop bioleaching fit within the industrial ecology framework, in that they create a means to convert waste into a resource while optimally utilizing system efficiencies (Fathima et al. 2022; Magoda et al. 2022).

Using bioleaching together with a closed-loop processing system creates many synergistic benefits. Reusing microbial cultures allows microbes to adapt and develop resistance against metal toxicity, thus increasing leaching efficiency in subsequent cycles (Hsu et al. 2021). The ability to recycle both leachates (during the leaching process) and process water, therefore, minimizes the need for additional freshwater and minimizes the potential for secondary pollution (Lee et al. 2025). Likewise, combining bioleaching with the use of downstream metal recovery methods – such as biosorption or through precipitation or electrochemical recovery – provides an overall higher recovery of metals and a more sustainable method of recovery when conducted in closed-loop form (Frazzoli et al. 2022; Srivastava et al. 2025). In addition to these considerations, improvements in the design of bioreactors, the engineering of microbial consortia, and the modeling of bioleaching processes have improved the feasibility of such closed-loop systems for processing complex matrices of e-waste (Golzar-Ahmadi et al. 2024; Soetrisno & Delgado-Saborit, 2020).

Closed-loop bioleaching directly contributes, from a global sustainability standpoint, to numerous UN Sustainable Development Goals (SDG) (Crowther et al. 2024), including Responsible Consumption & Production (SDG 12), Clean Water & Sanitation (SDG 6), Climate Action (SDG 13) and Industry, Innovation & Infrastructure (SDG 9) by decreasing our dependence on primary resource extraction, (Raman et al. 2024) and hence reducing environmental impacts, and also providing opportunities to transition to more sustainable and resilient materials delivery systems (Mondejar et al. 2021).

A comprehensive evaluation of closed-loop systems for bioleaching of e-waste is needed due to the limited understanding we have about these systems (e.g. operational stability, technoeconomic feasibility, scale-up potential) (Sadeghi et al. 2026) and the fact that most of the literature has primarily been focused on batch testing instead of the integrated closed-loop systems and on lifecycle environmental impacts of bioleaching (Yan et al. 2025). Since research has been conducted primarily at the laboratory scale, it is critical to conduct far-reaching assessments of bioleaching performance through a closed-loop approach that would bridge the gap between laboratory-scale research and industrial application and develop recommendations for industry (Danaee et al. 2025). In addition, the VOSViewer keyword co-occurrence network was developed to compare keyword co-occurrence patterns among the four major topic areas of e-waste, bioleaching, metal recovery, and recycling, which demonstrates the interdisciplinary nature of this field of research (Fig. 1). Overlap among the nodes associated with "e-waste," "metals," and "bioleaching" indicates their significant roles in connecting metallurgical processes with sustainable waste management (Rendón-Castrillón et al. 2023). The overlap and association between bioleaching, copper, and PCBs indicates a growing trend towards the bio-recovery of valuable metals from electronic components. Furthermore, the links to sustainability and waste management indicate a growing research emphasis on environmentally friendly and circular economy strategies (Grima-Carmena et al. 2025; Rendón-Castrillón et al. 2023). Consequently, the network shows the movement of metal manufacturing away from conventional metallurgical techniques and towards the development of integrated systems for the recovery of metal resources from e-waste (Tian et al. 2023).

The current research is focused on the assessment of the feasibility of closed-loop bioleaching as a method of retrieving metals sustainably from e-waste. It is based on exploring microbial mechanisms for bioleaching and optimising processes, along with examining the environmental consequences of closed-loop bioleaching, so that we may gain greater insight into the use of this technology in the future of closed-loop e-waste management systems. In addition, the research should provide important information concerning environmentally efficient, scalable and cost-effective solutions for future e-waste recycling efforts.

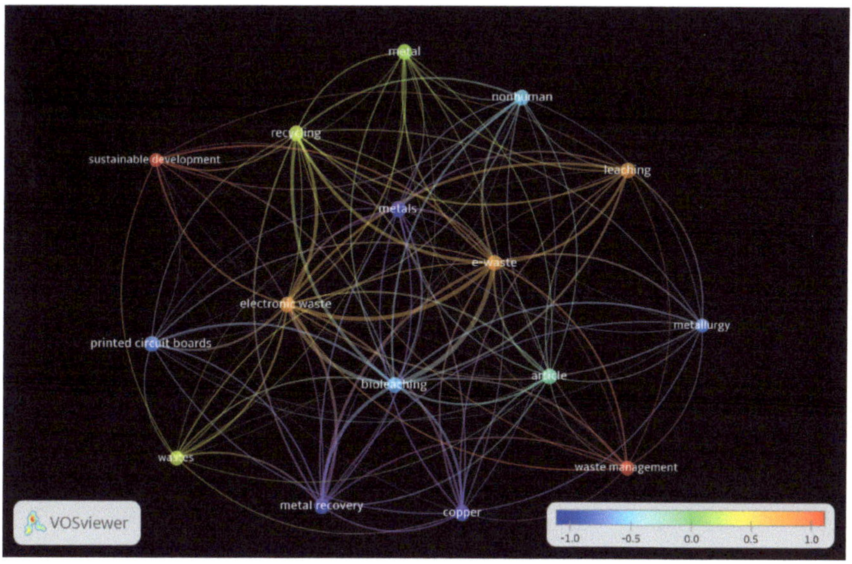

Fig. 1 VOSviewer keyword co-occurrence network illustrating research linkages among e-waste, bioleaching, metal recovery, recycling, and sustainability themes

2 Composition and Resource Potential of E-Waste

E-waste, characterised by its high heterogeneity, can be defined as a stream of materials that consists of various components, including PCBs, batteries, display panels, cables, plastic materials, and ceramics (Blumbergs et al. 2024). Among all of these materials, PCBs contain the most concentration of metals about 30% to 40% by weight, especially copper and precious metals such as gold, silver, and palladium, as well as other critical elements (Rivera et al. 2024). Enhancing the potential to extract natural resources from e-waste are lithium-ion batteries. These batteries serve as secondary sources for critical metals that are key for energy storage applications, including lithium, cobalt, nickel, and manganese (Rivera et al. 2024). The economic value associated with these metals is significant. However, the sourcing of these metals through primary mining raises environmental and geopolitical concerns (Gao et al. 2021). The complex composition of e-waste, including the presence of hazardous additions and the nature of the strongest form of metal to matrix associations, creates challenges for traditional recycling processes (Dash et al. 2025b). Therefore, to successfully recover these economically valuable metals and natural resources, new and environmentally sustainable recycling methods must be developed to effectively deal with this level of complexity (Dash et al. 2025b). As shown in Fig. 2, the large environmental and health risks are associated with the traditional approach to electronic waste management through uncontrolled disposal and landfill, The pollution of water and air is the result of this disposal method and leads to the potential for damage to human health through a number of pathways. The alternative approach highlighted is the recycling of electronic waste with recovery of metals through hydrometallurgical and pyrometallurgical processes as well as through bioleaching (Mvokwe et al. 2025). The bioleaching process is the most environmentally friendly of these processing methods. Also, this bioleaching process releases fewer pollutants into the environment and has less of a negative impact on the environment than the traditional high energy metallurgical processing methods (Junior et al. 2023; Magoda et al. 2022).

3 Conventional E-waste Recycling Technologies: Limitations

3.1 *Pyrometallurgical Processes*

Pyrometallurgy is an established method for recycling electronic waste (e-waste) that uses a combination of high-temperature smelting and refining to recover metals from complex electronic devices (Desmarais et al. 2020). This process has proven to be advantageous for bulk recovery of metals, especially copper and precious metals, and therefore is largely used by the e-waste recycling industry on an industrial scale (Minimol et al. 2021). Nevertheless, there are several major limitations associated

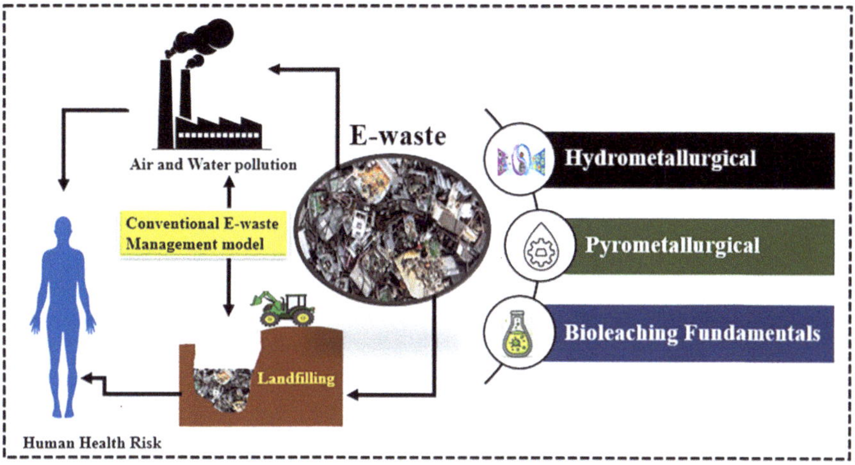

Fig. 2 Schematic comparison of conventional e-waste management impacts with hydrometallurgical, pyrometallurgical, and bioleaching-based recovery pathways

with the pyrometallurgical process (Desmarais et al. 2020). The need for extremely high temperatures means more energy is required to operate the process, resulting in increased levels of greenhouse gas emissions and a reduced sustainable environmental impact (Udayakumar et al. 2022). As a result of the high temperatures required, metals that have low vaporization points, such as mercury, cadmium and zinc, are frequently lost or require additional complex systems to capture them during the smelting process (Blumbergs et al. 2024). Toxic off-gases emitted during smelting also contain dioxins, compounds related to dioxins called furans and metal fumes, which mean that the gas generated during the process requires extensive additional investment in resources for advanced gas cleaning and emission control equipment, leading to much higher start-up and ongoing operational costs (Udayakumar et al. 2022). Because of these factors, pyrometallurgical recycling can only be economically viable for large centrally located facilities and is typically not well suited for smaller, decentralized e-waste recycling facilities or the recycling of low grade e-waste (Desmarais et al. 2020; Minimol et al. 2021).

3.2 Hydrometallurgical Processes

The hydrometallurgical process is a technique used to extract metals from electronic waste (Sikder et al. 2024). Hydrometallurgy is a method for obtaining metals using solutions made from strong acids or bases (hydrometals) followed by various ways of separating and recovering the metals (solvent extraction, ion exchange, precipitation, etc.) (Junior et al. 2023). The primary advantage of hydrometallurgy over pyrometallurgy (high temperature metal recovery) is that hydrometallurgy has a much greater

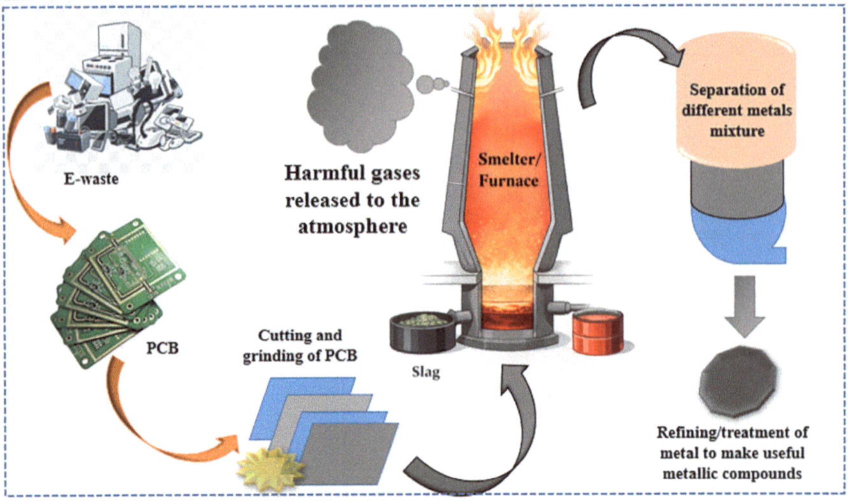

Fig. 3 Pyrometallurgical processing of e-waste showing PCB pre-treatment, smelting, gas emissions, slag formation, and downstream metal separation and refining

ability for selective recovery of a specific metal (i.e., lower amounts of metal loss) (Mvokwe et al. 2025). In addition, hydrometallurgy is more flexible in that it can process metals that are in complex matrices or were previously thought to be too difficult to recover using pyrometallurgy (Rivera et al. 2022). Although hydrometallurgical methods have these advantages over pyrometallurgy, they are associated with significant economic and environmental disadvantages because of the extensive use of harsh chemical reagents (sulfuric acid, nitric acid, cyanide, and ammonia), and thus generate large volumes of hazardous liquid waste that must be treated and disposed of at great expense (Mvokwe et al. 2025; Oke et al. 2025; Rivera et al. 2022). Corrosive effects on equipment, loss of chemicals, and increased need for safety measures add to operational costs of hydrometallurgy (Fig. 3). In addition, the high levels of chemicals used in hydrometallurgy conflict with sustainability objectives, and the use of hydrometallurgy to process low-value and heterogeneous fractions of e-waste decreases the long-term use of hydrometallurgical processes (Oke et al. 2025).

4 Bioleaching Fundamentals and Mechanisms

Metal extraction using bioleaching processes is performed using microorganisms to release metals from solid matrix materials while maintaining environmentally manageable process conditions (Gao et al. 2021). Examples of microorganisms used in E-Waste Recycling include *Acidithiobacillus Ferrooxidans* and *Leptospirillum Ferrooxidans*, as well as many genera of acid-loving fungi including *Aspergillus* and *Penicillium* species (Adetunji et al. 2023). Bioleaching using bacteria occurs through two types: a direct mechanism is where the bacteria oxidize ferrous iron and reduced sulphur compounds which will form ferric iron and sulphuric acid that can be used as powerful leaching agents to recover metals (Adetunji et al. 2023; Biswal et al. 2018); a second is indirect mechanism through oxidation of organic matter to produce carbon dioxide, nitrogen oxides, and water inhibit the formation of metals (Dusengemungu et al. 2021). Fungal bioleaching uses a different mechanism than bacterial bioleaching: here, fungi produce organic acids such as citric acid, oxalic acid and gluconic acid, which can either release metals directly from the matrix by chemical association or chelate the metal ions (Adetunji et al. 2023; Nkuna et al. 2024). The overall effectiveness of either of these processes depends on many different parameters such as pH, temperature, density of the pulp, particle size, redox potential, and the communities of the microorganisms present, so it is imperative that these factors be optimised to maximise the number of metals recovered from an extremely complex matrix that is in the form of electronic waste (Adetunji et al. 2023; Khan et al. 2024).

5 Closed-Loop Bioleaching Processing: Concept and Design

This approach combines the processes of bioleaching with the processes used to recycle waste products (electrolytic recovery) and/or produce trans-generation of glacial water (Harshan & Rajan, 2025). In addition, since all metals recovered will be used again in future operations, it reduces the need for additional water, chemicals, and trans-gre resources. Therefore, closed-loop operation provides substantial economic cost savings through lower materials usage and less environmental waste compared to other methods of electrolyte recycling and recycling of micro-organism-derived waste (Quinto et al. 2025).

Through these closed-loop systems of operations, microbes are continually adapting and evolving and therefore, will become greater leaching agents (Vera et al. 2022). This also means that they will be better capable of surviving with higher concentrations of toxic metals and therefore produce a higher efficiency of leaching over multiple cycles (Zhang et al. 2023). Closed-loop recycling methods are also highly consistent with the circular economy model as they reduce the amount of waste produced, allow for trans-generation of glacial water and offer a viable option for e-waste recycling that is both environmentally friendly and economically feasible (Fig. 4) (Tian et al. 2022).

5.1 *Advantages of Closed-Loop Systems*

Closed-loop bioleaching systems offer greater environmental, technical and economic benefits compared to conventional open-loop recycling systems (Morell

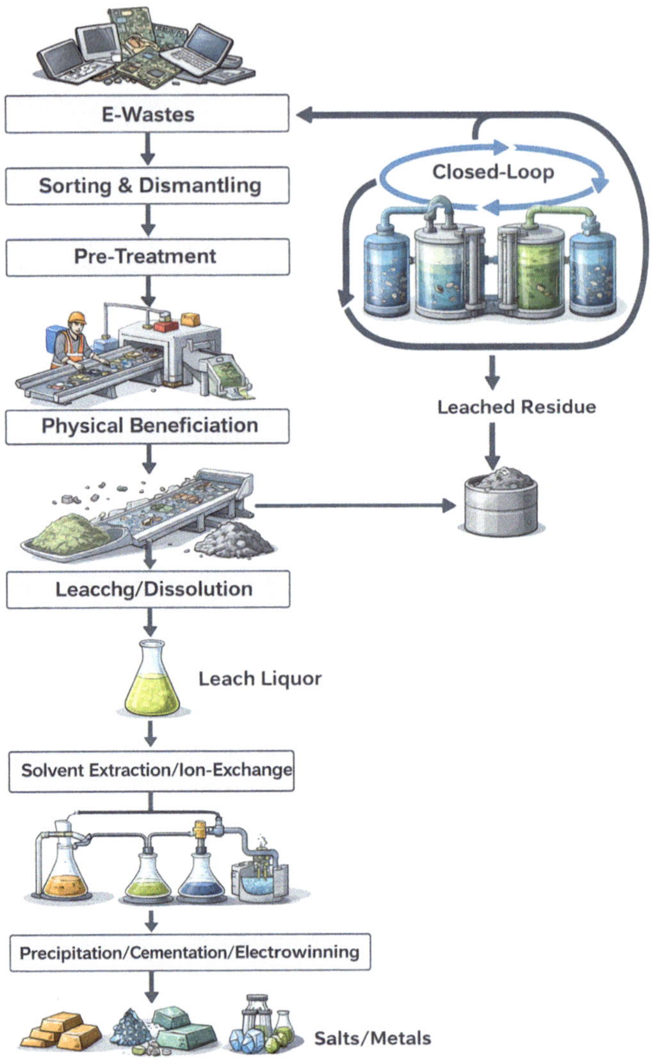

Fig. 4 Closed-loop bioleaching-based e-waste recycling process illustrating pre-treatment, metal dissolution, solution recycling, and final recovery of salts and metals

et al. 2025a). The most significant benefit of the use of closed-loop systems is the significant decrease in the amount of freshwater and chemical required to continuously recycle leachates and process solutions (Quinto et al. 2025). As a result, the amount of wastewater produced will be greatly reduced and the strain on effluent treatment facilities will be significantly lessened. Because closed-loop bioleaching systems also reduce the amount of secondary wastes produced, they promote environmental compliance and the sustainability of processes (Manikandan et al. 2023). The recycling of microbial cultures in the process of bioleaching promotes the development of physiological adaptations to improve resistance to metal toxicity (Dash et al. 2025c). Ultimately, by promoting physiological adaptation and improving the resistance to toxicity, the overall stability of the closed-loop system increases, allowing for the same level of operational stability to recover metals consistently (Morell et al. 2025a). At the same time, the reduced amount of reagent used, reduced costs associated with the disposal of waste products and the greater level of resource efficiency will ultimately increase the economic viability of the bioleaching process on a larger scale (Adetunji et al. 2023; Khan et al. 2024). Closed-loop bioleaching systems are also aligned with the concepts of circular economies and zero waste by turning electronic waste into a resource while minimizing negative environmental effects and promoting sustainable industrial practices.

5.2 *Bioreactor Configurations*

Numerous types of bioreactors have been tested in order to conduct closed-loop bioleaching of e-waste using the methods that are optimal for both the size and type of system required (Khan et al. 2024). Many researchers utilise the design of the stirred tank (also called stirred-tank) bioreactor as it allows for control over pH, temperature, aeration, mixing, and in fact, these controls help provide greater transfer of mass into the liquid and more interaction between bacterial growth and substrates (Harshan & Rajan, 2025). Airlift bioreactors are capable of transferring oxygen into liquid faster and with a lower energy consumption, reducing mechanical stress on microbes. In addition, due to the nature of the materials being processed, airlift bioreactors allow for continuous and closed-loop bioleaching of both individual electronic parts and electronic assemblies (Quinto et al. 2025). Bioleaching for e-waste using heap systems (previously used in mining) is beginning to show promise on low-grade or bulky fractions of the e-waste stream, primarily due to their simplicity and scalability (Wang et al. 2025). Several reactor designs have been optimised recently to improve the contact of solid and liquid phases, keep toxic compounds from accumulating, and help in recycling of the solution (Liu et al. 2022). To have a successful closed-loop bioleaching system, it is critical that the reactor design allows for maximal microbial growth efficiency, stabilisation of the bioleaching process and scalability of the system.

6 Microbial Consortia and Process Optimization

Because of their synergistic metabolic connections and greater tolerance to the toxic substances encountered in e-waste compared to pure cultures (Manikandan et al. 2023), mixed consortia of microorganisms are now thought to be superior to single species when it comes to the use of bioleaching systems (Table 1) (Khan et al. 2024).

Mixed consortia made up of a combination of iron-and sulfur-oxidizing bacteria, along with heterotrophic fungi, can generate acids, assist in redox reactions, and create metal chelates all at the same time, leading to increased metal solubilization efficiency (Harshan & Rajan, 2025; Khan et al. 2024). In addition, mixed consortia can withstand higher levels of metal toxicity, fluctuations in pH and inhibitory organics usually found in e-waste than pure cultures (Ankit et al. 2021). Advances in genetic and metabolic engineering will provide further opportunities to enhance the strength, specificity and leaching efficiency of mixed consortia through the optimization of specific metabolic pathways and through modifications to genes that regulate stress responses (Ankit et al. 2021; Harshan & Rajan, 2025). Strategies for optimizing the effectiveness of processes used to improve the performance of mixed microbial consortia include using adaptive laboratory evolution, using nutrients recycled from other processes to support the growth of mixed consortia and controlling the physicochemical parameters that influence microbial growth and metal recovery efficiencies (Manikandan et al. 2023). Integration of bioleaching with other pretreatment methods such as mechanical separation, size reduction, or mild thermal treatment increases the efficiency of the bioleaching process and improves metal recovery from complex e-waste materials (Quinto et al. 2025).

7 Downstream Metal Recovery and Purification

In a closed-loop bioleaching technology, effective downstream recovery and purification of solubilised metals are fundamental to the economic viability and sustainable nature of the process (Morell et al. 2025b). Once the metals are solubilised via bioleaching, they need to be recovered from the complex leachate by selective methods while allowing regeneration of the leaching solution (Restrepo Diaz & Amin, 2025). The most common methods for recovering the target metals at high purity are selective chemical precipitation, ion exchange, solvent extraction, and membrane separation (Sen & Chattoraj, 2021). New innovations like bioelectrochemical systems and electro-winning have several advantages over conventional methods because they allow simultaneous recovery of metals and regeneration of the leaching solution and use very little chemical input (Zhao et al. 2025). The use of functionalised biomaterial for biosorption has also been identified as an inexpensive and environmentally friendly option for metal recovery. The selection of appropriate downstream processes depends on the composition, concentration, and desired purity of the target metal (Adetunji et al. 2023; Morell et al. 2025a; Zhao et al. 2025). Integrating the recovery and purification processes in a closed-loop system allows for the least amount of reagents to be lost, the least amount of secondary waste generated, and longevity of performance from bioleaching (Singh et al. 2025; Villares et al. 2016).

Table 1 Microbial consortia and process optimization strategies in bioleaching of E-waste

Aspect	Description	Key Benefit for E-Waste Bioleaching	References
Microbial Consortia Type	Mixed cultures of iron-oxidizing (*Acidithiobacillus ferrooxidans*), sulfur-oxidizing (*Acidithiobacillus thiooxidans*), and heterotrophic fungi (*Aspergillus*, *Penicillium*)	Enhanced metal solubilization through synergistic metabolic pathways	(Singh et al. 2025)
Synergistic Interactions	Cooperative oxidation of Fe^{2+} and sulfur, organic acid production, chelation	Improved leaching efficiency and stability	(Kumar et al. 2026)
Metal Toxicity Tolerance	Consortia exhibit higher resistance to Cu, Ni, Co, Zn, and Pb toxicity	Sustained microbial activity in complex e-waste matrices	(Chu et al. 2025; Sharma et al. 2024)
Genetic & Metabolic Engineering	Stress-response enhancement, pathway optimization, metal resistance genes	Increased robustness, selectivity, and leaching rates	(Chu et al. 2025; Ramakrishnan et al. 2025)
Adaptive Evolution	Gradual exposure to increasing metal concentrations	Improved long-term microbial performance	(Villares et al. 2016)
Nutrient Recycling	Reuse of growth media and metabolites in closed-loop systems	Reduced operational cost and chemical input	(Mishra et al. 2023)
Process Monitoring	Real-time control of pH, redox potential, temperature, and dissolved oxygen	Enhanced process control and reproducibility	(Mathivanan et al. 2025)
Pre-treatment Integration	Mechanical separation, size reduction, mild thermal treatment	Increased metal accessibility and bioleaching efficiency	(Rath et al. 2012)
Closed-Loop Optimization	Reuse of microbes, leachates, and process water	Improved sustainability and circular economy compliance	(Morell et al. 2025b)

8 Environmental and Techno-Economic Assessment

Techno-economic and environmental assessment allow for a comprehensive feasibility assessment of the closed-loop bioleach method of recycling e-waste (de Oliveira Neto et al. 2017). Results from LCA studies continue to show that bioleach processes demonstrate significantly lower environmental impacts than those of the conventional pyrometallurgical and hydrometallurgical methods, specifically in regards to the amount of energy consumed, the amount of greenhouse gases emitted, and the amount of toxic waste produced (Cottes et al. 2023). Furthermore, using mild operating conditions and recyclable reagents increases the environmental benefits associated with bioleaching methods (Erkmen et al. 2025). The results of techno-economic analyses have demonstrated that closed-loop bioleaching can be commercially viable as the process will typically provide a better return on investment than traditional recycling methods, especially for e-waste streams that have low-grade materials, are heterogeneous in composition, and are complex in nature (Erkmen et al. 2025). While favourable economic benefits can be achieved through reduced chemical use, lower waste treatment costs and improved resource efficiency, capital investment, process scale-up and system integration ultimately dictate the final economic success of the bioleaching process (Wang et al. 2025). Therefore, pilot-scale validation and optimisation of the process before large-scale investments will be necessary to determine the economics of this technology.

9 Challenges and Future Perspectives

Several significant technical, operational and environmental constraints hinder the full-scale application of closed-loop bioleaching technologies. One of the major limitations is the relatively slow leaching kinetics as compared to chemical leaching, which can lead to delay in processing. High concentrations of metals, the presence of toxic additives, and the presence of organic contaminants from e-waste can reduce the ability of microorganisms to leach metals out of e-waste and, consequently, affect the stability and efficiency of the system. Structural and engineering challenges will arise as closed-loop bioleaching reactors are scaled up, in order to maintain the same level of microbial performance and integrity of the closed-loop. There is a need for additional development of resilient and metal-tolerant consortiums of microorganisms through adaptive evolution and synthetic biology; therefore, it is important that future research be conducted to explore these strategies. Combining advanced separation and purification technologies with closed-loop bioleaching can further improve the overall efficiency of bioleaching. The use of digitalization, real-time monitoring, and automation of processes will yield tools for enhancing control and scalability of closed-loop bioleaching technologies. Strong support from policymakers and regulatory incentives and furtherance of partnerships between industry and academic institutions are critical to enhancing the potential for large-scale commercialization and adoption of closed-loop bioleaching technologies.

10 Conclusions

The closed-loop bioleaching process is a groundbreaking new way to recycle electronic waste through environmentally friendly methods rather than traditional metal recovery methods. Through microbial processes, closed-loop bioleaching combines metal solubilization and metal recovery with the regeneration of reagents in a closed system to effectively recover valuable and critical metals from electronic waste, while simultaneously reducing energy use, chemical consumption, and generation of secondary waste. As a result, the use of closed-loop systems supports both circular economy and "hazard-free" concepts, meaning they address the dual concerns of resource depletion caused by rapid growth in e-waste and pollution caused by e-waste. While many challenges still exist with regards to kinetics, microbial inhibition, and scale-up, advancements are constantly being made in the areas of microbial consortia development, reactor design, and process integration that keep moving us toward making closed-loop bioleaching more feasible. Through continuous interdisciplinary research, with pilot-scale demonstrations and research-based techno-economic evaluations to validate pilot plant studies and industrial applications, closed-loop bioleaching will eventually become instrumental in creating sustainable and resilient e-waste management systems.

References

Abbadi A, Rácz Á, Bokányi L (2024) Exploring the comminution process of waste printed circuit boards in recycling: a review. J Mater Cycles Waste Manage 26(3):1326–1348. https://doi.org/10.1007/s10163-024-01945-3

Adetunji AI, Oberholster PJ, Erasmus M, Adetunji AI, Oberholster PJ, Erasmus M (2023) Bioleaching of metals from E-waste using microorganisms: a review. Minerals 13(6). https://doi.org/10.3390/min13060828

Ankit, Saha L, Kumar V, Tiwari J, Sweta, Rawat S, Singh J, Bauddh K (2021) Electronic waste and their leachates impact on human health and environment: global ecological threat and management. Environ Technol Innov 24:102049. https://doi.org/10.1016/j.eti.2021.102049

Beula D, Sureshkumar M (2021) A review on the toxic E-waste killing health and environment—Today's global scenario. Mater Today: Proc, Int Web Conf Adv Mater Sci Eng 47:2168–2174. https://doi.org/10.1016/j.matpr.2021.05.516

Biswal BK, Jadhav UU, Madhaiyan M, Ji L, Yang E-H, Cao B (2018) Biological leaching and chemical precipitation methods for recovery of Co and Li from spent lithium-ion batteries. ACS Sustain Chem Eng 6(9):12343–12352. https://doi.org/10.1021/acssuschemeng.8b02810

Blumbergs E, Shishkin A, Markus K, Serga V, Goljandin D, Klauson A, Abramovskis V, Baronins J, Zarkov A, Pankratov V (2024) Economic aspects of mechanical pre-treatment's role in precious metals recovery from electronic waste. Metals 14(1). https://doi.org/10.3390/met14010095

Castro G, Cobo M, Rodríguez I (2024) Identification of hazardous organic compounds in e-waste plastic using non-target and suspect screening approaches. Chemosphere 356:141946. https://doi.org/10.1016/j.chemosphere.2024.141946

Cayumil R, Khanna R, Rajarao R, Ikram-ul-Haq M, Mukherjee PS, Sahajwalla V, Cayumil R, Khanna R, Rajarao R, Ikram-ul-Haq M, Mukherjee PS, Sahajwalla V (2016) Environmental impact of processing electronic waste—Key issues and challenges. In: E-waste in transition—From pollution to resource. IntechOpen. https://doi.org/10.5772/64139

Chu X, Qu G, Zou H, Cheng M, Li J, Ren N (2025) Dehydration, heavy metal removal patterns, and microbial response in high moisture copper-containing sludge composting assisted by electromagnetic techniques. Int J Environ Sci Technol 22(4):2319–2332. https://doi.org/10.1007/s13762-024-05672-y

Cottes M, Mainardis M, Simeoni P, Cottes M, Mainardis M, Simeoni P (2023) Assessing the techno-economic feasibility of waste electric and electronic equipment treatment plant: a multi-decisional modeling approach. Sustainability 15(23). https://doi.org/10.3390/su152316248

Crowther TW, Rappuoli R, Corinaldesi C, Danovaro R, Donohue TJ, Huisman J, Stein LY, Timmis JK, Timmis K, Anderson MZ, Bakken LR, Baylis M, Behrenfeld MJ, Boyd PW, Brettell I, Cavicchioli R, Delavaux CS, Foreman CM, Jansson JK, van Galen LG (2024) Scientists' call to action: microbes, planetary health, and the sustainable development goals. Cell 187(19):5195–5216. https://doi.org/10.1016/j.cell.2024.07.051

Danaee S, Vo PHN, Malik S, Ralph PJ (2025) Can bioleaching of critical metals from secondary resources be a sustainable extraction solution? J Ind Eng Chem. https://doi.org/10.1016/j.jiec.2025.10.015

Dash J, Ojha R, Pradhan D (2025) Progress in bioleaching and its mechanism: a short review. Discover Environ 3(1):238. https://doi.org/10.1007/s44274-025-00454-w

Desmarais M, Pirade F, Zhang J, Rene ER (2020) Biohydrometallurgical processes for the recovery of precious and base metals from waste electrical and electronic equipments: current trends and perspectives. Biores Techn Reports 11:100526. https://doi.org/10.1016/j.biteb.2020.100526

Dusengemungu L, Kasali G, Gwanama C, Mubemba B (2021) Overview of fungal bioleaching of metals. Environ Adv 5:100083. https://doi.org/10.1016/j.envadv.2021.100083

Erkmen AN, Ulber R, Jüstel T, Altendorfner M (2025) Towards sustainable recycling of critical metals from e-waste: bioleaching and phytomining. Resour Conserv Recycl 215:108057. https://doi.org/10.1016/j.resconrec.2024.108057

Fatema K, Hassan MN, Hasan S, Roy H (2025) E-waste recycling in an optimized way for copper recovery by leaching and a case study on E-waste generation and management in Dhaka city. Heliyon 11(1):e41453. https://doi.org/10.1016/j.heliyon.2024.e41453

Fathima A, Tang JYB, Giannis A, Ilankoon IMSK, Chong MN (2022) Catalysing electrowinning of copper from E-waste: a critical review. Chemosphere 298:134340. https://doi.org/10.1016/j.chemosphere.2022.134340

Frazzoli C, Ruggieri F, Battistini B, Orisakwe OE, Igbo JK, Bocca B (2022) E-WASTE threatens health: the scientific solution adopts the one health strategy. Environ Res 212:113227. https://doi.org/10.1016/j.envres.2022.113227

Gao X, Jiang L, Mao Y, Yao B, Jiang P (2021) Progress, challenges, and perspectives of bioleaching for recovering heavy metals from mine tailings. Adsorpt Sci Technol 2021:9941979. https://doi.org/10.1155/2021/9941979

Goel A, Masurkar S, Pathade GR, Goel A, Masurkar S, Pathade GR (2024) An overview of digital transformation and environmental sustainability: threats, opportunities, and solutions. Sustainability 16(24). https://doi.org/10.3390/su162411079

Golzar-Ahmadi M, Bahaloo-Horeh N, Pourhossein F, Norouzi F, Schoenberger N, Hintersatz C, Chakankar M, Holuszko M, Kaksonen AH (2024) Pathway to industrial application of heterotrophic organisms in critical metals recycling from e-waste. Biotechnol Adv 77:108438. https://doi.org/10.1016/j.biotechadv.2024.108438

Gonzalez Baez A, Pantoja Muñoz L, Garelick H, Purchase D (2022) Characterization of industrially pre-treated waste printed circuit boards for the potential recovery of rare earth elements. Environ Technol Innov 27:102481. https://doi.org/10.1016/j.eti.2022.102481

Grandhi SP, Dagwar PP, Dutta D (2024) Policy pathways to sustainable E-waste management: a global review. J Hazard Mat Adv 16:100473. https://doi.org/10.1016/j.hazadv.2024.100473

Grima-Carmena L, Giner-Sanz JJ, Pocoví-Martínez S, Bosch-Mossi F, Pérez-Herranz V (2025) Study of the effect of the pretreatment and the concentrations of iron (II) and chloride, on the

electrochemical leaching of waste printed circuit boards using ANOVA analysis. J Ind Eng Chem 152:234–244. https://doi.org/10.1016/j.jiec.2025.04.054
Harshan K, Rajan AP (2025) Unveiling the potential of microbial biominers in bioleaching for heavy metal recovery from E-waste—A comprehensive review. J Hazard Mat Adv 18:100691. https://doi.org/10.1016/j.hazadv.2025.100691
Hsu E, Durning CJ, West AC, Park A-HA (2021) Enhanced extraction of copper from electronic waste via induced morphological changes using supercritical CO_2. Resour Conserv Recycl 168:105296. https://doi.org/10.1016/j.resconrec.2020.105296
Islam MdS, Hasan MdR, Mostakim K, Joarder MdSA, Hasan MdH, Ahmed MdR (2025) E-waste management in Bangladesh: Environmental impacts, health risks, and sustainable policy strategies. Cleaner Waste Syst 11:100297. https://doi.org/10.1016/j.clwas.2025.100297
Junior ABB, Sultana UK, Vaughan J (2023) Hydrometallurgical processing of E-waste and metal recovery. Wiley, management of electronic waste: resource recovery, technology and regulation, 234–288. https://doi.org/10.1002/9781119894360.ch11
Khan S, Alam FS, Borah D (2024) Microbial cell factories in the remediation of e-wastes: an insight. Biotechn Environ 1(1):4. https://doi.org/10.1186/s44314-024-00003-4
Kumar A, Lima AT, Kirkelund GM, Jensen PE, Ottosen LM, Funari V, Shemi A, Ndlovu S, Gomes HI (2026) Bioleaching: from natural ores to urban mines for sustainability, circularity, and carbon neutrality. Resour Conserv Recycl 227:108746. https://doi.org/10.1016/j.resconrec.2025.108746
Lee C-H, Lin Y-C, Li J-Y, Valix M, Wang DK (2025) The role of membrane separation in bioleaching process. Separat Purificat Rev 0(0):1–20. https://doi.org/10.1080/15422119.2025.2491085
Li Q, Wang Y, Li Y, Li L, Tang M, Hu W, Chen L, Ai S (2022) Speciation of heavy metals in soils and their immobilization at micro-scale interfaces among diverse soil components. Sci Total Environ 825:153862. https://doi.org/10.1016/j.scitotenv.2022.153862
Liu J, Ren N, Qu C, Lu S, Xiang Y, Liang D (2022) Recent advances in the reactor design for industrial wastewater treatment by electro-oxidation process. Water 14(22). https://doi.org/10.3390/w14223711
Magoda K, Mekuto L, Magoda K, Mekuto L (2022) Biohydrometallurgical recovery of metals from waste electronic equipment: current status and proposed process. Recycling 7(5). https://doi.org/10.3390/recycling7050067
Manikandan S, Inbakandan D, Valli Nachiyar C, Karthick Raja Namasivayam S (2023) Towards sustainable metal recovery from e-waste: a mini review. Sustain Chem Environ 2:100001. https://doi.org/10.1016/j.scenv.2023.100001
Mathivanan K, Zhang R, Chandirika JU, Mathimani T, Wang C, Duan J (2025) Bacterial biofilm-based bioleaching: sustainable mitigation and potential management of e-waste pollution. Waste Manage 193:221–236. https://doi.org/10.1016/j.wasman.2024.12.010
Minimol M, Shetty KV, Saidutta MB (2021) Biohydrometallurgical methods and the processes involved in the bioleaching of WEEE. In: Environmental management of waste electrical and electronic equipment, pp. 89–107. Elsevier. https://doi.org/10.1016/B978-0-12-822474-8.00005-2
Mishra S, Ghosh S, van Hullebusch ED, Singh S, Das AP (2023) A critical review on the recovery of base and critical elements from electronic waste-contaminated streams using microbial biotechnology. Appl Biochem Biotechnol 195(12):7859–7888. https://doi.org/10.1007/s12010-023-04440-x
Mondejar ME, Avtar R, Diaz HLB, Dubey RK, Esteban J, Gómez-Morales A, Hallam B, Mbungu NT, Okolo CC, Prasad KA, She Q, Garcia-Segura S (2021) Digitalization to achieve sustainable development goals: steps towards a smart green planet. Sci Total Environ 794:148539. https://doi.org/10.1016/j.scitotenv.2021.148539
Morell J, Lao-Luque C, Sole-Sardans M, Ribas D, Guimerà X, Dorado AD (2025) Closed-loop bioleaching for mobile phones from the bioregeneration of the leaching agent to the recovery of metal: a review. J Ind Eng Chem 151:90–111. https://doi.org/10.1016/j.jiec.2025.04.020

Mvokwe SA, Oyedeji OO, Agoro MA, Meyer EL, Rono N, Mvokwe SA, Oyedeji OO, Agoro MA, Meyer EL, Rono N (2025) A critical review of the hydrometallurgy and pyrometallurgical recovery processes of platinum group metals from end-of-life fuel cells. Membranes 15(1). https://doi.org/10.3390/membranes15010013

Nkuna R, Matambo TS, Nkuna R, Matambo TS (2024) Determining the metabolic processes of metal-tolerant fungi isolated from mine tailings for bioleaching. Minerals 14(3). https://doi.org/10.3390/min14030235

Oke EA, Fedai Y, Potgieter JH, Oke EA, Fedai Y, Potgieter JH (2025) Hydrometallurgical leaching of copper and cobalt from a copper–cobalt ore by aqueous choline chloride-based deep eutectic solvent solutions. Minerals 15(8). https://doi.org/10.3390/min15080815

Olaniran AO, Balgobind A, Pillay B, Olaniran AO, Balgobind A, Pillay B (2013) Bioavailability of heavy metals in soil: impact on microbial biodegradation of organic compounds and possible improvement strategies. Int J Mol Sci 14(5):10197–10228. https://doi.org/10.3390/ijms140510197

de Oliveira Neto GC, de Jesus Cardoso Correia A, Schroeder AM (2017) Economic and environmental assessment of recycling and reuse of electronic waste: multiple case studies in Brazil and Switzerland. Resources, Conservat Recycl 127:42–55. https://doi.org/10.1016/j.resconrec.2017.08.011

Quinto S, Law N, Fletcher C, Le J, Jose SA, Menezes PL, Quinto S, Law N, Fletcher C, Le J, Jose SA, Menezes PL (2025) Exploring the E-waste crisis: strategies for sustainable recycling and circular economy integration. Recycling 10(2). https://doi.org/10.3390/recycling10020072

Rajesh R, Kanakadhurga D, Prabaharan N (2022) Electronic waste: a critical assessment on the unimaginable growing pollutant, legislations and environmental impacts. Environ Chall 7:100507. https://doi.org/10.1016/j.envc.2022.100507

Rajpoot S, Thakur S (2026) Global ecological threat and management of electronic waste and its leachates' effects on human health and the environment. In: Das AP, Sirisena K, Bhanja D (Eds.), Sustainable municipal waste management with a zero waste approach, pp. 159–194. Springer Nature Switzerland. https://doi.org/10.1007/978-3-032-05950-5_7

Ramakrishnan P, Sundaram T, Lahiri D, Nag M, Bhattacharya D (2025) Genetic engineering and modulation of metabolic pathways. In: Lahiri D, Nag M, Bhattacharya D, Ghosh S (Eds.). Introduct Metabolic Eng Applicat, pp. 295–330. Springer Nature Switzerland. https://doi.org/10.1007/978-3-031-93189-5_12

Raman R, Lathabai HH, Nedungadi P (2024) Sustainable development goal 12 and its synergies with other SDGs: identification of key research contributions and policy insights. Discover Sustain 5(1):150. https://doi.org/10.1007/s43621-024-00289-0

Rath SS, Nayak P, Mukherjee PS, Roy Chaudhury G, Mishra BK (2012) Treatment of electronic waste to recover metal values using thermal plasma coupled with acid leaching—A response surface modeling approach. Waste Manage 32(3):575–583. https://doi.org/10.1016/j.wasman.2011.11.001

Rendón-Castrillón L, Ramírez-Carmona M, Ocampo-López C, Gómez-Arroyave L, Rendón-Castrillón L, Ramírez-Carmona M, Ocampo-López C, Gómez-Arroyave L (2023) Bioleaching techniques for sustainable recovery of metals from solid matrices. Sustainability 15(13). https://doi.org/10.3390/su151310222

Restrepo Diaz JD, Amin SH (2025) A multi-objective optimization approach for sustainable management of computers E-waste in a closed-loop supply chain network. J Clean Prod 506:145494. https://doi.org/10.1016/j.jclepro.2025.145494

Rivera RM, Zante G, Hartley JM, Ryder KS, Abbott AP (2022) Catalytic dissolution of metals from printed circuit boards using a calcium chloride–based deep eutectic solvent. Green Chem 24(7):3023–3034. https://doi.org/10.1039/D1GC04694B

Rivera RM, Elgar CE, Jacobson B, Feeney A, Prentice P, Ryder KP, Abbott A (2024) Ultra-fast extraction of metals from a printed circuit board using high power ultrasound in a calcium chloride-based deep eutectic solvent. RSC Sustain 2(2), 403–415. https://doi.org/10.1039/D3SU00147D

Sadeghi B, Allam SN, Khazaei A, Cavaliere P (2026) Closed-loop extraction of precious metals from e-waste via NH_4HCO_3 desulfurization and secondary lead smelting. Waste Manage 210:115228. https://doi.org/10.1016/j.wasman.2025.115228

Sen K, Chattoraj S (2021) 5–A comprehensive review of glyphosate adsorption with factors influencing mechanism: kinetics, isotherms, thermodynamics study. In: Bhattacharyya S, Mondal NK, Platos J, Snášel V, Krömer P (Eds.), Intelligent environmental data monitoring for pollution management, pp. 93–125. Academic Press. https://doi.org/10.1016/B978-0-12-819671-7.00005-1

Sharma S, Sharma M, Kumar R, Akhtar MS, Umar A, Alkhanjaf AAM, Baskoutas S (2024) Recent advances and mechanisms of microbial bioremediation of nickel from wastewater. Environ Sci Pollut Res 31(28):40224–40244. https://doi.org/10.1007/s11356-023-30556-y

Sikder S, Toha M, Rahman M (2024) Prospective energy production from incineration: a brief overview on different types of energy produced by incineration. In: Anouzla A, Souabi S (Eds.), Generation of energy from municipal solid waste: circular economy and sustainability, pp. 261–279. Springer Nature Switzerland. https://doi.org/10.1007/978-3-031-74334-4_12

Singh S, Santhiya D, Sharma JG (2025) An overview of bioprocess based recovery and recycling techniques for battery waste management: transitioning from environmental hazard to sustainability. Miner Eng 233:109598. https://doi.org/10.1016/j.mineng.2025.109598

Soetrisno FN, Delgado-Saborit JM (2020) Chronic exposure to heavy metals from informal e-waste recycling plants and children's attention, executive function and academic performance. Sci Total Environ 717:137099. https://doi.org/10.1016/j.scitotenv.2020.137099

Song C, Sun S, Wang J, Gao Y, Yu G, Li Y, Liu Z, Zhang W, Zhou L (2023) Applying fulvic acid for sediment metals remediation: mechanism, factors, and prospect. Front Microbio 13. https://doi.org/10.3389/fmicb.2022.1084097

Srivastava RR, Ilyas N, Chaerun SK, Engström F, Lennartsson A, Öqvist LS, Ilyas S (2025) Biological recycling of critical metals from spent hydrodesulfurization catalysts: a review. Environ Chem Lett 23(5):1381–1402. https://doi.org/10.1007/s10311-025-01849-0

Tian T, Liu G, Yasemi H, Liu Y (2022) Managing e-waste from a closed-loop lifecycle perspective: China's challenges and fund policy redesign. Environ Sci Pollut Res Int 29(31):47713–47724. https://doi.org/10.1007/s11356-022-19227-6

Tian H, Chen L, Wu J, Zheng D, Yang Q, Ji Z, Cai J, Chen Y, Li Z (2023) Global research into the relationship between electronic waste and health over the last 10 years: a scientometric analysis. Front Public Health 10:1069172. https://doi.org/10.3389/fpubh.2022.1069172

Udayakumar S, Razak MIBA, Ismail S (2022) Recovering valuable metals from waste printed circuit boards (WPCB): a short review. Materials today: proceedings, 14th AUN/SEED-net regional conference on materials and 4th international postgraduate conference on materials, minerals and polymer 66:3062–3070. https://doi.org/10.1016/j.matpr.2022.07.364

Vera M, Schippers A, Hedrich S, Sand W (2022) Progress in bioleaching: fundamentals and mechanisms of microbial metal sulfide oxidation—Part A. Appl Microbiol Biotechnol 106(21):6933–6952. https://doi.org/10.1007/s00253-022-12168-7

Villares M, Işıldar A, Mendoza Beltran A, Guinee J (2016) Applying an ex-ante life cycle perspective to metal recovery from e-waste using bioleaching. J Clean Prod 129:315–328. https://doi.org/10.1016/j.jclepro.2016.04.066

Wang X, Huang W, Yan B, Zhou S, Zhu X, Wang Z, Cheng Z, Chen G (2025) E-waste recycling: Integrated life cycle assessment and techno-economic analysis unravels pyrometallurgy's edge and delivers an optimization framework for recovering waste printed circuit boards. Waste Manage 207:115135. https://doi.org/10.1016/j.wasman.2025.115135

Yan J, Xie Z, Yan J, Xie Z (2025) Multilevel mechanisms of magnetic nanoparticles in enhancing dark fermentative hydrogen production: from pure to mixed cultures. Hydrogen 6(4). https://doi.org/10.3390/hydrogen6040120

Zhang X, Shi H, Tan N, Zhu M, Tan W, Daramola D, Gu T (2023) Advances in bioleaching of waste lithium batteries under metal ion stress. Bioresourc Bioprocess 10(1):19. https://doi.org/10.1186/s40643-023-00636-5

Zhao P, Qiu J, Liang H, Jiang Y, Huang F, Wang N, Zhu Y (2025) Latest research progress in high-purity material purification technology. ACS Omega 10(33):36804–36823. https://doi.org/10.1021/acsomega.5c03602

Bio Remediation Strategies for the Sustainable Management of Coal Mining Wastes

Madhusmita Mohanty and Alok Prasad Das

Abstract A large volume of waste is produced by coal mining, including both inorganic and organic pollutants that impose a severe threat to both the environment and health. The old management techniques, like physical and chemical treatments, have a lot of drawbacks, like high cost, incomplete removal, and secondary environmental implications. A more convenient technique, bioremediation using microorganisms to degrade, transform, or immobilise pollutants in coal mining wastes, including coal tailings, coal fly ash, acid mine drainage (AMD), and coal rock, offers a sustainable and green remediation way. Techniques like bioaugmentation, bio stimulation, phytoremediation, and especially bioleaching have given us fruitful results. Successful bioleaching requires specific microbial consortia, including iron and sulphur oxidising bacteria, for the recovery and solubilization of critical metals from the coal mining wastes. Investigations on pyrite coal waste have shown pyrite removal yield up to 80% with customised bioreactor conditions, which is a reason for the reduction of the risk of acid mine drainage along with environmental toxicity. The combination of the circular economy framework with bioremediation further increases the significance of these practices, enabling simultaneous pollution removal and production of bioenergy or recovery of critical elements. Although the bioremediation techniques are slower than the chemical approaches and often site-specific, their adaptation, economical application, minimal geo-environment impacts, and compatibility with other waste management strategies make them significant for sustainable mining. All these approaches hold great potential for the restoration of coal mining environments and the setup of sustainable and closed-loop waste management systems.

Keywords Coal mining wastes · Bioleaching · Critical metals · Circular economy

M. Mohanty · A. P. Das (✉)
Department of Life Sciences, Rama Devi Women's University, Bhubaneswar, India
e-mail: alok1503@gmail.com

A. P. Das and S. Priyadarsini (eds.), *Recovery and Recycling of Critical Metals from Electronic Waste for Environmental Sustainability*, Earth and Environmental Sciences Library, https://doi.org/10.1007/978-3-032-29064-9_6

1 Introduction

Globalisation and the industrial revolution have put great strain on energy producers to fulfil the expanding needs of the industrial and residential sectors. Particularly in coal-dependent economies like India, coal mining has become a vital sector to meet these rising energy demands. But unfortunately, the sudden growth of the mining industry has, regrettable, had detrimental effects on the use of land, degraded the environment, and posed a risk to the natural security and the way of life of people who reside near mining areas. Coal is an organic, combustible, black, sedimentary rock that, along with natural gas and oil, is recognised as a fossil fuel. Since it takes millions of years to develop and cannot be replenished during an individual's lifetime, it is considered a non-renewable natural resource (Priyadarsini and Das 2025). The combustibility attributes of coal have been known for approximately a thousand years, although it may have been used as an ornament in prehistoric times. In fact, coal use as fuel possibly began in China around 1000 BC. Varying its rank (degree of evolution) across its geological history, coal exhibits distinct physical, chemical, and technological features. Coal is mostly made up of organic matter, which comes mostly from diverse plant remains (higher plants, ferns, fungi, and algae) and various tissues (leaves, stalks, wooden trunks, bark, pollen, spores, sclerotia, resins, etc.). Alternatively, most rocks are inorganic. An ash yield (ISO 11760, 2005) of less than or equal to 50% on a dry mass basis matches up with coal's mineral matter amount. In the world, coal is the most frequently available, abundant, and accessible fossil fuel. It remains to be the primary source of electricity today, generating 41% of the world's energy production (Coal Council National 2015). Based on recent projections, coal will remain the globe's second-largest energy source until 2030–2040, despite the fact it has fallen somewhat recently (Tripathy et al. 2026). Energy Information Administration of the United States, 2016, 2017. Coal is primarily utilized in the steel sector, as well as in the manufacturing of cement, chemicals, and electricity. In the steel sector, coke is produced from coal and then used in blast furnaces to melt iron ore and make molten iron, which is the main part of steel. Coal is used to make 70% of the steel produced globally National Coal Council, 2015. In addition to its conventional use as an energy source, coal plays a critical role in various industrial mass production processes (Mishra et al. 2024). For evidence, it is integral to the manufacturing production; coal is employed as a primary fuel in kilns to heat raw materials, thereby producing clinker, a key intermediary material in cement manufacturing (Priyadarsini and Das 2024, 2026). Moreover, coal serves as an essential origin material for the synthesis of a range of carbon-based products, including high-surface-area activated carbon, which is widely utilized for air and water purification. It also provides the ground for the production of chemical derivatives such as oils, tar, and pitches, which are further processed into a variety of industrial chemicals and fertilizers. Through these diverse applications, coal continues to underpin critical sectors of the global industrial economy.

It is crucial to track land-use change in coal mining regions in order to comprehend the wider effects on biodiversity, ecosystem structures, surface energy balance, and biogeochemical cycles. In India, 44% of the landmass is degraded, mostly as a result of deforestation, excessive mining, fast industrialisation, and unsustainable land use practices. Degradation affects 23% of land resources worldwide (Swamy et al. 2023). Degradation affects 23% of land resources worldwide, but 44% of India's landmass is degraded, mainly as a result of deforestation, excessive mining, fast industrialisation, and unsustainable land use practices (Anthony et al. 2024).Coal mining is a key controller of the economic growth of the country, but it immensely contributes to environmental damage (Thakur et al. 2021).

One of the main outcomes of coal mining in India is deforestation, which adversely affects groups dependent on the forest for their livelihood. For a long time, India has banned cutting, but the demands of industries and coal mining continue to increase the forest loss. The abundantly seen grazing fields and agricultural zones are the consequences of the fragmentation of land by coal mining, which changes the dense forests into these open areas. The biodiversity and ecosystem health are going to face far-reaching consequences of this degradation.

2 Global Distribution of Coal

Coal is now considered a part of the GFP index according to its contribution to energy production in many countries. Given that peat formed in mainly tropical latitudes at different points in the geological record and the continents subsequently drifted to their current positions, the distribution of coal deposits around the world is uneven. Northern Hemisphere regions have the majority of coal deposits, while Southern Hemisphere regions—apart from Australia—have comparatively little of this kind of resource. Coal remains a cornerstone of the global energy mix and a critical resource for industries worldwide. The leading coal-producing nations dominate the global supply chain, meeting domestic and international energy demands. Among these, the top coal exporters have significantly contributed to industrial growth by supplying high-quality coal to energy energy-hungry market. China, being the largest coal producer in the world, contributes about 64.1% to the global coal production. India, ranking second in coal production, contributes 13.1% is followed by Indonesia with 8.8%, the United States of America with 7.3%, and Russia with 6.7% in the production market as of 2015. https://www.globalfirepower.com/coal-production-by-country.php (Fig. 1).

3 Coal Production in India

Coal mining, specifically open-cast mining, has grown to be a vital part of the Indian economy, helping to meet the nation's expanding energy needs. But there has been substantial environmental damage linked to this industrial prosperity. All India's coal production was able to reach a total of 997.83 million tonnes in 2023–2024,

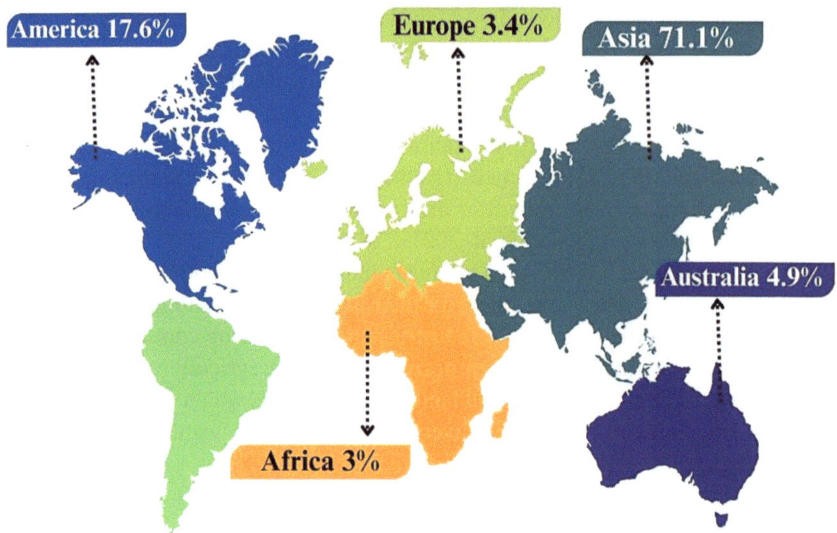

Fig. 1 Global distribution of coal

thanks to a consistent investment program and increased emphasis on the use of modern technologies. In 2024–2025, the total amount of coal produced in India was 1047.69 MT, representing an increase of 11.71%. Coal India Limited (CIL) and its subsidiaries produced 773.81 MT in 2023–2024, an increase of 10.04% over their production of 703.20 MT in 2022–2023. In 2024–2025, CIL produced 781.08 MT of coal, representing a positive growth of 0.94%. Coal is primarily supplied to the southern region by Singareni Collieries Company Limited (SCCL). Compared to 67.14 million tonnes of coal produced in the same period last year, the company produced 70.02 million tonnes in 2023–2024. The total SCCL coal production for 2024–2025 was 69.01 MT, representing a 1.45% increase. Additionally, TISCO, IISCO, DVC, and others produce small amounts of coal. Ministry of Coal.

The Government of India's Ministry of Coal claims that in recent years, India's coal output has been slowly increasing. Approximately 729 million tonnes (MT) of coal were produced in the nation during the 2019–20 fiscal year. Due to COVID-19 disruptions, production decreased slightly in FY 2020–21, totalling 716 MT. However, in FY 2021–2022, coal production reached 778 MT due to post-pandemic recovery and rising industrial demand. India crossed **893 MT** in net coal production in FY 2022–2023, for which Coal India Limited (CIL) alone contributed about **703 MT**. The remaining coal was produced by Singareni Collieries Company Limited (SCCL) and captive/private miners. For FY 2023–24, India showed a significant coal production of over **951 MT**, with a growth of about 6% from the previous year. The government has set an ambitious target to cross **1 billion tonnes** of coal production by 2025–26 to ensure energy security and reduce coal imports. The major coal-producing states of India are Odisha, Chhattisgarh, Jharkhand, and Madhya Pradesh, recording the production of 218.981 million tonnes MT, 184.895 MT, 156.445 MT, and 146.028 MT, respectively. The state of Odisha alone contributes 24.52% to the national coal production, followed by Chhattisgarh with 20.70%, Jharkhand with 17.54%, and Madhya Pradesh with 16.35%.

4 Coal Mining Wastes

During the extortion process of significant coal products, 0.4 tons of mining wastes warping coal washery products, tailing, and waste rocks, mine water, coal ash, spent heap leach material, sludge ponds, waste oil and other petroleum products is produced from 1 ton of hard coal (Dey et al. 2024). The tailings and waste rocks contain economically remediable coal. Runoff water pollution, spontaneous combustion, tailing storage facilities failures, and coal waste dumps are responsible for many detrimental environmental issues, for which there is an emerging demand for finding more responsible ways for the disposal of coal mining wastes. One of the convincing ways that has been shown is to manufacture secondary products from the wastes and generate revenue from that. This can be encountered by recovering fine coal from tailing streams, recovering lost coal from discarded plant materials, the extraction of rare earth elements, etc. Various low-cost building products, starting from roofing materials, road base, fabricated bricks and blocks, and agricultural products such as supplements for soil, along with moisture-retaining materials in various pellet forms, are included under the other secondary products (Jahandari et al. 2023a, b).

5 Coal Fly Ash

Tons of fly ash (FA), bottom ash, and polycyclic aromatic hydrocarbons are produced by coal-fed power plants (Harkness et al. 2015). The combustion of coal in thermal power plants is the source of a solid waste known as coal fly ash. It contributes 65–95% of the total ash generated. A range of oxides, including SiO2, Al2O3, and metal oxides that can be alkaline, alkaline earth, or transition metal oxides, make up this mineral mixture (Izquierdo et al. 2008). This solid waste covers a vast area of land, demanding a high maintenance cost, and also having ecological and environmental risks. By affecting the DNA repair mechanism with PAHs (from unburned organic matter) and inorganic elements (mineral fraction) through the creation of reactive organic species, fly ash discharge causes major health risks through inhalation of fine particulate matter (Matzenbacher et al. 2017). Apart from that, coal fly ash has plentiful valuable elements which make it a substance of potential. Knowing the limited supply of rare earth elements and the global decline in natural mineral resources, recycling valuable elements from coal fly ash is crucial to solving the problems of resource scarcity and large-scale solid waste disposal. All residues from coal combustion systems are referred to as coal ash (CA). Coal fly ash (CFA), which is usually collected from flue gas by precipitators, is the lighter portion, while coal bottom ash (CBA), which is heavier, is collected from the bottom of the combustion chamber. The residue that is produced when dust particles in flue gas from burning coal or lignite at 1100–1400 °C are precipitated mechanically or electrostatically is referred to as CFA by the European Coal Combustion Products Association (ECOBA) (Ju et al. 2021).

It is calculated that probably 1 ton of CFA is produced from burning 2 tons of coal, with a global total production of CFA projected at 1.143 billion tons. CFA is the most bountiful industrial residue for those countries that are more dependent on coal as an energy source, like India and China. Annually Coal coal-based power stations in India are generating approximately 131 million tons of CFA, while China exceeded this amount with 600 million tons in 2015, along with Brazil, which contributes 3 million tons of CFA every year (Mushtaq et al. 2019). CFA contains toxic trace elements in 4–10 times higher concentrations than those in coal (Neupane and Donahoe 2013). Severe waste management issues can be faced because of the continuous growth of the CFA in the future. For now, we are facing problems like stricter disposal site restrictions, diminishing landfill space, and rising disposal costs. This substantial amount of CFA production and its inappropriate handling is leading towards severe pollution of soil, water, and air, and are disrupting ecological cycles. The sustainable use of CFA by refining it into high-value items has gained attention recently as a way to reduce wastage of precious metals and environmental damage (Das et al. 2024).

6 Caol Tailing

In China and India, most extracted coal is beneficiated to boost its quality; coal tailings, a byproduct that constitutes about 6–8% of the initial coal output, are also a byproduct. Because of its tiny particle size, poor calorific value, high ash content, and high moisture content, a significant amount of coal tailings is thrown away, polluting the environment. Additionally, coal tailings frequently contain significant quantities of heavy metal elements, and food security is at risk from these heavy metals bioleaching into the soil. There are currently more than 5 million sites with 20 million hectares of heavy metal contamination globally (Filella et al. 2002, Ruiping et al.). In a nationwide soil quality survey conducted between 2005 and 2013, China's Ministry of Environmental Protection revealed a contamination rate that surpassed requirements at 16.1%, with heavy metals accounting for 19.4% of pollution sources.

Acid Mine Drainage (AMD) is considered to be one of the primary environmental issues related to coal mining. Iron sulphides, which are prevalent in waste and tailings, oxidize when exposed to air, water, and the bacteria *Thiobacillus ferroxidans*. This leads to the drainage process becoming acidic and metals dissolving. Mines that are still in action or have been abandoned might encounter these effects. Additionally, it is well recognized that the kind and number of sulphides, as well as the presence or lack of alkaline elements, define the kinds of metals, their concentration, and the degree of acidity that characterize AMD. In addition to the serious threat of contaminated surface and subsurface water supplies, this phenomenon also raises the possibility of threatening aquatic habitats. Recovery of these areas is both challenging, economically as well as in terms of reforestation (Milani et al. 2025).Approximately 60% of the quantity of coal extracted from open-pit and underground mines is thought to be solid waste following processing. Additionally, for every ton of coal dealt with, coal washing releases about 1.5 m3 of acidic effluents. Roughly 4,106 tons of waste were produced from coal processed in Brazilian coal plants in 2000 alone (Jahandari et al. 2023b). Due to the presence of potentially toxic metals and the high acidity of the AMD, it has significantly affected numerous regions of the world. AMD is responsible for long-term damage to waterways like river basins, agriculture, and biodiversity, with significant ecological consequences for human health. The released toxic substances are dangerous for living organisms, significantly affecting the ecosystem and communities (Akcil and Koldas 2006).

Substantial volumes of mine water (MW), which is made up of mine, pit, and drainage water, are produced by underground mining activities. Approximately one billion cubic meters of MW are generated annually by mining companies in Ukraine alone, using approximately 0.4 billion m^3 of drinking and technical water.

Since it requires typically 3–4 t of pumped water for every 1 t of extracted coal, treating these highly mineralized streams is a major barrier to meeting the SDGs. The quantity of MW and its mineralisation rise when coal seams' geology and mining conditions degrade. The challenge of converting MW into technical water is crucial, given the alarming global shortage of drinkable water. However, the majority of MW are still released into water bodies, with only a small portion being used in the mining sector's water supply. Likewise, useful chemical products, including sodium sulphate (Na2SO4), edible salt (NaCl), and different calcium and magnesium compounds, can be produced through MW processing. A fragment of rocks and coal brought to the surface during coal extraction processes is known as Coal mine waste (CMW). CMW contributes probably 20–40% of the mining-related wastes.

7 Sustainable Management of Coal Mining Wastes

Coal fly ash is a particle byproduct of burning coal, and its management has become critically needed due to environmental concerns about air, water, and soil pollution. Researchers have tried several times to turn coal fly ash into money-making goods while looking for feasible methods to use it ecologically. Coal fly ash wastewater remediation is one of such attempts to deal with waste management and water quality concerns. Coal fly ash has properties such as shape, surface area, porosity, and chemical composition (silica, alumina, iron oxide, titania, etc.) that make it a material that may be used in wastewater treatment. Notably, a substantial amount of CFA gets used for a variety of purposes, particularly in construction (Xu and Shi 2018)and civil engineering (Hoy et al. 2016)where it serves as a cement substitute in concrete (Ebrahimi et al. 2017; Wasserman and Bentur 1997).

It has been recognized for its use as a soil nutrient that improves texture, bulk density, water-holding capacity, and nutrients because of its greater surface area and multiple components (Bhattacharya et al. 2012). It also serves as a precursor to make mesoporous silica (Li and Qiao 2016) and zeolites (Fukasawa et al. 2017). It's applied as an adsorbent to eliminate mercury, SO2, CO2, and NOx from the air (Ji et al. 2017). It can also be used for making FA-based geopolymers, recovering important metals from CFA, and creating fly ash polymer composites for a variety of uses, such as creating materials to protect against electromagnetic interference.

One of the major pollutants of Coal mines is the mine water. The known methods for the mine water (MW) purification include distillation, Osmosis, electrodialysis, etc., which require a significant amount of energy expenditure and have challenges regarding the degree of mineralisation of desalinated water (Dastgheib et al. 2016). However, Desalination techniques are now much more advanced, but they are still not able to address the limitations on an industrial scale. For instance, reverse osmosis has $\eta E = 10\%$, electrodialysis has $\eta E = 3\%$, and evaporation desalination has $\eta E = 1\%$ for thermodynamic perfection (exergy efficiency). But there remains the problem of brine disposal. With an exergy efficiency of 12–26%, gas hydrate desalination is far more effective (Liang et al. 2024). Chlorofluorocarbons (CFCs), propane, and N-butane have all been utilised as gases that produce hydrates. However, the use of mine gases is a viable research route because of their high toxicity and limited availability.

The development of underground coal mines is currently accompanied by deeper depths and increasingly difficult geological conditions. Specifically, the amount of gas in mine workings rises to 25–30 m^3 for every ton of coal. Methane emissions into the mine environment have the potential to materially threaten miner safety, slow down the rate of coal output, and raise production costs. Depending on the degassing approach used, the extracted gas-air mixture's methane content varied greatly. The methane-air combination showed significant changes (ranging from 0.5% to 80%) during downhole degassing. Several countries currently only use just a portion of the mine gas that is recovered; the rest is released into the atmosphere, which causes severe air pollution and drives up the greenhouse effect.

Mine gases can be utilized in the following ways depending on the methane content: Synthesis of Methane(85–98%CH4), production of motor fuel ($CH4 \geq 95\%$), for the combustion process in boiler ($CH4 \geq 30\%$), combustion in steam generators ($CH4 \geq 9\%$), in the internal combustion engines ($CH4 \geq 6\%$), combustion in furnaces (methane-air mixture with 2% CH4).

The transformation of degassing techniques to the easy use of mine gas would improve the productivity and profitability of coal deposit extraction while significantly reducing environmental effects. The design of mine gas usage methods that are not solely dependent on variations in the flow rate and concentration of CH4 in the gaseous mixture, therefore, seems sensible. This makes the utilisation of the gas hydrate concentration process more hopeful. For the simultaneous utilisation of MW and methane, gas hydrate technology holds potential for the economic development of coal deposit resources while greatly minimising the harmful environmental effects of mining activities. Gas hydrate technology is therefore expected to reduce water body contamination and provide an extra freshwater supply. In areas where coal mining ended, as well as in areas where mining is still in progress, it may be feasible to further spread the suggested approach. Although the potential rewards extend to other mining industry sectors, stakeholders may include coal companies. The effective management of mining waste and the creation of additional revenue through the sale of valuable products (such as desalinated water, purified methane, dry salts, etc.) are two benefits of implementing the described technological process in any coal-mining region of the world.

Across the world, innovative approaches for managing and recycling coal mining waste are being developed. Using coal waste as a raw source for green construction supplies is one possible way to encourage the circular economy in coal mining communities. Other research suggests that this waste could be used to make aggregates for construction (Nixon and Gartner 1980), hot mix asphalt (Modarres and Rahmanzadeh 2014), and building ceramics materials. Lemeshev et al. (2004) demonstrated that coal mining waste is viable for producing construction ceramics. Different research recommends that coal waste can be used to make materials for construction (Nixon and Gartner 1980), building ceramic materials, hot mix asphalt (Modarres and Rahmanzadeh 2014)which suggests that coal mining wastes are significant for construction ceramics.

Coal mine waste (CMW) is currently being utilised in a number of agricultural processes, such as soil water retention, slow-release fertiliser, and mixed cropping substrate. As a result, CMW has come to be accepted as a useful farming resource. Its use in agricultural production is a topic of great ecological and economic value. It enables the large-scale use of CMW and satisfies the enormous agricultural demands of major producing nations (Zheng et al. 2024). Various minerals and elements like aluminium, silicon, iron, and some critical rare earth elements such as lithium, titanium, gallium, and germanium are present in CMW along with minerals like kaolin and coal. Metal recovery is the critical focus of secondary recycling of CMW. Cao et al. (2024), for example, assessed the recovery of lithium from coal mining waste. Pretreatment, lithium leaching, and selective removal from the leaching solution are the main steps in the recovery technology for these minerals. The extraction technique is currently in the laboratory stage, but it has promising future prospects and uses a similar enrichment and leaching procedure as lithium ore (Cao et al. 2024). CMW also has applications in the generation of power and environmental restoration. Areas like solar energy conversion, electrochemistry, and gas purification also have applications of CMW (Guo et al. 2024).

The discovery of sustainable and highly profitable coal and coal waste utilisation solutions has received a lot of attention. The recent discovery of new coal-based carbon nanomaterials, such as graphite (in 2003), carbon nanotubes (CNTs) (in 1999) (Williams et al. 1999), fullerene C60 and C70 (in 1991), and graphene (in 2012), mark critical turning points. There have been several reports of sophisticated techniques for creating carbon materials based on coal, such as graphene oxide (Savitskii 2017), carbon fibers (Mu et al. 2017), carbon foams (Calvo et al. 2005), and carbon dots (CDs) (Hu et al. 2016), all of which use various types of coal. Now there are also new innovative strategies where we can use coal waste as a raw material, like aerogels. Zeolites (reference) (Querol et al. 1995) and geopolymers.

Photocatalysis serves as a convincing solution to purify various pollutants, including the degradation of organic pollutants in wastewater, the Photocatalytic reduction of Cr (VI), photocatalytic disinfections, reduction and conservation of carbon dioxide. However, low stability, high cost, reduced catalysis efficiency, and post-separation issues are some of the technical obstacles that ordinary photocatalysts face in pollution reduction and other environmental protection activities (Liao et al. 2022). The direct use of photocatalytic materials in environmental protection is impacted by these issues, which stresses the significance of discovering economical and highly effective solutions. Researchers from a variety of domains are becoming more interested in the creation of photocatalytic nanocomposites using coal and coal waste as precursor materials in this context (Lum et al. 2020). As coal and coal waste consist of trace metals, photocatalytic nanocomposites made from such elements should have a lower band gap energy and be able to trap electrons similarly to metal-doped photocatalysts. The oxides in coal waste can also prevent photogenerated charges from recombining through coupling mechanisms. Coal-based and coal waste-based photocatalytic substances, therefore, hold a lot of possibilities to mitigate pollution and other connected uses. It is also feasible to significantly decrease the cost of manufacturing photocatalytic elements by using coal and coal waste as raw materials. The cost of coal and coal waste (such as coal fly ash and coal gangue) is much lower than that of typical raw materials used in the making of conventional catalytic materials (Sawunyama et al. 2023). The broad application of photocatalytic composites based on coal and coal waste in environmental protection applications is additionally supported by their cost-effectiveness (Lv et al. 2020) (Table 1).

8 Conclusion

Coal and coal waste-based supplies have received a lot of scientific attention as an innovative strategy for using coal and coal waste and safeguarding the environment. For the management of coal mining wastes, bioremediation strategies, particularly including bioleaching, offer a promising pathway. To detoxify, transform, or remove hazardous substances and recover critical elements, microorganisms are utilised as an eco-friendly approach for coal mine tailings and waste rock. By understanding the natural metabolic processes of microorganisms, the mobility and toxicity of pollutants can be reduced, and the impact on the environment can be minimised using bioremediation. Bioleaching is one of those strategies that contribute to the extraction of valuable metals along with the remediation of acid mine drainage (AMD) by microbially driven redox reactions. This brings both economic benefit and environmental protection. In comparison to traditional physicochemical techniques, bioremediation strategies require less energy, fewer secondary pollutants are produced, and can be tailored to site-specific conditions, increasing their effectiveness and adaptability. For maintained success factors like survival of microorganisms, the availability of substrate, temperature, and pH must be carefully managed. The blending of bioremediation into coal mine waste management plans can be a part of green technologies moving towards ecological restoration and circular economic goals.

Table 1 Bioremediation techniques for the sustainable management of coal mine wastes

Sl. no	Coal product used	Bioremediation techniques used	Product obtained	Reference
1	Coal tailing	Bioleaching using a mutated strain of *Bacillus mucilaginosus*	Silicon	(Zhang et al. 2025)
2	Coal tailing	Bioleaching using *Bacillus inaquosorum B.4* strain	Lead, Zinc, and Copper	(Zhang et al. 2024)
3	Coal fly ash (CFA)	Bioleaching (Acid and Alkaline leaching)	Aluminum	(Tian et al. 2025)
4	Stone coal	Bioleaching using *Aspergillus niger*	Vanadium	(Dong et al. 2024)
5	Acid mine drainage (AMD)	Bioleaching using *Leptospirillum, Ferrimicrobium, Sulfobacillus* and *Acidiferrobacter* at 30 °C, and *Acidithiomicrobium, Sulfobacillus* and *Acidithiobacillus* at 48 °C	Pyrite	(Joulian et al. 2020)
6	Coal gangue	Bioleaching using *A. ferrooxidans*	Desulfurization	(Liu et al. 2025)
7	Coal fly ash	Bioleaching using *A. niger*	REE	(Ma et al. 2023)
8	Stone coal	Bioleaching using biochar and plasma pretreatment with *Bacillus mucilaginosus*	Vanadium	(Dong et al. 2023)
9	Coal fly ash	Bioleaching using *Thiobacillus thiooxidans*	Aluminum and Iron	(Seidel et al. 2001)

Lastly, bioremediation using bioleaching shows a significant sustainable strategy for coal mining waste management. It decreases environmental liability, fosters resource recovery, and copes with the principles of green chemistry and sustainable development, subsequently paving the path for stronger and environmentally responsible mining practices.

References

Akcil A, Koldas S (2006) Acid Mine Drainage (AMD): causes, treatment and case studies. J Clean Prod 14(12–13):1139–1145. https://doi.org/10.1016/J.JCLEPRO.2004.09.006

Anthony T, Shohan AAA, Oludare A, Alsulamy S, Kafy AA, Khedher KM (2024) Spatial analysis of land cover changes for detecting environmental degradation and promoting sustainability. Kuwait J Sci 51(2):100197. https://doi.org/10.1016/J.KJS.2024.100197

Bhattacharya SS, Iftikar W, Sahariah B, Chattopadhyay GN (2012) Vermicomposting converts fly ash to enrich soil fertility and sustain crop growth in red and lateritic soils. Resour Conserv Recycl 65:100–106. https://doi.org/10.1016/J.RESCONREC.2012.05.008

Calvo M, García R, Arenillas A, Suárez I, Moinelo S R. (2005) Carbon foams from coals. A preliminary study. Fuel84(17):2184–2189. https://doi.org/10.1016/J.FUEL.2005.06.008

Cao Y, Zhou C, Gao F, Huang Y, Zhu W, Liu G, Wang J (2024) Lithium recovery from typical coal-based solid wastes: critical technologies, challenges, and prospects. Chem Eng J 498:155121. https://doi.org/10.1016/J.CEJ.2024.155121

Das A, Dey S, Das AP (2024) Ecotoxicological impacts of synthetic microfiber pollutants and development of sustainable mitigation strategies. Environ Chem Ecotoxicol. https://doi.org/10.1016/J.ENCECO.2024.12.003

Dastgheib SA, Knutson C, Yang Y, Salih HH (2016) Treatment of produced water from an oilfield and selected coal mines in the Illinois Basin. Int J Greenhouse Gas Control 54:513–523. https://doi.org/10.1016/J.IJGGC.2016.05.002

Dey S, Dash A, Das AP (2024) Bioremediation as a sustainable and circular solution for mining waste management and bioenergy generation. In: Sustainable management of mining waste and tailings: a circular economy approach, pp 107–119. https://doi.org/10.1201/9781003442455-6

Dong Y, Zan J, Lin H (2023) Enhanced bioleaching efficiency of vanadium from stone coal vanadium ore by adding biochar and plasma treatment. Miner Eng 196:108059. https://doi.org/10.1016/J.MINENG.2023.108059

Dong Y, Zan J, Lin H (2024) Enhanced vanadium bioleaching from stone coal employing Aspergillus Niger: optimization of stone coal pretreatment method. Miner Eng 215. https://doi.org/10.1016/j.mineng.2024.108792

Ebrahimi A, Saffari M, Milani D, Montoya A, Valix M, Abbas A (2017) Sustainable transformation of fly ash industrial waste into a construction cement blend via CO_2 carbonation. J Clean Prod 156:660–669. https://doi.org/10.1016/J.JCLEPRO.2017.04.037

Filella M, Belzile N, Chen YW (2002) Antimony in the environment: a review focused on natural waters: II. Relevant solution chemistry. Earth-Sci Rev 59(1–4):265–285. https://doi.org/10.1016/S0012-8252(02)00089-2

Fukasawa T, Karisma AD, Shibata D, Huang AN, Fukui K (2017) Synthesis of zeolite from coal fly ash by microwave hydrothermal treatment with pulverization process. Adv Powder Technol 28(3):798–804. https://doi.org/10.1016/J.APT.2016.12.006

Guo J, Fan Y, Qiao C, Ma X, Dong X, Zeng H (2024) Harnessing coal and coal waste for environmental conservation: a review of photocatalytic materials. Sci Total Environ 946:174437. https://doi.org/10.1016/J.SCITOTENV.2024.174437

Harkness JS, Ruhl LS, Millot R, Kloppman W, Hower JC, Hsu-Kim H, Vengosh A (2015) Lithium isotope fingerprints in coal and coal combustion residuals from the United States. Procedia Earth Planet Sci 13:134–137. https://doi.org/10.1016/J.PROEPS.2015.07.032

Hoy M, Horpibulsuk S, Rachan R, Chinkulkijniwat A, Arulrajah A (2016) Recycled asphalt pavement—fly ash geopolymers as a sustainable pavement base material: strength and toxic leaching investigations. Sci Total Environ 573:19–26. https://doi.org/10.1016/J.SCITOTENV.2016.08.078

Hu S, Wei Z, Chang Q, Trinchi A, Yang J (2016) A facile and green method towards coal-based fluorescent carbon dots with photocatalytic activity. Appl Surf Sci 378:402–407. https://doi.org/10.1016/J.APSUSC.2016.04.038

Izquierdo M, Moreno N, Font O, Querol X, Alvarez E, Antenucci D, Nugteren H, Luna Y, Fernández-Pereira C (2008) Influence of the co-firing on the leaching of trace pollutants from coal fly ash. Fuel 87(10–11):1958–1966. https://doi.org/10.1016/J.FUEL.2007.11.002

Jahandari S, Tao Z, Chen Z, Osborne D, Rahme M (2023a) Coal wastes: handling, pollution, impacts, and utilization. The coal handbook: Volume 2: Towards cleaner coal utilization, 2nd Edition, 2, 97–163. https://doi.org/10.1016/B978-0-12-824327-5.00001-6

Jahandari S, Tao Z, Chen Z, Osborne D, Rahme M (2023b) Coal wastes: handling, pollution, impacts, and utilization. The coal handbook: Volume 2: Towards cleaner coal utilization, 2nd Edition, 2, 97–163. https://doi.org/10.1016/B978-0-12-824327-5.00001-6

Ji L, Yu H, Wang X, Grigore M, French D, Gözükara YM, Yu J, Zeng M (2017) CO2 sequestration by direct mineralisation using fly ash from Chinese Shenfu coal. Fuel Process Technol 156:429–437. https://doi.org/10.1016/J.FUPROC.2016.10.004

Joulian C, Fonti V, Chapron S, Bryan CG, Guezennec AG (2020) Bioleaching of pyritic coal wastes: bioprospecting and efficiency of selected consortia. Res Microbiol 171(7):260–270. https://doi.org/10.1016/J.RESMIC.2020.08.002

Ju T, Meng Y, Han S, Lin L, Jiang J (2021) On the state of the art of crystalline structure reconstruction of coal fly ash: a focus on zeolites. Chemosphere 283:131010. https://doi.org/10.1016/J.CHEMOSPHERE.2021.131010

Li CC, Qiao XC (2016) A new approach to prepare mesoporous silica using coal fly ash. Chem Eng J 302:388–394. https://doi.org/10.1016/J.CEJ.2016.05.029

Liang W, Liu G, Han C, Xia L, Zhu W, Yang J, Han J (2024) Techno-economic analysis and fuel applicability study of a novel gasifier-SOFC distributed energy system with CO_2 capture and freshwater production. Desalination 586:117901. https://doi.org/10.1016/J.DESAL.2024.117901

Liao W, Zhao M, Rong H, Jiang P, Liao Q, Zhang C, Chen Y (2022) Photocatalyst immobilized by hydrogel, efficient degradation and self regeneration: a review. Mater Sci Semicond Process 150:106929. https://doi.org/10.1016/J.MSSP.2022.106929

Liu W, Liu Y, Yan Y, Jia G, Zhang Y, Lei Y, Li Z, Ma G, Li Z (2025) Application of A. ferrooxidans bioleaching on large-scale desulfurization of high-sulfur coal gangue sands and assessment of sulfur oxidation risk in cement-based materials. J Environ Chem Eng 13(5):117941. https://doi.org/10.1016/J.JECE.2025.117941

Lum PT, Foo KY, Zakaria NA, Palaniandy P (2020) Ash based nanocomposites for photocatalytic degradation of textile dye pollutants: a review. Mater Chem Phys 241:122405. https://doi.org/10.1016/J.MATCHEMPHYS.2019.122405

Lv Y, Xing B, Yi G, Guo H, Huang G, Zhang C, Liu Q, Zhang X, Cao Y, Li L (2020) Synthesis of oxygen-rich TiO_2/coal-based graphene aerogel for enhanced photocatalytic activities. Mat Sci Semicond Process 117:105169. https://doi.org/10.1016/J.MSSP.2020.105169

Ma J, Li S, Wang J, Jiang S, Panchal B, Sun Y (2023) Bioleaching rare earth elements from coal fly ash by Aspergillus niger. Fuel 354:129387. https://doi.org/10.1016/J.FUEL.2023.129387

Matzenbacher CA, Garcia ALH, dos Santos MS, Nicolau CC, Premoli S, Corrêa DS, de Souza CT, Niekraszewicz L, Dias JF, Delgado TV, Kalkreuth W, Grivicich I, da Silva J (2017) DNA damage induced by coal dust, fly and bottom ash from coal combustion evaluated using the micronucleus test and comet assay in vitro. J Hazard Mater 324:781–788. https://doi.org/10.1016/J.JHAZMAT.2016.11.062

Milani LAP, Fernandes L, Ribeiro EA, Battisti R, Bristot VM, De Noni A, Hotza D, Dal Bó M (2025) Synthesis and characterization of a clay-coal waste-based adsorbent for contaminant removal from acid mine drainage. Miner Eng 233:109616. https://doi.org/10.1016/J.MINENG.2025.109616

Mishra S, Dash D, Das AP (2024) Isolation and molecular characterization of polyester-degrading bacterial strains from municipal wastewater and their application in microplastic degradation in contaminated wastewater. Geomicrobiol J. https://doi.org/10.1080/01490451.2024.2376101

Modarres A, Rahmanzadeh M (2014) Application of coal waste powder as filler in hot mix asphalt. Constr Build Mater 66:476–483. https://doi.org/10.1016/J.CONBUILDMAT.2014.06.002

Mu X, Xu Z, Xie Y, Mi H, Ma J (2017) Pt nanoparticles supported on Co embedded coal-based carbon nanofiber for enhanced electrocatalytic activity towards methanol electro-oxidation. J Alloy Compd 711:374–380. https://doi.org/10.1016/J.JALLCOM.2017.04.008

Mushtaq F, Zahid M, Bhatti IA, Nasir S, Hussain T (2019) Possible applications of coal fly ash in wastewater treatment. J Environ Manage 240:27–46. https://doi.org/10.1016/J.JENVMAN.2019.03.054

Neupane G, Donahoe RJ (2013) Leachability of elements in alkaline and acidic coal fly ash samples during batch and column leaching tests. Fuel 104:758–770. https://doi.org/10.1016/J.FUEL.2012.06.013

Nixon PJ, Gartner EM (1980) An assessment of processes for the manufacture of synthetic aggregates from colliery spoil. Int J Cem Compos Lightweight Concrete 2(3):141–164. https://doi.org/10.1016/0262-5075(80)90016-0

Priyadarsini S, Das AP (2024) Analytical and structural characterization of waste lithium-ion batteries for their effective recycling strategy. Environ Chem Ecotoxicol. https://doi.org/10.1016/j.enceco.2024.12.004
Priyadarsini S, Das AP (2025) Lithium bioleaching: a review on microbial-assisted sustainable technology for lithium bio-circularity. J Water Process Eng 69:106744. https://doi.org/10.1016/j.jwpe.2024.106744
Priyadarsini S, Das AP (2026) Lithium in waste water: Impacts on pollution and human health. Decarbonization of Wastewater Pollutants as a Sustainable Solution, 353–364. https://doi.org/10.1016/B978-0-443-33062-9.00010-2
Querol X, Alastuey A, Fernández-Turiel JL, López-Soler A (1995) Synthesis of zeolites by alkaline activation of ferro-aluminous fly ash. Fuel 74(8):1226–1231. https://doi.org/10.1016/0016-2361(95)00044-6
Savitskii DP (2017) Preparation and characterization of colloidal dispersions of graphene-like structures from different ranks of coals. J Fuel Chem Techn 45(8):897–907. https://doi.org/10.1016/S1872-5813(17)30043-9
Sawunyama L, Oyewo OA, Seheri N, Onjefu SA, Onwudiwe DC (2023) Metal oxide functionalized ceramic membranes for the removal of pharmaceuticals in wastewater. Surf Interf 38:102787. https://doi.org/10.1016/J.SURFIN.2023.102787
Seidel A, Zimmels Y, Armon R (2001) Mechanism of bioleaching of coal fly ash by Thiobacillus thiooxidans. Chem Eng J 83(2):123–130. https://doi.org/10.1016/S1385-8947(00)00256-4
Swamy SL, Darro H, Mishra A, Lal R, Kumar A, Kumar Thakur T (2023) Carbon stock dynamics in a disturbed tropical forest ecosystem of Central India: Strategies for achieving carbon neutrality. Ecol Ind 154:110775. https://doi.org/10.1016/J.ECOLIND.2023.110775
Thakur TK, Patel DK, Dutta J, Kumar A, Kaushik S, Bijalwan A, Fnais MS, Abdelrahman K, Javed Ansari M (2021) Assessment of decadal land use dynamics of upper catchment area of Narmada River, the lifeline of Central India. J King Saud University–sci 33(2):101322. https://doi.org/10.1016/J.JKSUS.2020.101322
Tian X, Guo Z, Zhu D, Pan J, Yang C, Li S (2025) Recovery of valuable elements from coal fly ash: a review. Environ Res 282:121928. https://doi.org/10.1016/J.ENVRES.2025.121928
Tripathy B, Singh S, Behera ID, Mishra S, Das AP (2026) Metagenomic profiling of diversified marine microbiome across microplastic contaminated niches of Bay of Bengal, India. Mar Pollut Bull 222:118872. https://doi.org/10.1016/J.MARPOLBUL.2025.118872
Wasserman R, Bentur A (1997) Effect of lightweight fly ash aggregate microstructure on the strength of concretes. Cem Concr Res 27(4):525–537. https://doi.org/10.1016/S0008-8846(97)00019-7
Williams KA, Tachibana M, Allen JL, Grigorian L, Cheng SC, Fang SL, Sumanasekera GU, Loper AL, Williams JH, Eklund PC (1999) Single-wall carbon nanotubes from coal. Chem Phys Lett 310(1–2):31–37. https://doi.org/10.1016/S0009-2614(99)00725-3
Xu G, Shi X (2018) Characteristics and applications of fly ash as a sustainable construction material: a state-of-the-art review. Resour Conserv Recycl 136:95–109. https://doi.org/10.1016/J.RESCONREC.2018.04.010
Zhang Q, Ma L, Peng Y, Yan X (2024) Sustainable bioleaching of heavy metals from coal tailings using Bacillus inaquosorum B.4: mechanistic insights and environmental implications. J Environ Chem Eng 12(5):113400. https://doi.org/10.1016/J.JECE.2024.113400
Zhang Q, Yan X, Peng Y (2025) Feasibility of bioleaching for extracting effective silicon from coal tailings: optimization and modeling. J Environ Chem Eng 13(2). https://doi.org/10.1016/j.jece.2025.115964
Zheng Q, Zhou Y, Liu X, Liu M, Liao L, Lv G (2024) Environmental hazards and comprehensive utilization of solid waste coal gangue. Progress Natural Sci Mater Int 34(2):223–239. https://doi.org/10.1016/J.PNSC.2024.02.012

Transforming Lithium Slag into Sustainable Construction Materials: Pathways for Circularity and Environmental Sustainability

Salmabanu Luhar, Ashraf Ashour, and Ismail Luhar

Abstract The rapid global transition toward renewable energy and electric mobility has significantly increased the demand for lithium, which is predominantly extracted from spodumene ore. Lithium production generates substantial quantities of lithium refinery residue (LRR), commonly referred to as lithium slag (LS), which poses environmental hazards due to its leachable components and high disposal cost. This chapter presents a review on the utilization of LRR as a sustainable construction material, focusing on its application as a supplementary cementitious material (SCM) and its performance in concrete and mortar. The chapter covers detailed evaluations of workability, compressive strength, tensile and flexural strength, wear loss, impact resistance, reaction of alkali with silica, chloride (Cl^{-1}) ion infiltration, attacks of sulphate (SO_4^{2-}) ions and carbonation; water (H_2O) loss, shrinkage, creep, and sorptivity. Across multiple studies, LRR content ranging from 2 to 60% demonstrates notable influence on mechanical and durability properties, with optimal performance generally observed below 20–30% replacement. Incorporation of LRR improves long-term strength, reduces shrinkage and water loss, enhances wear resistance, and mitigates chloride penetration, supporting both environmental sustainability and circular economy objectives. The findings emphasize the potential of LRR valorization in producing low-carbon, durable, and sustainable construction materials while reducing industrial waste and carbon emissions.

Keywords Lithium slag · Lithium refinery residue · Sustainable construction materials · Supplementary cementitious material (SCM) · Circular economy

---

S. Luhar (✉) · A. Ashour
Faculty of Engineering and Digital Technologies, University of Bradford, Bradford, UK
e-mail: dr.salmabanuluhar@gmail.com

I. Luhar
Shri Jagdishprasad Jhabarmal Tibrewala University, Jhunjhunu, Rajasthan, India

A. P. Das and S. Priyadarsini (eds.), *Recovery and Recycling of Critical Metals from Electronic Waste for Environmental Sustainability*, Earth and Environmental Sciences Library, https://doi.org/10.1007/978-3-032-29064-9_7

1 Introduction

The global demand for reliable, efficient, and sustainable energy solutions has reached unprecedented levels in today's rapidly evolving world. Energy storage technologies have emerged as critical enablers of the ongoing energy revolution, bridging the gap between energy generation and consumption (McIlwaine et al. 2021). As the global economy expands and environmental awareness increases, the development of new energy industries is advancing dynamically across the world. Notably, the global transition toward renewable energy and electric mobility has sharply increased the demand for lithium. Lithium is a soft, silvery-white alkali metal valued for its unique properties and wide-ranging applications (Tang et al. 2021). As the least dense and lightest solid element, it exhibits a high electrochemical potential, making it essential for electric vehicles, rechargeable batteries, energy storage systems, and electronic devices. Beyond battery applications, lithium compounds also play important roles in the glass, ceramics, and pharmaceutical industries (Dong et al. 2021). Its significance in advancing modern technologies and environmentally friendly energy systems underpins the rapidly increasing global demand and its growing economic importance. Lithium (Li) is typically extracted from lithium-rich ores and subsequently converted into lithium carbonate (Li_2CO_3), from which lithium is recovered for use in battery manufacturing. Although lithium resources are abundant worldwide, major deposits are found in Bolivia, the United States, Argentina, Chile, China, and Australia (Jaskula 2020). According to the United States Geological Survey (USGS), global lithium consumption was approximately 170,000 tons in 2023 and increased by 29% to 220,000 tons in 2024. Projections suggest that global demand may exceed 2 million tons by 2030 and potentially reach 5 million tons by 2050 (USGS 2022). Similarly, global lithium salt production now exceeds 1.2 million tons annually. This surge is driven by the accelerating clean energy transition (Asif et al. 2024). Currently, high-purity lithium compounds are predominantly extracted from spodumene, a Li–Al silicate ore, and to a lesser extent from lepidolite, a lithium-bearing mica. The processing of spodumene generates LRR, commonly referred to as LS, along with a variety of metal oxides as by-products of lithium carbonate production. Lithium production yields approximately 9–10 tons of LRR for every ton of lithium carbonate produced (Lemougna et al. 2019; Rahman et al. 2022). Unfortunately, the disposal of LRR is costly, hazardous, and poses significant environmental risks. Improper disposal can lead to the leaching of fluoride (F^-) and sulphate (SO_4^{2-}) ions, requiring careful and systematic waste management (Liu et al. 2019a, b). Conventional disposal methods—such as open-air dumping or landfilling—not only consume vast areas of fertile land but also increase ecological hazards. High sulphur levels in LS can contaminate soil and groundwater, making safe disposal an important research priority.

To address these challenges, researchers are exploring the valorization of LS in fields such as construction materials (Rahman et al. 2023, 2024), ceramics, molecular sieves, and bio-electrode materials. LS also has potential applications in producing silica, alumina, aerogels, and glass ceramics. Rich in silica (SiO_2), alumina (Al_2O_3), and other oxides, LS represents a valuable resource for use in construction and infrastructure industries (Gu et al. 2024; Zhai et al. 2024). Given its scalable generation, the identification of economically viable and sustainable reuse pathways is crucial. In particular, the construction industry can effectively utilize LS as an aggregate in cement or as a supplementary cementitious material (SCM) in concrete. LS has a particle size distribution similar to that of ordinary Portland cement (OPC), the primary binder in concrete (Lu et al. 2019). However, LS produced through sulphuric acid (H_2SO_4) extraction often contains elevated levels of sulphur trioxide (SO_3), necessitating careful control to maintain cement matrix stability (Li et al. 2021; Gou et al. 2023; Rupasinghe et al. 2017). Thus, a cautious and informed approach is required to achieve the sustainability benefits of LS without compromising material performance.

OPC production accounts for approximately 7–9% of global carbon dioxide (CO_2) emissions—the primary greenhouse gas contributing to climate change (Luhar and Luhar 2019). Incorporating LS into cement and concrete formulations can reduce clinker consumption, enhance the sustainability of construction materials, and minimize waste burden from the lithium industry. This approach aligns with global circular economy principles by reintegrating industrial by-products into high-value applications, thereby improving resource efficiency, reducing landfill demand, and lowering carbon footprints (Priyadarsini and Das 2024, 2026). Beyond environmental benefits, the reuse of LRR offers economic advantages by reducing raw material and waste management costs for both lithium production and construction sectors. Incorporating LRR into construction materials can also improve performance properties, including durability, strength, chemical resistance, microstructural densification, and the formation of additional C–A–S–H gel (Gu et al. 2024; Luo et al. 2024). Not only does this strategy serve the dual purpose of waste valorisation and the promotion of sustainable construction, but it also aligns with global circular economy principles. The circular economy framework emphasizes product design and processes that minimize waste while maximizing resource efficiency. Reusing LRR in the manufacture of construction materials such as concrete offers numerous economic and environmental advantages. Importantly, this approach reduces the negative impacts associated with landfill disposal and mitigates the carbon footprint of construction-material production by partially replacing conventional raw materials that are typically associated with high energy consumption and greenhouse gas emissions (Chaudhury et al. 2023; Cheng et al. 2023). From an economic perspective, incorporating LRR can lower the costs of both lithium production and construction by reducing expenditures related to raw materials and waste management, making this approach highly cost-effective. Moreover, blending LRR into construction materials can improve their performance characteristics—including enhanced durability, strength, and resistance to chemical attacks—by promoting microstructural densification and the formation of additional C–A–S–H gel (Gu et al. 2024; Luo et al. 2024). In light of the dual global challenges posed by carbon- and energy-intensive construction processes and the growing need for effective industrial waste management, LS valorisation offers a highly impactful and economically advantageous solution. It reduces disposal costs for the lithium industry while simultaneously lowering raw material consumption and production costs in the construction sector for manufacturing building materials (Chaudhury et al. 2023; Luo et al. 2024). Table 1 provides an overview of progress in LS research, its applications, and the performance enhancements reported in recent literature.

Critically, this chapter examines the potential of converting LS into sustainable building materials, highlighting pathways toward circularity, decarbonization, and enhanced environmental sustainability. The primary objectives are to analyze the origin, chemical composition, and physicochemical characteristics of LS relevant to eco-construction; evaluate its performance as a sustainable building material; assess its environmental and economic benefits; identify the challenges hindering large-scale industrial adoption; and propose future research directions. Ultimately, this chapter supports informed decision-making aimed at transforming LS into viable construction and infrastructure materials, advancing circularity and environmental sustainability while contributing to global "net-zero" and "waste-to-resource" goals.

2 Tendency in Generation of LRR

Industrially, approximately 8–10 tons of LRR are generated for every ton of lithium carbonate produced (Shah et al. 2021; Zhai et al. 2024). The increasing trend in annual lithium carbonate production indicates that LRR is also being produced in substantial quantities. For example, in 2016 alone, an estimated 500,000 tons of LRR were generated solely from lithium carbonate production, and this figure is expected to continue rising in the future. The rapid accumulation of unrecycled or unutilized LRR has created significant challenges for lithium salt producers, particularly with respect to storage, handling, and environmental compliance. Therefore, establishing systematic and sustainable LRR management strategies—rather than continuing reliance on landfilling or stockpiling—is essential for the long-term, environmentally responsible operation of the lithium salt industry.

3 Mineralogy and Chemistry of LS

LS is characterized by the presence of several key mineral phases, including quartz (SiO_2), calcite ($CaCO_3$), gypsum ($CaSO_4$), ruby (Al_2O_3)—a red variety of corundum—trisodium aluminate ($Al_2O_3 \cdot 3H_2O$), and corundum (Al_2O_3), along with smaller amounts of kaolinite, lithium carbonate (Li_2CO_3), and glassy phases (Duan et al. 2022; Shi et al. 2020; Wu et al. 2012; Chen et al. 2012). The silica (SiO_2) content primarily originates from the co-magmatic mineral lepidolite, while gypsum ($CaSO_4$) is formed through reactions between limestone or calcium carbonate powder and sulphuric acid (H_2SO_4) during lithium extraction.

The relatively low content of glassy phases in LS is attributed to the low roasting temperatures (250–300 °C) used during the sulphuric acid processing stage. These

Table 1 Overview of LS research progress, applications, and performance enhancements reported in recent literature

Research area	Key scientific findings	References
SCM and geopolymer precursor	LS is an aluminosilicate by-product with pozzolanic characteristics, suitable as a supplementary cementitious material and as an alkali-activated precursor. It contributes to pore structure refinement and densification in cementitious matrices	Shah et al. (2021), Guo and Wang (2023), Rahman et al. (2023), Javed et al. (2022a, b, 2024a, b), Luo et al. (2023a, b)
Broader material utilization	Demonstrated in structural and manufacturing uses such as alkali-activated binders, mine backfill, lightweight ceramsite, and bricks, indicating its versatile resource potential	He et al. (2019, 2021), Li et al (2023a, b), Borges et al. (2016)
Mechanical property enhancements	Promotes secondary C–S–H/C–A–S–H gels and reinforces the ITZ, resulting in higher compressive, tensile, and flexural strength	Zhai et al. (2021), Yang et al. (2024), Rahman et al. (2023), Zhang et al. (2022), Gu et al. (2024), Rahman et al (2024), Amin et al. (2024), Chen et al. (2024), Wang et al. (2024a, b, c), Li et al. (2019a, b, c, 2020a, b), Luo et al. (2017a; b)
Durability improvements	Reduces permeability-related deterioration mechanisms including chloride ingress, sulfate attack, drying shrinkage, and freeze–thaw damage by refining pore network and improving impermeability	Zhai et al. (2021), Yang et al. (2024), Rahman et al. (2023), Gou et al. (2023), Wen (2013), Wu et al. (2014a, b, c), Gu et al. (2024), Amin et al. (2024), Li et al. (2019a, b, c), Luo et al. (2017a, b), He et al. (2021), Foghi et al. (2024), Silva et al. (2025), Li et al. (2024)
Synergistic blended systems	When used with slag, FA, metakaolin, fibers, or recycled aggregates, LS improves rheology, microstructure, and cost-performance balance	Zhai et al. (2021), Tabelin et al. (2021), Liu et al. (2019a, b), Gao et al. (2022), Gou et al. (2023)
Use in Alkali-activated and 3D-printed materials	Enhances polymerization, setting kinetics, chemical resistance, and mechanical stability in emerging additive manufacturing and geopolymer systems	de Souza et al. (2024), Mim et al. (2025)

(continued)

Table 1 (continued)

Research area	Key scientific findings	References
Microstructural evidence	SEM/XRD confirm denser matrices, increased gel volume, reduced pore connectivity, and stronger ITZ interfaces that support durability enhancement	Luo et al. (2017a, b), He et al. (2018), Gou et al. (2023), Zhang et al. (2022), Rahman et al. (2024), Amin et al. (2024), Wu et al. (2014a, b, c), Vigneshwari and Jayaprakash (2024)
Environmental sustainability benefits	LS utilization reduces CO_2 emissions, landfill disposal, and natural resource extraction; however, quantitative LCA data remain insufficient and future priority	Zhai et al. (2021), Tabelin et al. (2021), Yang et al. (2024), Gao et al. (2022)
Challenges to large-scale adoption	Barriers include absence of standards, limited economic advantage, low early-age reactivity, and reduced market willingness for industrial implementation	Zhai et al. (2024)
Activation strategies for improved reactivity	Physical, chemical, and combined activation technologies can improve early hydration and pozzolanic efficiency; systematic performance benchmarking remains limited	Zhang et al. (2022), Gou et al. (2023)
LS as fine aggregate substitute	Similar particle size and SiO_2 content to sand provide a viable opportunity for fine aggregate replacement, enhancing strength and increasing LS consumption rate	Zhu et al. (2024), Dong et al. (2023), Luo et al. (2021)
Relation to other industrial residues	Findings align with circular utilization of steel slag, high-titanium slag, phosphorous slag, and gold tailings sludge in low-carbon construction materials	Zhan et al. (2022), Chen et al. (2023), Wang et al. (2023a, b)
Emerging focus on s LRR	Rising LRR quantity from lithium production drives its incorporation into concrete and geopolymer applications as a sustainable waste solution	Javed et al. (2024a, b), Rahman et al. (2023, 2024), Zhai et al. (2024)

temperatures are insufficient to convert SiO_2 and Al_2O_3 present in lepidolite into amorphous glass phases. As a result, the silica and alumina in LS exhibit limited reactivity, reducing their ability to directly participate in cement hydration reactions and consequently impacting the strength development of LS-based concrete. Importantly, the combined SiO_2 and Al_2O_3 content typically exceeds 60% of the total mass of LS, indicating substantial potential for pozzolanic reactions under appropriate processing or activation conditions (He et al. 2017a, b; Wang et al. 2020, 2023a, b). The main chemical components of LS are SiO_2, Al_2O_3, CaO, and Fe_2O_3. These compositional variations significantly influence the pozzolanic activity of LS and its suitability for use in concrete. For example, the use of sulphuric acid during lithium extraction increases the SO_3 content, which can inhibit the formation of C–S–H gel in concrete, leading to a reduced rate of hydration (Zhou et al. 2022). In contrast, when K_2SO_4 is used in the sulphate process, the resulting K_2O reacts with sulphate in LS during the early hydration stage to produce temporary hydration products such as potassium gypsum or potassium sulphate (Zhou et al. 2022). Additionally, the presence of Na_2O formed during the leaching stage—due to the use of sodium hydroxide—facilitates cement hydration. Conversely, limestone-based processing increases the CaO content, creating an alkaline environment that further promotes hydration reactions (Zhou et al. 2022). Among these methods, the sulphuric acid process has emerged as the predominant commercial technique.

4 Utilization of LRR or LS in Construction Materials Production

LRR, also known as LS, is a by-product generated during lithium extraction and refining. Owing to its high silica (SiO_2) and alumina (Al_2O_3) content, LS exhibits strong potential as a supplementary cementitious material (SCM) for sustainable construction applications. Its incorporation into cement, concrete, geopolymers, and other construction materials not only mitigates environmental impacts but also contributes to improved mechanical performance and enhanced durability. Moreover, the use of LS reduces CO_2 emissions and lowers the demand for natural raw materials, thereby decreasing overall production costs (Rahman et al. 2023; Gu et al. 2024; Gou et al. 2023).

4.1 Utilization of LRR or LS as Cement or SCM

LRR, or LS, contains substantial amounts of SiO_2 and Al_2O_3, making it suitable for use as a partial replacement for ordinary Portland cement (OPC) or as a supplementary cementitious material (SCM) (Rahman et al. 2023; Gu et al. 2024; Gou et al. 2023). The pozzolanic activity of LS can be further enhanced through mechanical, thermal, or chemical treatments (Tan et al. 2020, 2021; Wang et al. 2024a, b, c; He et al. 2023; Li et al. 2019a, b, c). LS can also be synergistically combined with industrial by-products such as fly ash (FA), blast furnace slag (BFS), and slags from steel and carbide plants. Such combinations improve compressive strength, refine microstructure, and enhance CO_2 sequestration in OPC-based systems (Gu et al. 2023a, b; Wang et al. 2024a, b, c; Zhou et al. 2023). Additionally, LS can be incorporated into specialized cement types, including white Portland cement (PC), magnesium oxychloride cement (MOC), magnesium ammonium phosphate cement (MAPC), and cemented paste backfill (Li et al. 2020a, b, 2023a, b; Dong et al. 2021; He et al. 2019, 2021). Overall, LS proves to be an effective SCM, improving mechanical properties, durability, and sustainability while simultaneously reducing industrial waste and CO_2 emissions.

4.2 Utilization of LRR or LS in Cement Concrete

LRR, or LS, has been extensively studied as a partial replacement for cement in concrete. Its incorporation improves mechanical properties, microstructure, and durability. Optimal replacement levels generally range from 20 to 40%, enhancing compressive and tensile strength as well as the elastic modulus (Ec) (Rahman et al. 2024; Amin et al. 2024). Microstructural analyses indicate that LS promotes a denser interfacial transition zone (ITZ) and contributes to the formation of additional C–S–H and C–A–S–H gels. In addition, LS enhances resistance to environmental attacks such as acid rain (Zhang et al. 2022). In high-performance and ultra-high-performance concretes (HPC and UHPC), LS acts both as a nucleation site and as a pozzolanic material, promoting secondary hydration reactions and reducing the formation of weaker hydration products (Yang et al. 2024). Optimal LS addition of 10–20% improves long-term strength, and partial substitution of silica fume with LS can reduce costs while promoting sustainability. Synergistic use with other industrial by-products—such as fly ash, blast furnace slag (BFS), and iron ore tailings—further enhances compressive strength, workability, and durability (Zhang et al. 2023a, b; Zhu et al. 2024). Overall, incorporating LS in concrete improves mechanical performance, microstructure, and environmental sustainability. However, replacement levels above 30% may reduce strength due to the presence of unreacted particles.

4.3 Utilization of LRR or LS in Geopolymer Concrete

LRR or LS is a promising precursor for production of geopolymer concrete, owing to its alumina-silicate content. Its pozzolanic activity can be enhanced through thermal and mechanical activation, increasing amorphous content and Si/Al dissolution (Dong et al. 2024; Javed et al. 2022a, b; Luo et al. 2024; Wang et al. 2020). Calcination at ~ 700°C improves compressive strength significantly, with N(C)-A-S–H gel formation facilitates for densification of the micro-structure (Luo et al. 2024). Activator type and dosage significantly influence geopolymer performance. The potassium (K) enclosing activators increase the compressive strength as well as resistance against acids inferior to those of Sodium (Na) based, while hybrid activators NaOH + $CaCO_3$ + $Ca(OH)_2$ enhance C-A-S–H gel formation and overall strength of roughly 35.6 MPa for 28 days (Javed et al. 2024a, b; Luo et al. 2023a, b). Also, LRR or LS can be combined with mineral additives or industrial wastes such as FA, silica fume (SF), blast furnace slag (BFS), and metakaolin (MK) to improve mechanical performance, durability, and thermal stability (Li et al. 2021; Shah et al. 2021; Guo and Wang 2023; Luo et al. 2021). Optimal LRR content ranges from 10–30%, with higher proportions potentially reducing long-term strength. LRR or LS-based geopolymers offer high performance, sustainability, and inferior emissions of carbon dioxide (CO_2) in comparison with contemporary OPC production process. Their mechanical as well as durability performances can be optimized through activator selection, thermal activation, and synergistic use of mineral additives.

4.4 Utilization of LRR or LS in Miscellaneous Applications in Construction

LRR or LS is explored in foamed ceramics and lightweight aggregates (LWA) for sustainable construction: (i) Foamed ceramics: LRR blended with albite and kaolin using high-temperature foaming produces materials with compressive strength ~ 5.7 MPa, porosity ~ 73.7%, and water absorption ~ 2.24% at 40% LRR content (Xiong et al. 2023). Excessive LRR reduces mechanical strength due to particle agglomeration. (ii) Lightweight aggregates (LWA): LRR combined with contaminated soil or FA can produce aggregates with compressive strength > 1.5 MPa, stabilized heavy metals, and lower costing for manufacturing in the range of 0.3 to 0.6 US $ per one ton of production against the present 7 or more US $ for each ton (Li et al. 2023a, b). (iii) Lightweight ceramsite: Optimal mixes in the ratio of FA: Li-residue from refinery: albite: kaolin:: 4:4:1:1 to produce ceramsite presenting lesser water absorption of 0.96%; strength for compression as 1.76 MPa as well as 0.46 kg/m^3 symbolizing for bulk and 0.86 kg/m^3 figuring for apparent densities, respectively (Li et al. 2023a, b). These applications demonstrate that LRR or LS can be valorised beyond traditional cement and concrete, providing environmentally safe, cost-effective, and sustainable materials for construction. Table 2 shows the effects, applications, and activation of LS or LRR in conventional Cement and Concrete, as well as in Geopolymer systems.

Table 2 Effects, applications, and activation of LS or LRR in cement, concrete, and geopolymer systems

Application	Material	Optimal LS/LRR replacement (%)	Mechanical properties	Microstructural/ durability benefits	Key findings and notes	References
SCM/cement	OPC, special cements	2.5–20	Compressive strength up to 70 MPa (thermally activated)	Enhanced pozzolanic activity, formation of C-S–H/C-A-S–H gels, refined pore structure	LS improves ITZ, reduces harmful voids, and contributes to early and late-age strength at optimal doses	Gou et al. (2023), Gu et al. (2024), He et al. (2023), Li et al. (2020a, b)
SCM/cement	OPC + industrial wastes (SS, FA, BFS)	10–30	Strength ~ 50–65 MPa	Synergistic effects, improved hydration, refined microstructure, CO_2 sequestration	LS combined with other SCMs enhances durability and strength, reduces carbon footprint	Gu et al. (2023a, b), Wang et al. (2024a, b, c), Zhou et al. (2023)
Concrete	Normal concrete	10–40	Compressive 48–52 MPa, improved tensile strength	Denser ITZ, additional C-S–H/ C-A-S–H gels, reduced drying shrinkage and creep, acid resistance	Optimal replacement (10–20%) refines microstructure; > 30% may reduce flowability and early strength	Rahman et al. (2024), He et al. (2017a, b), Amin et al. (2024), Zhang et al. (2022)
Concrete	HPC/UHPC	10–20	Compressive strength up to 100 MPa	Filler and nucleation effects, secondary hydration, refined microstructure, reduced dependency on silica fume	Improves mechanical performance while maintaining workability; excessive LS can cause pore imbalance	He et al. (2018), Wu et al. (2014a, b, c), Yang et al. (2024), Zhang et al. (2023a, b)

(continued)

Table 2 (continued)

Application	Material	Optimal LS/LRR replacement (%)	Mechanical properties	Microstructural/durability benefits	Key findings and notes	References
Concrete	Synergistic mixes (IOT, BFS, grain slag)	10–30	Compressive strength up to 65 MPa	Optimized workability, chloride and carbonation resistance, densified microstructure	Combining LS with other industrial residues improves durability and cost-efficiency	Zhang et al. (2023a, b), Luo et al. (2023a, b), Zhu et al. (2024)
Geopolymer	LRR-based	10–30	Compressive 36–50 MPa (subjected to activation)	Formation of N-(C)-A-S-H/C-(N)-A-S-H gels, denser matrix, room-temperature curing possible	Activation essential due to low amorphous content (<20%) in unactivated LS; improves silica-alumina reactivity	Liu et al. (2019a, b), Luo et al. (2024), Wang et al. (2020)
Geopolymer	LRR + FA/SF/BFS/MK	10–50	Compressive 35–77 MPa	Enhanced durability, thermal stability, densified gel formation, reduced porosity	Synergistic blends enhance polymerization, gel formation, and microstructural density	Javed et al. (2022a, b, 2024a, b), Li et al. (2021), Shah et al. (2021), Luo et al. (2021)
Miscellaneous	Foamed ceramics	40	Compressive 5.7 MPa	High porosity (~73.7%), low water absorption (~2.2%)	LS/LRR used as filler; improves lightweight and insulating properties	Xiong et al. (2023)
Miscellaneous	Lightweight aggregate/ceramsite	10–40	Compressive > 1.5 MPa (LWA), 1.76 MPa (ceramsite)	Stabilization of heavy metals, reduced density, low water absorption, cost-effective	LS/LRR incorporation reduces resource usage and enhances microstructure	Gao et al. (2022), Li et al. (2023a, b)

(continued)

Table 2 (continued)

Application	Material	Optimal LS/LRR replacement (%)	Mechanical properties	Microstructural/ durability benefits	Key findings and notes	References
Hydration and reactivity notes	–	10–20 (optimal)	Early and late-age strength improvement	Refinement of ITZ, C-S–H/C-A-S–H formation, reduced harmful voids	Excessive LS (>30%) reduces flowability, pore structure, and mechanical performance; activation needed to enhance amorphous SiO_2/Al_2O_3 content	Li et al. (2020a, b), Zhou et al. (2023), Zhang et al. (2023a, b), Liu et al. (2019a, b), Rahman et al. (2023), Wang et al. (2024a, b, c)
Activation requirement	Thermal/ chemical/hybrid	—	Improved strength and pozzolanic activity	Higher amorphous phase (~50%), enhanced N-(C)-A-S–H gel formation	Activation critical for high-level cement replacement and large-scale utilization	Tan et al. (2020), Tan et al. (2021), Liu et al. (2021), Rahman et al. (2023)

5 Setting Time of LRR-Based Cementitious Systems

Several studies have examined how lithium residue resources LRR influence the setting behavior of cement-based materials. Haigh et al. (2013) investigated two concrete mixtures with target strengths of 25 MPa and 40 MPa, incorporating 25% of LRR as a supplementary cementitious material. For the 25 MPa mixture, the initial and final setting times were measured at 580 min and 690 min, respectively. In comparison, a mixture containing 25% FA exhibited shorter such as 4 h 40 min and 7 h for early and final setting timings correspondingly. An inclusion of 25% LRR led to setting times that were 64–78% longer than the corresponding control specimens with respect to 40 MPa mixtures. The authors suggested that variations in superplasticizer content may have contributed to these observations. Tan et al. (2018) explored the use of sulfoaluminate cement (SAC) without ordinary Portland cement (OPC) and reported significantly shorter setting times. The study further indicated that increasing the LRR content generally reduced the setting duration. These findings are consistent with previous observations by Tan et al. (2021). For example, in systems with 10% of LRR addition, initial and final setting times were eight minutes and 10 and half minutes in that order, compared to 8.7 and 11.35 min in the control mixtures. Two main mechanisms were proposed to explain this acceleration. First, SAC contains a higher proportion of corundum than OPC (Li et al. 2019a, b, c), which promotes rapid formation of ettringite (AFt) in the presence of dissolved lithium salts from LRR, effectively eliminating the induction period. Second, the fine particle size of LRR serves as nucleation sites that further enhance AFt formation and overall hydration kinetics (Han and Yan 2010). A comparison for the setting behaviour of concrete mixtures incorporating LRR (calcined as well as non-calcined) with FA-based concretes was performed by Munn et al. (2019). Initial and final setting timings for mixtures containing 25 percentages LRR (calcined) were found slightly lower than those for FA and LRR (non-calcined) mixtures amounting to 5 h 55 min and 7 h 30 min. Similarly, Tan et al. (2021) evaluated cement pastes with 0.5% to 4% nano-LRR as additive material and observed that increasing nano-LRR content led to faster setting. In particular, pastes containing 4% nano-LRR exhibited 16.34 min and 21.93 min as early and final setting times, representing reductions of roughly 33 and 37 percentages respectively compared to controlled samples. For paste and mortar specimens, setting times below 20 min were primarily governed by the combined use of SAC and nano-LRR. Concrete mixtures, due to their larger aggregate content and lower water-to-cement ratios, displayed longer setting times ranging from 290 to 690 min. Overall; the setting behaviour was strongly influenced by both the percentage of LRR incorporated and the type of cementitious system.

6 Workability of LRR-Based Cementitious Systems

The workability of cementitious materials incorporating LRR has been widely studied, with effects observed on fresh density, quantity of air, slump, as well as flow characteristics. Huang et al. (2011), reported that LRR-based aerated mortar combined with FA exhibited a lower fresh density compared to FA-only mixtures. The maximum density recorded for 100% LRR mortar was 918 kg/m^3, roughly 4 percentages greater than corresponding FA mixture. Similarly, Zhang and Wang (2011), evaluated ternarily as well as quaternarily groupings of LRR, GGBFS plus FA for concrete production. They observed that a binary blend of 80% slag of GGBF and 20% LRR reached a maximum density of 2532 kg/m^3, whereas a ternary mixture containing slag of of GGBF and FA in equal quantity of 40 percentage each, plus 20% LRR led to density of 2493 kg/m^3. The LRR-concrete's density examination by Haigh et al. (2013), setting the target strengths of 25 and 40 MPa, was then compared to FA and controlled specimens. Part cement substitution with 25% LRR put on the show the concrete density faintly declining in both mix designs. At a steady ratio of water/cement as well as dosage of the superplasticizer, a boost in slump noted up to 20 percentage quantity of LRR, reaching 196 mm compared to 186 mm for the control as reported by Shi and Zhang (2011). Beyond 20%, slump decreased sharply. Haigh et al. (2013), maintained constant slump values for 25 and 40 MPa concrete blends with ratios of 0.65 and 0.5 for water/cement, respectively, by adjusting high-range water-reducing admixtures. Wen (2013), reported that increasing lime powder content in 10% LRR concrete decreased workability. A research study by Wu et al. (2014a, b, c), on ternary mixes containing OPC, LRR, and steel slag with water-cement ratios between 0.27 and 0.35 revealed that the slump showed no consistent trend with LRR content, increasing at lower water-cement ratios of 0.27 to 0.30 but decreasing at 0.35, likely due to reduced cement content and higher admixture usage. Further investigations on high-performance concrete (HPC) by Wu et al. (2014a, b, c), found an increasing LRR content led to a 6% slump reduction at 30% LRR and a 0.30 water-cement ratio, contrasting with earlier studies (Wen 2013). A report of He et al. (2017a, b), showed middling slump values of 150 to 180 mm for mixes containing ten to thirty percentages of pozzolanic supplement like LRR. An observation of Munn et al. (2019), for twenty-five percentages of LRR (calcined and non-calcined) showed a drop in slump by four percent in calcined and twelve percent in non-calcined, when a comparison made with twenty-five percent FA-concrete, corroborating the trends reported by Wu et al. (2014a, b, c). At higher replacement levels of 40 to 60%, He et al. (2020), observed slump values ranging from 180 to 200 mm, indicating that workability generally decreases as LRR content rises. Flow characteristics of ultra-HPC containing 5–15% LRR with 5–20% silica fume (SF) was reported by He et al. (2019), with flow values ranging from 180 to 200 mm. Application of "Minimum mortar flow" technique by Li and Huang (2020), to determine the greatest usable quantity of LRR was performed, adding-up 2 to 11% LRR. Their results indicated boosting LRR has elevated water demand and reduced slurry fluidity, leading to a recommendation of a maximum 11% LRR for optimal mortar performance. Air content measurements for 25 and 40 MPa LRR-concrete mixes containing 25% LRR showed slightly higher values like 2% to 2.4% than FA and control specimens (Tan et al. 2020). Munn et al. (2019), reported a 14% reduction in air content for both types of LRR-calcined as well as non-calcined were compared with FA based concrete. Differing trends were attributed to water-cement ratio variations: higher ratios increase free water and entrained air. Both studies also found that LRR incorporation reduced concrete bleeding relative to control and FA specimens. The electrical properties of LRR-modified mortar were explored by Tan et al. (2018, 2020). For wet-ground 2.5–10% LRR mortar, conductivity decreased with increasing LRR content, with maximum of 636 μS/cm when 10% LRR added, which stands roughly fifty-six percentage inferior to the controlled samples. Similarly, micro-LRR (4%) cement pastes exhibited a maximum conductivity of 5783.3 μS/c.m. at 4 and half hours, lower than 6799.5 μS/cm for the control, reflecting decreased dissolved ion concentrations and enhanced hydration product formation (e.g., AFt and C-S-H gel). Resistivity measurements also indicated that LRR additions accelerated the early development of resistivity, with 10% LRR samples showing 28% lower resistivity than control after 8 h (Tan et al. 2018). Table 3 displays the workability examination data for LRR-based cement and concrete mixtures.

7 Mechanical Strength of LRR-Based Cementitious Systems

The effect of LRR on the compressive strength of cementitious systems has been extensively investigated under various combinations and curing conditions. An evaluation of concrete by Li et al. (2012) with cement replacement by LRR in proportion of ten, twenty and forty percentages together with 10% lime powder reported that a 10% LRR substitution achieved highest SAI—strength activity index; of 121% on 28th day and a the greatest 139 MPa on 180th day in context of compressive strength. An examination on 25% addition of LRR by Haigh et al. (2013) with concrete blends of 25 and 40 MPa strength, has noted that no active pozzolanic action found till 28th day. The values for SAI on 28th day were found 113% and 99% for the respective concrete mixes. Wen (2013) studied ternary blends containing 10–40% limestone powder and 10% LRR, reporting reductions in 28- and 60-day compressive strengths by 24–54% and 2.5–29%, respectively, compared to control specimens. Interestingly, early-age strengths at 3 and 7 days for 10% limestone + 10% LRR exceeded those of 10% LRR alone, with SAI values of 131% and 118%, respectively. Wu et al. (2014a, b, c), investigated partial cement replacement using 10–20% FA combined with 10–30% LRR. For the case of equal FA and LRR as ten percent each, the SAI recorded somewhat greater like 1% to 3.6% than for 10% LRR alone at 3, 7, 28, and 90 days. Combinations of 20% FA + 30% LRR maintained SAI values near 100% on 28th and 90th day relative to the controlled specimens. Tan et al. (2015) disclosed that

Table 3 Workability test of LRR-based cement/concrete mixes

LRR %	SCM combination	Slump/flow (mm)	Fresh density (kg/m^3)	Air content (%)	Observations	References
0	–	186	918	2.0	Baseline slump and density	Huang et al. (2011)
20	FA ternary combination	196	–	–	Slump increased up to 20% LRR, sharp decrease beyond 20%	Shi and Zhang (2011)
25	Partial cement replacement	–	–	2.4 (25 MPa), 2.0 (40 MPa)	Air content higher than FA samples; constant slump with variable admixtures	Haigh et al. (2013)
10	LRR + 10–40% LP	–	–	–	Increased lime powder decreased workability	Wen (2013)
10–30	25–65% OPC + 20–40% steel slag	0.27–0.35 w/c	–	–	Slump not following trend; inversely proportional at lower w/c	Wu et al. (2014a, b, c)
10–30	Calcined and non-calcined LRR	–	–	–	25% calcined-LRR plunged four percent slump and non-calcined-LRR by 12% when compared to FA based concrete	Munn et al. (2019)
5–15	LRR + 5–20% SF	180–200	–	–	Flow value consistent for all combinations	He et al. (2018)
2–11	High strength mortar	172–225	–	–	Increase in LRR increased water demand, reducing fluidity; maximum 11% recommended	Li and Huang (2020)
4	Micro-LRR	–	–	–	Conductivity 5783.3 μS/cm vs 6799.5 μS/cm control	Tan et al. (2020)
2.5–10	Wet grinded LRR	–	–	–	Electrical resistivity inversely proportional to conductivity; 10% LRR lowest resistivity 48 Ω·m	Tan et al. (2018)

both dry grinding duration and LRR content significantly influenced the compressive strengths of early and long-standing time strengths. Specifically, 50% LRR with 10 min of dry grinding exhibited a 28-day SAI of 85% compared to ungrounded samples. When using accelerators, aluminum silicate slightly outperformed sodium silicate for enhancing compressive strength in 20% LRR mixes. He et al. (2017a, b), reported that partial replacement of cement with 10–30% LRR improved long-term compressive strength. A 20% replacement increased 60- and 90-day strengths by 7.12% and 9.57%, in that order, when comparing with controlled concrete. Similarly, investigations of He et al. (2018), on the ultra-HPC with ternarily blends of LRR and silica fume (SF), finding that equal proportion of ten percent each for LS as well as SF led to the greatest 28- and 90-day compressive strengths, with increases of 4% and 7% relative to control mixtures. Wet grinding of LRR further enhanced pozzolanic activity and compressive performance. For instance, 30-min wet grinding of 2.5–10% LRR in cement mortar produced the highest 28-day compressive strength at 5% LRR (32.3 MPa) (Qi et al. 2017). A 120 min wet grinding increased the SAI to 101–128% for 0.5–4% LRR addition (Tan et al. 2020). Lu et al. (2019), investigated ternary blends of industrial slag (3–50%) and 3–30% LRR, identifying 12% LRR + 18% slag as the optimal combination, achieving thirty-six and 60.8 MPa on 7th and 28th day, correspondingly. A comparison of calcined as well as non-calcined LRR in twenty-five and 12.5% together in that order with 12.5% FA was made by Munn et al. (2019) and they noted the 28th day compressive strengths as 62.3 MPa and 34.4 MPa, respectively. Equal LRR and FA in ten percent content of each have produced the greatest strength among ternary combinations as revealed by Wu et al. (2014a, b, c), highlighting the influence of LRR origin and composition. A comparison performed by Li et al. (2019a, b, c) on wet and steam curing at 80°C, for mortar enclosing twenty percent LRR and noted that steam-cured samples exhibited a 35.6% lower compressive strength than control specimens after 720 days due to over-steaming effects. Qin et al. (2019) explored the use of LRR in recycled coarse aggregate (RCA) concrete, observing a 28-day compressive strength of 37.65 MPa and an SAI of 115.14% for 10% LRR + 50% RCA. Zhang et al. (2020), demonstrated that the addition of 0.03–0.10% TIPA with 30% LRR improved paste compressive strength, with 0.10% TIPA achieving 60th day strength as 67.4 MPa representing 118.7% SAI. The 40–60% LRR as SCM remained largely inactive until 180 days, with only a 3.66% strength gain after 60 months for 40% LRR as revealed by He et al. (2020). Li and Huang (2020), prepared high-strength mortars using ternary combinations of 10% SF and 2–11% LRR, with the maximum SAI (99%) and compressive strength (112.2 MPa) achieved at 11% LRR. Li et al. (2020a, b), reported similar results for mortar prepared using white coloured PC containing 2.5 to ten percent of LRR, where found compressive strength as 48 MPa with SAI as 171% on 28th day. The influence of LRR on tensile and flexural properties of cementitious materials has been studied under various mix designs and curing regimes. Splitting tensile strength generally varies with the LRR content, showing both enhancements and reductions depending on the replacement level and combination with other materials. Shi and Zhang (2011), reported the percentage use from 15 to 45 of additive LRR for HP-concrete. Their results put on the show that splitting tensile strength decreased with increasing cement changeover on 7th and 14th day; however, the uppermost strengths observed were 6.0 and 6.5 MPa, corresponding to 3.45 and 4.84 percentages greater than the controlled mix. The binary and ternary groupings enclosing ten to thirty percent LRR and ten to twenty percent FA examined by Wu et al. (2014a, b, c). The binary mix with 10% LRR achieved the highest 28-day splitting tensile strength of 5.6 MPa, while the ternary combination with 10% LRR + 10% FA showed an approximately 9% increase compared to the binary mix. A study of Qin et al. (2019) on RCA-concrete using 10–100% recycled coarse aggregates along with 10–25% LRR and found that 70% RCA + 15% LRR yielded a maximum splitting tensile strength of 5.04 MPa, which was 57.5% greater than the controlled samples. Collectively, these studies suggest that an LRR replacement of 10–15% optimizes splitting tensile strength, though the inclusion of RCA can slightly increase the recommended LRR dosage. Flexural strength similarly depends on LRR content and complementary materials. Wen (2013), reported that 10% LRR combined with 10–40% limestone powder (LP) produced flexural strengths of 11–11.1 MPa at 28 days, suggesting a moderate LP content in LRR mixes achieves optimal performance. Tan et al. (2015), evaluated cement pastes with varying LRR content and grinding durations. They observed that 50% LRR had the slowest early flexural development, while the addition of aluminum sulfate accelerated strength gain. The maximum 28-day flexural strengths were 10 MPa (30% LRR), 9.2 MPa (10 min grinding), 10 MPa (0.8% sodium silicate), and 10.3 MPa (0.4% aluminium sulphate). Li et al. (2019a, b, c), investigated 20% LRR in wet-cured and steam-cured mortars. High-temperature steam curing has reduced 28-day flexural strength by 12.2% for 20% LRR and 7.8% for control specimens. Qin et al. (2019) reported that across different RCA replacement levels (0% to 100%), 20% LRR consistently produced the highest 28-day flexural strengths, with increases ranging from 17.8% to 46.1% over respective controls. Li and Huang (2020) also demonstrated that in high-strength mortars containing 10% silica fume, 8% LRR yielded the highest 28-day flexural strength of 27.2 MPa, indicating the importance of optimizing LRR content to avoid strength reductions. Elastic modulus (EC) follows trends similar to compressive strength. He et al. (2017a, b) found that at early ages, 30% cement replacement with LRR decreased EC from 33.23 to 28 GPa at 7 days. However, at 28, 60, and 90 days, the elastic modulus increased due to pozzolanic activity, with 20% LRR achieving a maximum EC of 38.69 GPa at 90 days. Qin et al. (2019) corroborated that 20% LRR provides an optimal elastic modulus in RCA-containing concrete at 28 days, while higher LRR content (>20%) results in a rapid decline in stiffness, regardless of curing age. Overall, tensile, flexural, and elastic modulus tests indicate that moderate LRR replacement levels of typically 10% to 20% optimize mechanical performance, whereas higher dosages can compromise early-age and long-term properties. The combination with supplementary materials such as FA, silica fume, or limestone powder, as well as appropriate curing regimes, further enhances performance and mitigates potential drawbacks. The influence of LRR on the wear and impact resistance of concrete and mortar has been evaluated in several studies. Qi et al. (2017), investigated ten to thirty percent quantity of LRR as a supplementary cementitious material in concrete mixes and cement-concrete testing standards on 28th day. Their results indicated that replacing cement with up to 20%

LRR minimized wear loss to 3.4 kg/m^2, representing a 43.33% reduction compared to controlled concrete. In the same way, samples of mortar with high-strength mortar sizing 150 × 150 × 150 mm^3 studied by Li and Huang (2020), with two to eleven percent changeover by LRR. They reported that 11% LRR produced a wear loss of 0.82 kg/m^2, approximately 44.6% lower than the control. These findings suggest that moderate LRR replacement effectively enhances resistance to surface wear, though performance depends on both the proportion of LRR and specimen dimensions. Impact resistance of LRR-containing mortars has also been assessed. Li and Huang (2020), applied a 5.2 kg weight from a 1 m drop height on 28-day cured specimens. The results showed that increasing LRR content generally improved impact resistance up to a certain threshold. Specifically, 5% LRR as a partial cement replacement required seven impacts to cause failure, which was 40% higher than the control specimen. However, LRR content above 5% tended to slightly reduce impact resistance, indicating the need for careful optimization. Overall, these studies demonstrate that moderate LRR replacement around 5% to 20%) improves both wear and impact resistance of concrete and mortar, providing a sustainable strategy to enhance durability while utilizing industrial waste. The mechanical characteristics of LRR-based Cementitious Systems are portrayed by Table 4.

8 Durability Properties of LRR-Based Cementitious Systems

The influence of LRR on durability-related deterioration mechanisms in cementitious systems has been widely investigated. Munn et al. (2019) evaluated alkali-silica reaction (ASR) using mortar bars in a sample size of 25 by 25 by 285 mm immersed in 1 M sodium hydroxide (NaOH) solution keeping the temperature of 80 °C. This study compared control specimens made with general-purpose (GP) cement to mixes containing FA, and LRR—calcined as well as non-calcined. Results indicated that LRR (calcined type) has slimmed down the expansion to 0.05 and 0.14% on 10th and 21th day, in that order, whereas LRR (non-calcined type) exhibited higher expansions of 0.10 and 0.39%. Calcined LRR also outperformed FA, showing 3.96% and 11.76% lower expansion on the same test days. These findings were confirming that LRR, particularly in calcined form, effectively mitigates ASR. Chloride migration in LRR-containing concretes has also been studied. A partial replacing of cement with 10–30% LRR by Qi et al. (2017), disclosed measurement of electric flux over 6 h. Increasing LRR content reduced the flux, with 30% LRR achieving the minimum value, roughly 43% inferior to the controlled. In the high-strength white Portland cement mortars enclosing two to eleven percent LRR, Li and Huang (2020), observed that 8% LRR offered the most effective reduction in chloride migration, decreasing penetration by approximately 22%. Further, Wu et al. (2014a, b, c), highlighted water/cement ratio's influence on ClDC, i.e., chloride diffusion coefficient on 28th and 84th day. Their results indicated that lower water-cement ratios reduced ClDC, with the

Table 4 Mechanical properties of LRR-based cementitious systems

LRR %	SCM/combination	Compressive strength (MPa)	SAI (%)	Splitting tensile strength (MPa)	Flexural strength (MPa)	Elastic modulus (GPa)	Notes/ observations	References
10	LRR + 10% Lime Powder	–	121 (28 d), 139 (180 d)	6.0 (7 d), 6.5 (14 d)	–	–	Max early SAI and tensile strength; pozzolanic activity enhanced with lime powder	Li et al. (2012), Shi and Zhang (2011)
25	Partial cement replacement	–	113 (25 MPa mix), 99 (40 MPa mix, 28 d)	–	–	–	Pozzolanic activity dormant till 28 d	Haigh et al. (2013)
10–30	LRR + 10–20% FA	–	100 (28 and 90 d)	5.6–6.1	–	–	Ternary combination higher tensile strength than binary	Wu et al. (2014a, b, c)
20	LRR	–	–	–	10.0	–	Early flexural strength development moderate	Tan et al. (2015)
20	LRR + 0.4–0.8% Al2(SO4)3	–	–	–	10.3	–	Flexural strength accelerated with aluminum sulfate	Tan et al. (2015)

(continued)

Table 4 (continued)

LRR %	SCM/combination	Compressive strength (MPa)	SAI (%)	Splitting tensile strength (MPa)	Flexural strength (MPa)	Elastic modulus (GPa)	Notes/ observations	References
20	LRR + 0.4–0.8% Sodium Silicate	–	–	–	10.0	–	Slightly lower flexural strength than aluminum sulfate	Tan et al. (2015)
2–11	LRR + 10% SF	112.2 (28 d)	99	27.2	–	–	Maximum compressive strength at 11% LRR	Li and Huang (2020)
5	LRR + wet grinding 30 min	32.3 (28 d)	–	–	–	–	Amorphousness enhanced early and long-term strength	Tan et al. (2020)
12% LRR + 18% Slag	Ternary combination	36 (7 d), 60.8 (28 d)	–	–	–	–	Highest compressive strength among ternary combinations	Lu et al. (2019)
25	Calcined LRR	62.3 (28 d)	–	–	–	–	Non-calcined LRR ternary mix reduces compressive strength	Munn et al. (2019)

(continued)

Table 4 (continued)

LRR %	SCM/combination	Compressive strength (MPa)	SAI (%)	Splitting tensile strength (MPa)	Flexural strength (MPa)	Elastic modulus (GPa)	Notes/observations	References
20	LRR	–	–	–	27.2 (28 d)	38.69 (90 d)	Optimum elastic modulus and flexural strength; pozzolanic activity increases with time	He et al. (2017a, b), Qin et al. (2019)
40–60	LRR	–	–	–	–	–	SCM dormant until 180 d; slow strength gain	He et al. (2020)
10%	LRR + 10% FA	–	–	–	–	–	Maximum splitting tensile strength in binary and ternary mixes	Wu et al. (2014a, b, c), Shi and Zhang (2011)
15%	LRR + RCA	–	–	5.04	46.1	38.0	Coarse aggregate replacement slightly increased optimum LRR usage	Qin et al. (2019)

minimum value of 1.77×10^{-12} m^2/s observed at 0.27 water-cement ratio after 28 days—about 33% lower than the control. Additionally, the primary as well as the secondary hydration product's formulations over time densified interfacial transition zone (ITZ), further limiting chloride penetration.

Sulphate resistance of LRR mortars has been assessed under long-term exposure. An exposure of 20% LRR enclosing mortar specimens to 99% concentrated solution of sodium sulphate (Na_2SO_4) made for 1 year 355 days by Li et al. (2019a, b, c), and comparison made of wet-cured and steam-cured specimens sizing 40 by 40 by 160 mm. The samples which were steam-cured have put on the show a maximum mass change of 4%, which was found 2.67 times greater than controlled, highlighting the pessimistic impact of overheating. XRD analysis revealed the formation of thenardite (efflorescence) in both the parts of the up and the down sides of the partly submerged specimens, indicating that excessive curing temperature can accelerate sulfate-induced deterioration by limiting AFt and AFm formation.

Overall, these studies suggest that LRR incorporation, particularly when optimized in content and processing (e.g., calcination), improves resistance to ASR and chloride ingress, while careful curing is critical to mitigating sulfate-related degradation in cementitious systems. The effects of LRR on shrinkage and creep of concrete have been systematically studied. Haigh et al. (2013), evaluated 25 and 40 MPa concrete grades incorporating 25% LRR as a pozzolanic additive. For the 25 MPa mix, LRR concrete exhibited 26.5 percentage greater shrinkage than the FA based concrete, but it was 2.3% inferior to controlled samples on 56th day. With regard to 40 MPa possessing blend, the shrinkage of LRR enclosing concrete matched to that of FA concrete and was 12% lower than the control. The report of He et al. (2017a, b), revealed increasing quantity of LRR generally reduced shrinkage, although contents above 30% led to higher shrinkage at all ages. Specifically, 30% LRR replacement reduced 28- and 90-day shrinkage by roughly 30% and 27% compared to the control. Similarly, an evaluation by He et al. (2017a, b) and Qi et al. (2017) for ten to thirty percent LRR changeover and recommended 20% and 30% LRR, respectively, for optimal shrinkage reduction. He et al. (2017a, b), observed shrinkage reductions of 42% and 27% at 28 and 180 days with 20% LRR, whereas Qi et al. (2017) reported reductions of 37% and 24% at 28 and 180 days with 30% LRR. Differences in optimum LRR content may arise from variations in coarser sorts of aggregates, surface area of LRR, and blend proportions. The investigations on binarily and ternarily blending of FA with LRR enclosing calcined as well as non-calcined were carried out by Munn et al. (2019). They revealed that the demonstration of superior shrinkage control was put on the show by calcined ternary blends to those non-calcined ternary, whereas non-calcined binarily blends have shown more better-quality than their calcined counterparts. Although, LRR plus FA in equal quantity as 12.5% in a blend was identified as an optimized mix, shrinkage recorded 9.3 and 4.3% superior to 25% FA controlled on 28th and 56th day. A study by Li and Huang (2020) on the high-strength mortars enclosing within LRR in the proportion of 2 to 11% as a partial cement replacement. The results indicated that LRR, due to its porous nature, retained free water within the mortar matrix, slowing water evaporation and thereby reducing shrinkage. The optimal composition was 11% LRR, which lowered shrinkage by factors of 0.83 and 0.76 at 28 and 56 days, respectively. Data from various studies show that shrinkage remains below the maximum allowable value of 600 με specified by AS 3972, demonstrating compliance with standard requirements.

Creep behaviour has also been investigated. He et al. (2017a, b), evaluated concrete with 10–30% LRR under a sustained dead or static load of twenty-five percent for the compressive strength of the 7th day (sample size: 100 × 100 × 300 mm^3). The 7-day compressive strength was selected as it reflects the timing of early construction activities. The minimum creep was observed for 20% LRR, reducing creep by 30% and 29% at 90 and 180 days compared to the control. Corresponding creep coefficients were 1.31 and 1.38, indicating enhanced resistance to long-term deformation.

The sorptivity behaviour of LRR concrete has been evaluated using different LRR types, including those bagged, damp, and waxed. Munn et al. (2019) prepared samples of 50 × 50 × 100 mm^3 and cured them at 50 °C with 50–70% relative humidity for 7 days. The study revealed that non-calcined LRR specimens exhibited better resistance to sorptivity compared to FA and calcined LRR-concrete. Specifically, the sorptivity of 25% wetted, waxed and bagged non-calcined LRR based concretes found roughly 18, 17, and 25% inferior to the correlating calcined LRR-blends. Each value found to be satisfactory, with H_2O penetration remaining well under 25 mm.

Carbonation resistance was investigated by Qi et al. (2017) for concrete containing LRR in proportions of ten to thirty percent. Investigations recorded carbonation depths at 3, 7, 14, and 28 days. Results showed that higher LRR contents reduced carbonation resistance. At 28 days, the carbonation depth ranged from 5.2 mm to 10.99 mm. This behaviour was attributed to the reduction in cementitious material, which decreased the concrete's alkalinity and neutralization resistance, thus accelerating CO_2 penetration.

Water retention, which directly affects moisture migration and drying shrinkage, was studied by He et al. (2017a, b), for 15–45% LRR concrete. Using samples of 40 × 40 × 160 mm^3 possessing a rate of 2 to 3.4 g/h for H2O evaporation, the investigations revealed that LRR in thirty percent proportion has exhibited the lowest water loss, approximately 46% inferior to the controlled on 360th day. Increasing LRR content beyond 30% caused a sudden rise in water loss (up to 170 g at 45% LRR) due to the excessive unhydrated material and insufficient binding. The examination by Tan et al. (2020) carried out on the loss in the weight of specimens of cement-pastes enclosing four percent micro-LRR and sizing 40 by 40 by 40 mm^3 on 16 and 24 h as well as on 28th day under 50°–200 °C. Weight loss increased by 39%, 19%, and 26%, respectively, compared to the controlled, highlighting significance of non-evaporating H2O in evaluating hydration characteristics.

Finally, He et al. (2020), verified the non-evaporating H2O quantity of concrete enclosing LRR in the proportion of forty to sixty percent on 7th, 28th, 90th, and 180th day. Samples (40 × 40 × 40 mm^3) were dried at 105 °C, ground, sieved (80 μm), and further heated at 1000 °C. A greatest quantity of non-evaporating H2O was found with samples enclosing forty percent LRR on 180th day, slightly exceeding controlled by 2.4%, indicating enhanced hydration in these mixes. Table 5 stands for data of durability attributes of LRR-based cement and concrete blends.

Table 5 Durability properties of LRR-based cement/concrete mixes

LRR %	SCM/ Combination	Test type	Result	Notes/ observations	References
10–30	LRR	Wear loss	Min: 3.4 kg/ m^2 at < 20% LRR	~ 43% lower than control; minimum wear loss achieved at < 20% LRR	Qi et al. (2017)
2–11	LRR	Wear loss	0.82 kg/m^2 at 11% LRR	~ 44.6% lower than control; specimen size affects wear loss	Li and Huang (2020)
5	LRR	Impact resistance	7 hits from 1 m height	~ 40% higher than control; optimal at 5% LRR	Li and Huang (2020)
10–30	LRR	ASR expansion	Calcined: 0.05–0.14%; Non-calcined: 0.10–0.39%	ASR within acceptable limits; calcined LRR shows less expansion than non-calcined	Munn et al. (2019)
10–30	LRR	Chloride migration	Min: 1333 C at 30% LRR	~ 43% lower than control; electric flux decreases with LRR content	Qi et al. (2017)
2–11	LRR + 10% SF	Chloride migration	~ 22% lower than control at 8% LRR	High-strength mortar; optimal LRR content 8%	Li and Huang (2020)
0.27 w/c	LRR	Chloride diffusion coefficient (CIDC)	1.77×10^{-12} m^2/s at 28 d	~ 33% lower than control; improved ITZ densification reduces chloride penetration	Wu et al. (2014a, b, c)
20	LRR	Sulfate attack (mass change)	Max 4% after 720 d	Wet cured vs. 7 h steam cured; steam curing reduced AFt/ AFm formation	Li et al. (2019a, b, c)
10–30	LRR	Shrinkage	28 d: 37–42% reduction; 180 d: 24–27%	Optimal shrinkage reduction at 20–30% LRR; higher percentages increase shrinkage	He et al. (2017a, b), Qi et al. (2017), Haigh et al. (2013), Munn et al. (2019), Li and Huang (2020)

(continued)

Table 5 (continued)

LRR %	SCM/ Combination	Test type	Result	Notes/ observations	References
10–30	LRR	Creep	Min creep coefficient: 1.31 (90 d), 1.38 (180 d)	Dead load 25% of 7 d compressive strength; lowest for 20% LRR	He et al. (2017a, b)
25	LRR (wet, bagged, waxed)	Sorptivity	18–25% lower than calcined LRR	Non-calcined LRR improves water penetration resistance; ASTM C1585 compliant	Munn et al. (2019)
10–30	LRR	Carbonation depth	5.2–10.99 mm at 28 d	Higher LRR content reduces cementitious material; increased CO_2 infiltration	Qi et al. (2017)
15–45	LRR	Water loss	Min 46% lower than control at 30% LRR; abruptly increased at 45%	Excessive LRR cannot hydrate; retains water in unhydrated zones	He et al. (2017a, b)
4	Micro-LRR	Weight loss	16 h: + 39%, 24 h: + 19%, 28 d: + 26%	Non-evaporable water used to assess hydration characteristics	Tan et al. (2020)
40–60	LRR	Non-evaporable water	180 d: slightly higher (2.4%) than control	Indicates hydration progression and water retention in LRR concrete	He et al. (2020)

9 Analysis of Pore Size Distribution of LS

The porous structure of LS, pigeonholed with a huge inner surface region, demonstrates higher absorption competence for water and brilliant adsorption attributes (Gu et al. 2023a, b). An analysis of distribution of pore size discloses a bimodal allocation having somewhat greater percentage of meso-pores in comparison with macro-pores. The entire pore volume of LS is more or less fourfold to the cement, whereas, its surface area is roughly five times to cement. Nevertheless, the average diameter of LS's pore is yet inferior to cement (Li et al. 2019a, b, c). Substituteing part of the cement with LS can optimize the structure of the pore of composite paste piloting to enhanced mechanical attributes of the concrete. Moreover, the tiny pore diameter can augment, overall, the resistance against the infiltration of chloride ions and carbonation of concrete (Zheng et al. 2016; Wu et al. 2016), thus, enhancing the durability of concrete structures. Nonetheless, the adsorption characteristics of LS can escort to a drop in the quantity of free water of the concrete mix, ensuing in the amplified viscosity and reduced flowability of the paste (Luo et al. 2023a, b; Zhai et al. 2020) posing a challenge in preparing and transporting the LS-based concrete.

10 Pros of Usage of Li-Residue from Refinery

A consumption of Li-Residue from Refinery in manufacturing of structural and building materials represents numerous environmental and economical benefits as follows:

1. Reuse of LRR facilitates the reduction in the volume of generation of wastes from the Li-extraction industries, lessening the stress on the landfills and minimizing eco-pollution.
2. Preservation of resources: Substitution of conventional raw materials with LRR will decrease to consume limited natural resources of limestone and pyrophyllite rocks, endorsing the sustainable management of resources.
3. The contemporary process of OPC production is highly energy-intensive and generates momentous emissions of primary green house gas—carbon dioxide fuelling earth-heating. LRR reuse as partly substitution for OPC can lessen prerequisites for energy and CO_2 footprints from constructions helping decarbonisation targets to achieve from this industry, e.g., the entire CO_2 emissions and consumption of energy concerning to one ton production of LRR enclosing cement (activated using alkali) composites approximated about sixty-one and sixty-five percents of 1 ton of OPC production. For now, in manufacturing of ultra-HPC, 40% integration of LRR can slim down the earth-heating potential of ultra-HPC by thirty-six percent (Dong et al. 2024; Yang et al. 2024).
4. The integration of Li-Residue from refineries for production of materials to use in constructions can boost mechanical attributes, resistance against degrading eco-system as well as durability, piloting towards long-standing and superior resilient infrastructural practices.
5. A reuse of Li-Residue from refineries can decrease costing for materials to use in infrastructures, offering economical encouragements to adopt sustainable approaches. The information on the cost of production of Li-Residue from refineries induced cement (activated using alkali) composites was found more or less 70% to that of one ton of ordinary PC-production (Dong et al. 2024).

11 Challenges Reusing Li-Residues from Refineries to Manufacture Edifice Materials

Number of crucial challenges are standing in the pathway of widespread adoption of LRR to be promisingly reuse for manufacturing building materials. The key challenges can be described as:

a. The variable chemistry relying on Li-ore's source, the precise kind of processings and physical attributes of LRR escort to the inconsistent performances in constructions set hurdles in optimization as well as standardization (Zhai et al. 2024).
b. Financial viability is also an obstacle. When LRR represents potential as an adding-up material in constructions, the LRR processing cost, including the energy and chemicals needed for the treatment. It may bind its economical feasibility. In addition, the market competition of LRR with other waste materials can prospectively craft it with lesser attraction in support of scalable usage except incentivized by strategy or market exigency, which needs for economical appraisals as well as costing gain analysing to assess its longer feasibility.
c. A challenge of presence of impurities with LRR, viz., sulphate in spodumene LRR, sodium, lepidolite enclosing metals those are regarded as heavy, posing health as well as eco-dangers provided unless managed correctly. The leaching behaviour of the listed LRR enclosing substances requires methodical examination making sure to remain within regulatory frontiers posing no hazard to the eco-system as well as human health (Zhai et al. 2024).
d. Authoritarian barriers exist because till today, only a little harmonized guiding principles or set of laws existing meant for valuable reusing of Li-Residues from Refineries. Therefore, the improvement of vigorous strategy frame-works and industrial benchmarks are highly crucial for directing to safer as well as eco-friendly usage of Li-residues from refineries for production of building materials. The clearer standards are very much needed since the regulatory bodies more and more prioritize the sustainability factor so that promotion and the adoption of LRR can have no compromise on safety or eco-protection.
e. Integration of LRR with building material productions necessitates cautious thoughtfulness for mixture designing as well as conditions for processes (Zhai et al. 2024). A highly fine as well as precise surface area of refinery's Li-residues pilot to improved water exigency as well as influence concrete mixture's workability. To develop correct admixtures and optimize mixture proportions are vital to lessen these types of hurdles and attain the performance characteristics looked-for.
f. The deficient comprehensive information on the long-standing performances of LRR-integrated materials is an obstruction to extensive acceptance of LRR reuse. The stakeholders may be indecisive whether to go for reusing it or not for high-performance utilizations devoid of in-depth thoughtfulness for its stability and durability over the time.

g. Eco-trade-offs such as supplementary energy consuming linked to LRR process must also be considered for its advantageous reuse.

12 Eco-Influences of Mining and Extractions of Lithium

While lithium consumption contributes to energy savings by reducing reliance on fossil fuels, its mining, extraction, and use can have significant environmental impacts. Recent years have witnessed a 256% increase in lithium production due to record industrial demand, and it is projected to reach 740,000 tons by 2041 (Adeel et al. 2023). Improper disposal of lithium-containing waste poses threats to both aquatic and human health. For instance, water samples near central European lithium deposits recorded lithium concentrations of 0.7 ppm, exceeding U.S. EPA limits (Toupal et al. 2022). Lithium also occurs naturally in surface and subsurface waters through contact with lithium-bearing minerals. Certain lithium compounds, such as lithium aluminium hydride ($LiAlH_4$) and lithium methanolate (CH_3LiO), are listed as hazardous substances in Denmark (Kjølholt et al. 2003).

Large-scale lithium mining and extraction can lead to habitat destruction, biodiversity loss, and chemical contamination, endangering ecosystems. Disposal of electronic waste in landfills further exacerbates the problem, as battery electrolytes decompose and contaminate soil and water. Long-term open-brine evaporation processes require extensive land and water use, contributing to soil, water, and air pollution. For example, higher lithium concentrations were reported in water around Shanghai, China, due to extensive lithium battery disposal (Shen et al. 2020). In the arid Salar de Atacama region, lithium mining has negatively affected indigenous wildlife and their habitats (Moran et al. 2022). Lithium exposure through drinking water has been associated with potential thyroid disorders in humans (Broberg et al. 2011), although lithium is also used therapeutically for mental health, bipolar disorder, suicide prevention, mania, and short-term mortality reduction (Adeel et al. 2023). Generally, lithium does not bioaccumulate, and its human and environmental toxicity is relatively low (Aral and Vecchio-Sadus 2008). Three main lithium extraction techniques are currently employed: surface mining, evaporative concentration from brine, and direct lithium extraction (DLE) from brine, with DLE having comparatively lower environmental impacts. However, DLE still requires substantial freshwater and energy and generates chemical wastes (Vera et al. 2023). A greener approach involves extracting lithium from Li-ion batteries using reduced glutathione (GSH) as a reductant, which lowers leaching temperatures, reduces chemical use, and improves efficiency and safety (Gu et al. 2023a, b). Lithium mining and processing can also adversely affect agriculture. Lithium uptake by crops, due to high plant accumulation coefficients, can result in the entry of lithium into grains and vegetables, posing potential health risks (Bolan et al. 2021). Studies on leafy vegetables and Amaranthus species reported lithium accumulation primarily in above-ground plant parts, indicating potential risks to human consumers (Gayathri et al. 2022). Lithium exposure also affects plant physiological processes, ultimately influencing growth (Shahzad et al. 2016). Controlled studies on agromining or phytomining using rapeseed (Brassica napus) and sunflower (Helianthus annuus) demonstrated lithium accumulation exceeding 4000 ppm in some species, highlighting the potential for lithium recovery from plants (Kavanagh et al. 2018). Environmental studies have also addressed lithium pollution in aquatic ecosystems. Barbosa et al. (2023) emphasized the need for mitigation strategies to protect marine life and promote sustainable development. Thibon et al. (2021) analyzed lithium concentrations in over 400 specimens, including bivalves, cephalopods, crustaceans, and fish, across the Atlantic, Pacific, and Indian Oceans, providing a baseline for future eco-toxicological studies of lithium in marine environments.

13 Conclusion

LRR or LS is a promising industrial by-product with substantial potential for sustainable construction applications. The integration of LRR as a supplementary cementitious material demonstrates improvements in long-term compressive, tensile, and flexural strength, particularly at replacement levels of 10 to 30%. LRR enhances durability characteristics, including wear resistance, impact resistance, reduced chloride penetration, and controlled shrinkage and creep. Non-calcined LRR performs better in resisting sorptivity, while calcined LRR contributes to improved pozzolanic activity. Optimal LRR contents vary depending on mix composition, curing regime, and specific performance metrics. Excessive LRR (>40%) can negatively impact workability, early-age strength, carbonation resistance, and water loss due to incomplete hydration and higher content of pore water. Valorising LRR in concrete aligns with circular economy principles by reducing industrial waste, lowering the carbon footprints in form of carbon dioxide emissions, and decreasing reliance on natural raw materials. This approach offers both environmental and economic benefits, supporting decarbonisation and sustainable infrastructure development.

References

Adeel M, Zain M, Shakoor N, Ahmad MA, Azeem I, Aziz MA, Tulcan RXS, Rathore A, Tahir M, Horton R, Xu M, Rui YK (2023) Global navigation of Lithium in water bodies and emerging human health crisis. NPJ Clean Water 6:33

Amin MTE, Sarker PK, Shaikh FUA (2024) Transport properties of concrete containing lithium slag. Constr Build Mater 416:135073

Aral H, Vecchio-Sadus A (2008) Toxicity of lithium to humans and the environment—a literature review. Ecotoxicol Environ Safety 70:349–356

Asif AH, Li C, Lim H, Sun H (2024) Australia's spodumene: advances in lithium extraction technologies, decarbonization, and circular economy. Ind Eng Chem Res 63(5):2073–2086

Barbosa H, Soares AMVM, Pereira E, Freitas R (2023) Lithium: a review on concentrations and impacts in marine and coastal systems. Sci Total Environ 857:159374

Bolan N, Hoang AS, Tanveer M et al (2021) From mine to mind and mobiles lithium contamination and its risk management. Environ Pollut 290(18):118067
Borges PH, Santos FA, Milikic N, Barsante AC (2016) Lithium aluminosilicate residue as raw material in the production of sustainable concrete masonry units: a Brazilian case. Open Constr Build Technol J 10:418–430
Broberg K, Concha G, Engström K, Lindvall M, Grandér M, Vahter M (2011) Lithium in drinking water and thyroid function. Environ Health Perspect 119(6):827–830
Chaudhury R, Sharma U, Thapliyal P, Singh L (2023) Low-CO_2 emission strategies to achieve net zero target in cement sector. J Clean Prod 417:137466
Chen D, Hu X, Shi L, Cui Q, Wang H, Yao H (2012) Synthesis and characterization of zeolite X from lithium slag. Appl Clay Sci 59–60:148–151
Chen G, Huang Y, Yang R, Yu R, Xiao R, Wang Z, Ke X, Xie G, Cheng J, Bao M (2023) Comparative study on mechanical properties and microstructure development of ultra-high performance concrete incorporating phosphorous slag under different curing regimes. Constr Build Mater 392:131963
Chen X-B, Liang J-F, Li W (2024) Compression stress-strain curve of lithium slag recycled fine aggregate concrete. PLoS ONE 19:e0302176
Cheng D, Reiner DM, Yang F, Cui C, Meng J, Shan Y, Liu Y, Tao S, Guan D (2023) Projecting future carbon emissions from cement production in developing countries. Nat Commu 14(1):8213
de Souza EA, de Kassia Rodrigues e Silva R, Borges PHR (2024) Overburden materials from the iron mining as raw materials for AAM: preliminary assessment of mixes for 3D concrete printing. In: Proceedings of the FIB international conference on concrete sustainability, Guimar.es, Portugal, 11–13 September 2024. Springer, Berlin/Heidelberg, Germany, pp 414–421
Dong P, Ahmad MR, Chen B, Munir MJ, Kazmi SMS (2021) Preparation and study of magnesium ammonium phosphate cement from waste lithium slag. J Clean Prod 316:128371
Dong SK, Tu SW, Chen LL, Chen LL, Wu FF, Xie LL, Zhuo Q, Yu SH (2023) Investigation of the performance of cement mortar incorporating lithium slag as a super-fine aggregate. Front Mater 10:129979
Dong J, Chen L, Li L, Zhou P, Shi Z, Cai J, Zhang T (2024) Investigation into the alkali-activation of lithium slag: a sustainable alternative to conventional cement with optimized mechanical properties. Constr Build Mater 416:135022
Duan Y, Xiang C, Wang Y, Wang W, Xia F, Cheng L (2022) Research and application of lithium slag as cement mixing material. China Cem 07:69–71
Foghi EJ, Vo T, Rezania M, Nezhad MM, Ferrara L (2024) Early age hydration behaviour of foam concrete containing a coal mining waste: novel experimental procedures and effects of capillary pressure. Constr Build Mater 414:134811
Gao W, Jian S, Li X, Tan H, Li B, Lv Y, Huang J (2022) The use of contaminated soil and lithium slag for the production of sustainable lightweight aggregate. J Clean Prod 348:131361
Gayathri N, Sailesh AR, Srinivas N (2022) Effect of lithium on seed germination and plant growth of Amaranthus viridis. J Appl Natural Sci 14(1):133–139
Gou H, Rupasinghe M, Sofi M, Sharma R, Ranzi G, Mendis P, Zhang Z (2023) A review on cementitious and geopolymer composites with lithium slag incorporation. Materials 17(1):142
Gu K, Gu X, Wang Y, Qin W, Han J (2023a) A green strategy for recycling cathode materials from spent lithium-ion batteries using glutathione. Green Chem 25:4362–4374
Gu X, Wang H, Zhu Z, Liu J, Xu X, Wang Q (2023b) Synergistic effect and mechanism of lithium slag on mechanical properties and microstructure of steel slag-cement system. Constr Build Mater 396:131768
Gu T, Zhang G, Wang Z, Liu L, Zhang L, Wang W, Huang Y, Dan Y, Zhao P, He Y (2024) The formation, characteristics, and resource utilization of lithium slag. Constr Build Mater 432:136648
Guo C, Wang R (2023) Utilizing lithium slag to improve the physical-chemical properties of alkali-activated metakaolin-slag pastes: cost and energy analysis. Constr Build Mater 403:133164

Haigh M et al (2013) Development of new high performance supplementary cementitious material—a lithium production by-product. In: CIA biannual conference. Concrete Institute of Australia (CIA), Australia
Han J, Yan P (2010) Influence of lithium compound on sulphoaluminate cement hydration process. Guisuanyan Xuebao (j Chin Ceram Soc) 38(4):608–614
He Z, Li L, Du S (2017a) Mechanical properties, drying shrinkage, and creep of concrete containing lithium slag. Constr Build Mater 147:296–304
He Z et al (2017b) Effect of lithium slag on drying shrinkage of concrete with manufactured-sand. J Residuals Sci Technol 14(1):171–176
He Z, Du S, Chen D (2018) Microstructure of ultra high performance concrete containing lithium slag. J Hazard Mater 353:35–43
He Y, Chen Q, Qi C, Zhang Q, Xiao C (2019) Lithium slag and fly ash-based binder for cemented fine tailings backfill. J Environ Manage 248:109282
He Z et al (2020) Hydration and microstructure of concrete containing high volume lithium slag. Mater Express 10(3):430–436
He Y, Zhang Q, Chen Q, Bian J, Qi C, Kang Q, Feng Y (2021) Mechanical and environmental characteristics of cemented paste backfill containing lithium slag-blended binder. Constr Build Mater 271:121567
He Y, Kang Q, Lan M, Peng H (2023) Mechanism and assessment of the pozzolanic activity of melting-quenching lithium slag modified with MgO. Constr Build Mater 363:129692
Huang SW et al (2011) Study on preparing aero-concrete using leaching residual slag of lepidolite ore. Appl Mech Mater 99–100:375–378
Jaskula BW (2020) Lithium production. USGS Online Publ Dir 703:648–4908
Javed U, Shaikh FUA, Sarker PK (2022a) Microstructural investigation of thermo-mechanically processed lithium slag for geopolymer precursor using various characterization techniques. Constr Build Mater 342:127952
Javed U, Shaikh FUA, Sarker PK (2022b) Microstructural investigation of lithium slag geopolymer pastes containing silica fume and fly ash as additive chemical modifiers. Cem Concr Compos 134:104736
Javed U, Shaikh FUA, Sarker PK (2024a) A comprehensive micro-nano investigative approach to study the development of aluminosilicate gel in binary blends of lithium slag geopolymer. Cem Concr Compos 145:105338
Javed U, Shaikh FUA, Sarker PK (2024b) Corrosive effect of HCl and H_2SO_4 exposure on the strength and microstructure of lithium slag geopolymer mortars. Constr Build Mater 411:134588
Kavanagh L, Keohane J, Cabellos GG, Lloyd A, Cleary J (2018) Induced plant accumulation of lithium. Geosciences 8:56
Kjølholt J, Stuer-Lauridsen F, Skibsted Mogensen A, Havelund S (2003) The elements in the second rank—lithium. Miljoministeriet, Copenhagen, Denmar
Lemougna PN, Yliniemi J, Ismailov A, Levanen E, Tanskanen P, Kinnunen P, Roning J, Illikainen M (2019) Recycling lithium mine tailings in the production of low temperature (700–900 °C) ceramics: effect of ladle slag and sodium compounds on the processing and final properties. Constr Build Mater 221:332–344
Li J, Huang S (2020) Recycling of lithium slag as a green admixture for white reactive powder concrete. J Mater Cycles Waste Manage 22(6):1818–1827
Li HF, Guo L, Xia Y (2012) Mechanical properties of concretes containing super-fine mineral admixtures. Appl Mech Mater 174–177:1406–1409
Li B et al (2019a) Products and properties of steam cured cement mortar containing lithium slag under partial immersion in sulfate solution. Constr Build Mater 220:596–606
Li B, Cao R, You N, Chen C, Zhang Y (2019b) Products and properties of steam cured cement mortar containing lithium slag under partial immersion in sulfate solution. Constr Build Mater 220:596–606
Li B, You N, Cao R, Huo B, Chen C, Zhang Y (2019c) Composition of lithium slag powder and its physical and chemical reaction characteristics in cement paste. Mater Rep 34:10046–10051

Li J et al (2020a) Recycling of lithium slag extracted from lithium mica by preparing white Portland cement. J Environ Manage 265:110551
Li J, Lian P, Huang S, Huang L (2020b) Recycling of lithium slag extracted from lithium mica by preparing white Portland cement. J Environ Manage 265:110551
Li M, Liu H, Duan P, Ruan S, Zhang Z, Ge W (2021) The effects of lithium slag on microstructure and mechanical performance of metakaolin-based geopolymers designed by response surface method (RSM). Constr Build Mater 299:123950
Li C, Zhang G, Liu D, Wang M (2023a) Preparation of lightweight ceramsite from solid waste lithium slag and fly ash. Constr Build Mater 398:132419
Li S, Liu S, Du Y, Huang Q, Qu W, Zheng W (2023b) Effect of mineral-generated lithium slag on the properties of magnesium oxychloride cement. Crystals 13(3):513
Li T, Tang X, Xia J (2024) Functional characteristics of sustainable pervious cement concrete pavement modified by silica fume and travertine waste. Ceram Silikaty 68:582–597
Liu Z, Wang J, Jiang Q, Cheng G, Li L, Kang Y, Wang D (2019a) A green route to sustainable alkali-activated materials by heat and chemical activation of lithium slag. J Clean Prod 225:1184–1193
Liu Z, Wang J, Li L, Wang D (2019b) Characteristics of Alkali-activated lithium slag at early reaction age. J Mater Civ Eng 31:4019312
Lu J, Yu Z, Zhu Y, Huang S, Luo Q, Zhang S (2019) Effect of lithium-slag in the performance of slag cement mortar based on least-squares support vector machine prediction. Materials 12:1652
Luhar S, Luhar I (2019) Application of artificial neural network for predicting compressive strength of geopolymer concrete
Luo Q, Wang Y, Hong S, Xing F, Dong B (2021) Properties and microstructure of lithium-slag-based geopolymer by one-part mixing method. Constr Build Mater 273:121723
Luo Q, Liu Y, Dong B, Ren J, He Y, Wu K, Wang Y (2023a) Lithium slag-based geopolymer synthesized with hybrid solid activators. Constr Build Mater 365:130070
Luo X, Huang X, Liu Y, Tao J, Xiao S, Peng B (2023b) Performance characterization and optimization of cement-lithium powder-grain slag composite cementitious materials. Constr Build Mater 409:133851
Luo X, Huang L, Yan L, Li Y, Wei L, Chen Z, Qu Y (2024) Preparation of geopolymers from thermally activated lithium slag: activity enhancement and microstructure. J Build Eng 88:109256
Luo Q, Huang S, Zhou Y, Li J, Peng W, Wen Y (2017a) Influence of lithium slag from lepidolite on the durability of concrete. In: Proceedings of the IOP conference series: earth and environmental science, vol 61. IOP Publishing, Bristol, UK, p 12151
Luo Q, Wen YF, Huang SW, Peng WL, Li JY, Zhou YX (2017b) Effects of lithium slag from lepidolite on Portland cement concrete. In: Luo Q, Wen YF, Huang S, Peng W, Li J, Zhou Y (eds) Civil, architecture and environmental engineering. CRC Press, Boca Raton, FL, USA, pp 620–623
McIlwaine N, Foley AM, Morrow DJ, Al Kez D, Zhang C, Lu X et al (2021) A state of-the-art techno-economic review of distributed and embedded energy storage for energy systems. Energy 229:120461
Mim NJ, Shaikh FUA, Sarker PK (2025) Sustainable 3D printed concrete incorporating alternative fine aggregates: a review. Case Stud Constr Mater 22:e04570
Moran BJ, Boutt DF, McKnight SV, Jenckes J, Munk LA, Corkran D, Kirshen A (2022) Relic groundwater and prolonged drought confound interpretations of water sustainability and lithium extraction in arid lands. Earth's Future 10:7
Munn B, Dumitru I, Maree D (2019) Assessment of the performance of new supplementary cementitious materials from lithium production residues. In: CIA Biannual conference. Concrete Institute of Australia (CIA), Melbourne, Australia
Priyadarsini S, Das AP (2024) Analytical and structural characterization of waste lithium-ion batteries for their effective recycling strategy. Environ Chem Ecotoxicol. https://doi.org/10.1016/j.enceco.2024.12.004

Priyadarsini S, Das AP (2026) Lithium in waste water: Impacts on pollution and human health. Decarbonization of Wastewater Pollutants as a Sustain Solution, 353–364. https://doi.org/10.1016/B978-0-443-33062-9.00010-2
Qi L et al (2017) Influence of lithium slag from lepidolite on the durability of concrete. IOP Conf Ser Earth Environ Sci 61:012151
Qin Y et al (2019) The mechanical properties of recycled coarse aggregate concrete with lithium slag. Adv Mater Sci Eng 2019:1–12
Rahman SA, Shaikh FUA, Sarker PK (2022) A comprehensive review of properties of concrete containing lithium refinery residue as partial replacement of cement. Constr Build Mater 328:127053
Rahman S, Dodd A, Khair S, Shaikh F, Sarker P, Hosan MA (2023) Assessment of lithium slag as a supplementary cementitious material: pozzolanic activity and microstructure development. Cem Concr Compos 143:105262
Rahman SA, Shaikh FUA, Sarker PK (2024) Fresh, mechanical, and microstructural properties of lithium slag concretes. Cem Concr Compos 148:105469
Rupasinghe M, San Nicolas R, Mendis P, Sofi M, Ngo T (2017) Investigation of strength and hydration characteristics in nano-silica incorporated cement paste. Cem Concr Compos 80:17–30
Shah SFA, Chen B, Ahmad MR, Haque MA (2021) Development of cleaner one-part geopolymer from lithium slag. J Clean Prod 291:125241
Shahzad B, Tanveer M, Hassan W, Shah AN, Anjum SA, Cheema SA, Ali I (2016) Lithium toxicity in plants: reasons, mechanisms and remediation possibilities—a review. Plant Physiol Biochem 107:104–115
Shen J, Li X, Shi X, Wang W, Zhou H, Wu JW, Wang X, Li J (2020) The toxicity of lithium to human cardiomyocytes. Environ Sci Eur 32:59
Shi KB, Zhang S (2011) Ring method test on the early-age anti-cracking capability of high-performance lithium slag concrete. Appl Mech Mater 94–96:782–785
Shi Q, Huang S, Liang J, Cai H (2020) Research on the influence of different kinds of lithium slag powder on concrete performance. Jiangxi Build Mater S1:31–33
Silva YF, Burbano-Garcia C, Araya-Letelier G, Izquierdo S (2025) Sulfate attack performance of concrete mixtures with use of copper slag as supplementary cementitious material. Case Stud Constr Mater 22:e04846
Tabelin CB, Dallas J, Casanova S, Pelech T, Bournival G, Saydam S, Canbulat I (2021) Towards a low-carbon society: a review of lithium resource availability, challenges and innovations in mining, extraction and recycling, and future perspectives. Miner Eng 163:106743
Tan H et al (2015) Utilization of lithium slag as an admixture in blended cements: physico-mechanical and hydration characteristics. J Wuhan Univ Technol Mater Sci Ed 30(1):129–133
Tan H et al (2018) Utilization of lithium slag by wet-grinding process to improve the early strength of sulphoaluminate cement paste. J Cleaner Prod 205:536–551
Tan H, Li M, He X, Su Y, Zhang J, Pan H, Yang J, Wang Y (2020) Preparation for micro-lithium slag via wet grinding and its application as accelerator in Portland cement. J Clean Prod 250:119528
Tan H et al (2020) Preparation for micro-lithium slag via wet grinding and its application as accelerator in Portland cement. J Cleaner Prod 250:119528
Tan H, Li M, He X, Su Y, Yang J, Zhao H (2021) Effect of wet grinded lithium slag on compressive strength and hydration of sulphoaluminate cement system. Constr Build Mater 267:120465
Tang Y, Zhang L, Chen J, Sun H, Yang T, Liu Q, Huang Q, Zhu T, Huang J (2021) Electro-chemo-mechanics of lithium in solid state lithium metal batteries. Energy Environ Sci 14(2):602–642
Thibon F, Weppe L, Vigier N, Churlaud C, Lacoue-Labarthe T, Metian M, Cherel Y, Bustamante P (2021) Large-scale survey of lithium concentrations in marine organisms. Sci Total Environ 751:141453
Toupal J, Vann DR, Zhu C, Gieré R (2022) Geochemistry of surface waters around four hard-rock lithium deposits in Central Europe. J Geochem Explor 234:106937
USGS (2022) Mineral commodity summaries—lithium

Vera ML, Torres WR, Galli CI, Chagnes A, Flexer V (2023) Environmental impact of direct lithium extraction from brines. Nature Rev Earth Environ 4:149–165

Vigneshwari A, Jayaprakash J (2024) A review on 3D printable cementitious material containing copper and iron ore tailings: material characterization, activation methods, engineering properties, durability, and microstructure behavior. Innov Infrastruct Solut 9:74

Wang J, Han L, Liu Z, Wang D (2020) Setting controlling of lithium slag-based geopolymer by activator and sodium tetraborate as a retarder and its effects on mortar properties. Cem Concr Compos 110:103598

Wang H, Liu X, Zhang Z (2023a) Pozzolanic activity evaluation methods of solid waste: a review. J Clean Prod 402:136783

Wang W, Wang J, Shen J, Guo P (2023b) High titanium heavy slag powder as a sustainability filler and its influence on the performance of asphalt mortar. J Mater Res Technol 25:5586–5599

Wang H, Gu X, Liu J, Zhu Z, Wang S, Xu X, Nehdi ML (2024a) Synergistic effects of steel slag and lithium slag in carbonation-cured cement pastes: carbonation degree, strength and microstructure. J Build Eng 85:108706

Wang S, Gu X, Liu J, Zhu Z, Wang H, Ge X, Hu Z, Xu X, Nehdi ML (2024b) Assessment of lithium slag as a supplementary siliceous material in autoclaved aerated concrete: physical properties and hydration characteristics. Constr Build Mater 442:137621

Wang Y, Wang D, Zheng Y, Hua K, Liu J (2024c) Thermal activation mechanism and activity evaluation of lithium slag: insights from simulated hydration. Constr Build Mater 411:134615

Wen H (2013) Property research of green concrete mixed with lithium slag and limestone flour. Adv Mat Res 765–767:3120–3124

Wu F, Wang G, Shi K, Hao J, Zhai C (2012) Comprehensive utilization of lithium slag. Fly Ash Compr Util 3:46–50

Wu FF, Shi KB, Dong SK (2014a) Influence of concrete with lithium-slag and steel slag by early curing conditions. Key Eng Mater 599:52–55

Wu FF, Shi KB, Dong SK (2014b) Properties and microstructure of HPC with lithium-slag and fly ash. Key Eng Mater 599:70–73

Wu F, Wang X, Chen L, Shi K, Zhang K (2016) Activity evaluation and pore distribution of lithium slag concrete under alkaliactivated. Water Resour Power 34:119–122

Xiong H, Sun J, Liu C, Xian W, Wang W, Luo L, Li S (2023) Study on sintering behavior and properties of lithium slag-based foamed ceramics. J Non-Cryst Solids 617:122499

Yang B, Zhang Y, Zhang W, Sun H, Wang Q, Han D (2024) Recycling lithium slag into eco-friendly ultra-high performance concrete: hydration process, microstructure development, and environmental benefits. J Build Eng 91:109563

Zhai Y, Miao M, Xiao L (2020) Effect of lithium slag fineness on rheological properties of cement paste mixed with superplasticizers. Mater Rep 34:18056–18059

Zhai M, Zhao J, Wang D, Wang Y, Wang Q (2021) Hydration properties and kinetic characteristics of blended cement containing lithium slag powder. J. Build. Eng. 39:102287

Zhai J, Chen P, Long J, Fan C, Chen Z, Sun W (2024) Recent advances on beneficial management of lithium refinery residue in China. Miner Eng 208:108586

Zhan PM, Zhang XX, He ZH, Shi JY, Gencel O, Hai-Yen NT, Wang GC (2022) Strength, microstructure and nanomechanical properties of recycled aggregate concrete containing waste glass powder and steel slag powder. J Clean Prod 341:130892

Zhang LF, Wang RY (2011) Experimental study on alkali-activated slag-lithium slag-fly ash environmental concrete. Adv Mat Res 287–290:1237–1240

Zhang T et al (2020) Effect of TIPA on mechanical properties and hydration properties of cement-lithium slag system. J Environ Manage 276:111274

Zhang L, Pan Y, Xu K, Bi L, Chen M, Han B (2022) Corrosion behavior of concrete fabricated with lithium slag as corrosion inhibitor under simulated acid rain corrosion action. J Clean Prod 377:134300

Zhang Y, Yang B, Gu X, Han D, Wang Q (2023a) Improving the performance of ultra-high performance concrete containing lithium slag by incorporating limestone powder. J Build Eng 72:106610

Zhang Y, Zhang L, Wang Q, Han D, Li Z (2023b) Iron ore tailings, phosphate slags, and lithium slags as ternary supplementary cementitious materials for concrete: study on compression strength and microstructure. Mater Today Commun 36:106644

Zheng D-D, Ji T, Wang C-Q, Sun C-J, Lin X-J, Hossain KMA (2016) Effect of the combination of fly ash and silica fume on water resistance of magnesium-potassium phosphate cement. Constr Build Mater 106:415 421

Zhou J, Ai X, Shi J, Deng T, Fu Y (2022) Research progress of lithium extraction technology and resource utilization of lithium slag from lithium ore. Energy Res Manag 14:71–77

Zhou M, Yan J, Fan J, Xu Y, Lu Y, Duan P, Zhu Y, Zhang Z, Lu Z (2023) Insight to workability, compressive strength and microstructure of lithium slag-steel slag based cement under standard condition. J Build Eng 75:107076

Zhu Z, Gu X, Wang S, Xu X, Wang H, Liu J, Zhan K (2024) Investigation of synergistic effects of lithium slag and granulated blast furnace slag from the perspectives of physics and hydration. J Build Eng 89:109337

Capture Study of Lead Salts with Single-Walled Carbon Nanotubes Using Molecular Dynamics Simulations

Padmesh Medesety

Abstract Single-walled carbon nanotubes (SWCNTs) represent a remarkable breakthrough in nanotechnology, demonstrating their exceptional ability to capture lead salts from polar aqueous mixtures. In our recent exploratory simulation study, we gained significant insights into the absorption of lead salts, particularly Pb $(NO_3)_2$, into these exceptional nanotubes through molecular dynamics simulations. Our research highlights the essential role of Van der Waals interactions with the nanotube walls, illustrating their impact in both smaller and larger carbon nanotubes immersed in liquid. We meticulously designed our simulation model with the carbon nanotube positioned centrally, setting the stage for innovative exploration. In one of our simulations, we investigated the SWCNT (13, 0) configuration twice initially with 40 lead nitrate molecules and then with 10. We also examined the SWCNT (20, 0) configuration with a customized number of lead nitrate molecules. The outcomes from all three simulations were promising, demonstrating the effective separation and capture of lead salts from the aqueous solution. These lead salts offer tremendous potential across a wide range of applications, including mineral processing, medical imaging, nuclear facilities, ammunition manufacturing, and pigment production. Our findings open up new possibilities for these nanomaterials in various industries, paving the way for future innovations.

Keywords Single-walled carbon nanotube SWCNTs · Lead nitrate (Pb $(NO_3)_2$) · Molecular dynamic simulations · tip4pEw polar aqueous water molecules · Hydrophobic nature · Interaction potential · Van der Waals attraction

P. Medesety (✉)
Department of Chemical Engineering, V.S.B. Engineering College (Autonomous), Karur, Tamil Nadu, India
e-mail: padmeshmedesety@gmail.com

A. P. Das and S. Priyadarsini (eds.), *Recovery and Recycling of Critical Metals from Electronic Waste for Environmental Sustainability*, Earth and Environmental Sciences Library, https://doi.org/10.1007/978-3-032-29064-9_8

1 Introduction

Due to rapid industrialization, heavy metals (An Overview of Heavy Metal Pollution and Control | ACS Symposium Series, n.d.; Zhou et al. 2024) have contaminated water bodies. The increase in industrial activities over the past few decades has resulted in the release of substantial volumes of industrial wastewater (Shrestha et al. 2021), which is a significant contributor to environmental degradation (Swain 2024). While some research papers address this trend, we focus on ab initio molecular dynamics simulations (Hansson et al. 2002; Molecular Dynamics Simulations—ScienceDirect, n.d.) aimed at understanding the interaction between methyl ammonium lead iodide ($MAPbI_3$) perovskite surfaces and a liquid water environment (Mosconi et al. 2015). Our computations indicate that MAI-terminated surfaces undergo a swift solvation process, driven by the interaction of water molecules with lead (Pb) atoms. Moreover, in both aerobic and anaerobic processes, lead and mercury can be removed through calcium carbonate precipitation, which is mediated by statistically optimized nitrate reductases (Aerobic and Anaerobic Removal of Lead and Mercury via Calcium Carbonate Precipitation Mediated by Statistically Optimized Nitrate Reductases | Scientific Reports, n.d.). The available freshwater resources have been progressively more contaminated due to accelerated industrialization and urbanization. Numerous techniques have been developed for wastewater treatment, including physical, thermal, chemical, membrane-based, and electrochemical methods. However, finding a well-ordered and efficient wastewater treatment approach has been challenging for researchers over the years. In our study, we utilize a system designed to remove monovalent cations (Na^+ and K^+), divalent cations (Ca^{2+}), and heavy metals such as lead (Pb^{2+}) from water (Datar et al. 2023). Notably, PC-ZIF-67 demonstrates a sorption capacity of 117.2 mg/g for Pb^{2+}, which is greater than that for other saline cations. The mechanism of Pb^{2+} removal is attributed to electrostatic and π-π interactions between the surface of PC-ZIF-67 and Pb^{2+}.

This research encourages the future use of MOF-derived porous carbon for the efficient removal of dissolved Na^+, K^+, Ca^{2+}, and Pb^{2+} from aqueous sources using capacitive deionization (CDI). Additionally, another study investigates a novel approach to effectively eliminate Pb(II) ions (Namkoong et al. 2022) from water using immobilized beads made from Undaria pinnatifida and sodium alginate in conjunction with calcium chloride. The developed biosorbent was characterized using Fourier-transform infrared spectroscopy (FTIR), scanning electron microscopy (SEM), and energy-dispersive X-ray spectroscopy (EDS) (Axon and Klinowski 1990). Our studies on Pb(II) ion biosorption were analyzed using the Langmuir and the Freundlich adsorption models. Furthermore, a thermodynamic analysis indicates that the biosorption process is endothermic, suggesting that increasing the temperature of the Pb (II) ion solution enhances the biosorption reaction. The presence of lead (Pb (II)) in drinking water is a significant concern due to its harmful health effects on human physiology. A commercially available filter utilizes the adsorption potential of activated carbon to remove heavy metals, such as Pb (II) (Kumkum and Kumar 2020). However, this method has limitations, as it relies solely on surface area for adsorbing these contaminants. Biochar, produced through the slow pyrolysis of biomass, contains oxygen-bearing functional groups on its surface, which enhances its adsorption potential compared to activated carbon. This study investigates the adsorption kinetics and mechanisms of Pb (II) removal from water using biochar in a fixed-bed continuous-flow adsorption system. We analyzed the effects of initial Pb (II) concentration, mass of adsorbent, and flow rate on the adsorption efficiency. Various models, including the Adams-Bohart, Thomas, and Yoon-Nelson models, were employed to interpret the adsorption data. The maximum removal efficiency recorded for Pb (II) was 88.86 mg/g.

Our results suggest that the Yoon-Nelson model is the most suitable for describing the adsorption phenomenon of Pb (II) in a fixed-bed biochar column. The insights gained from this research can be applied to develop practical filters capable of efficiently removing Pb (II) from drinking water. Furthermore, this work explores the removal of Pb(II), Cd(II), Co(II), Zn(II), and Sr(II) using polyacrylic acid-acrylonitrile talc nanocomposites (Abass et al. 2022). Various analytical techniques were used to investigate the functional groups, morphology, particle size, and structure of these composites. The capacity of the P(AA-AN)-talc composite to capture Pb^{2+}, Cd^{2+}, Co^{2+}, Zn^{2+} and Sr^{2+} from mixed aqueous solutions was assessed using a batch method, resulting in the following distribution coefficients at different pH levels: $Pb^{2+} > Cd^{2+} > Co^{2+} > Zn^{2+} > Sr^{2+}$ Finally, the removal of lead (II) from water using magnetic nanocomposites (Magnetic Nanocomposite for Lead (II) Removal from Water | Scientific Reports, n.d.) has also been studied. Additionally, this research discusses the adsorption processes for heavy metal ions, specifically lead (Pb), using an artificial intelligence simulation model to predict the adsorption behavior of these ions.

This work explores the promising potential of using artificial intelligence simulation-based models to predict the adsorption process of heavy metal ions, specifically lead (Pb), through the innovative application of magnetic ash/graphene oxide (GO) nanocomposites (Zeng et al. 2022). The results revealed that the adsorption of Pb ions on the adsorbent is driven by electrostatic interactions and van der Waals bonding, with the lead-water-GO configuration exhibiting the highest adsorption affinity, as indicated by the ΔE gap computations. The equilibrium concentration of Pb^{2+} ions, a crucial factor in predicting the adsorption capacity of the adsorbent, was evaluated in conjunction with the model inputs of initial Pb^{2+} ion concentration and solution temperature. The association of molecular dynamics computations and artificial intelligence simulation methods presents a promising avenue for accurately predicting the adsorption performance of nanocomposites in removing heavy metal ions from aqueous solutions. Furthermore, the molecular dynamics (MD) simulation of the $Na_{0.98}Pb_{0.02}$ alloy (Shimkevich and Shimkevich 2011) reveals the profound impact of minor lead additives on the liquid sodium microstructure, highlighting the intricate beauty of materials at the atomic level.

The observed outcomes highlight the remarkable stability of lead-sodium clusters in liquid sodium, driven by the ionic bonds (Na^+–Pb^-) formed due to the significant difference in electronegativity between the alloy components. Building on this foundation, we delve into molecular dynamics (MD) simulations of the $Na_{0.98}Pb_{0.02}$ alloy within a three-component bipolar replica. Our exploration utilizes these simulations to envision the potential of employing the UiO-66 metal–organic framework (MOF) as a powerful agent for removing lead ions (Pb^{2+}) from water (Wehbe et al. 2020).

We investigated the concentration of Pb (II) in water in the context of a Metal–Organic Framework (MOF) under varying conditions, including particle number density, metal cation concentration, and the presence of NO^{3-} ions. Radial distribution functions indicated that the optimal adsorption sites for Pb^{2+} in UiO-66 correspond to the oxygen atoms within the Zr–O–Zr bridges. This study suggests that UiO-66 could be an effective candidate for removing Pb(II) from water. An article titled "NHMRC Information for Evidence on the Effects of Lead on Human Health" (NHMRC Information Paper: Evidence on the Effects of Lead on Human Health, n.d.) reported similar findings. Additionally, hydroxyapatite (HAP), prepared through a wet chemical process, was used to remove lead (II) from an aqueous solution. HAP (El Hammari et al. 2023) was characterized using Fourier Transform Infrared Spectroscopy, X-ray Diffraction, Brunauer–Emmett–Teller analysis, and Scanning Electron Microscopy. The maximum Pb removal rate of 98.94% was achieved at an initial concentration of 50 mg/L, a contact time of 15 min, and a pH of 8. At 323 K, isothermal adsorption was matched by the Langmuir isotherm model, with a regression coefficient of 0.96. Furthermore, the maximum adsorption capacity at equilibrium was determined to be 90.18 mg/g, and a pseudo-second-order kinetic model provided the best description of the adsorption kinetics.

The presence of lead compounds in our environment presents a challenge we can overcome. Notably, water consumption has been identified as a significant source of human exposure to lead, posing a serious health risk. Due to its toxicity, both the International Agency for Research on Cancer and the US Environmental Protection Agency (USEPA) have classified lead (Pb) and its compounds as potential human carcinogens. With the European Community and World Health Organization establishing a maximum allowable lead limit of 10 μg/L in tap water and the USEPA setting a guideline of 15 μg/L in drinking water, the urgency to eliminate lead ions from our water sources is apparent. This work highlights the remarkable research on the cost-effective removal of lead ions using a variety of innovative adsorbents (Aranda et al. 2022; Chowdhury et al. 2022; Zhou et al. 2006), paving the way for a safer tomorrow.

The adsorption capabilities of various natural materials (Abdel-Halim et al. 2003) for lead have been examined, showcasing the remarkable potential of resources such as animal bone powder, activated carbon, Nile rose plant powder, commercial carbon, and ceramics. In this exploration, lead absorption rates reached equilibrium at impressive intervals: 15 min for bone powder, 30 min for activated carbon, 45 min for plant powder, and 120 min for commercial carbon. We find that as the pH value rises, the uptake of lead flourishes. The ranking of lead uptake percentage (% adsorption) at a constant pH from a given concentration of lead nitrate solution inspires us: bone powder leads the way, followed by activated carbon, plant powder, and commercial carbon. Additionally, by reducing the concentration of lead while maintaining a constant pH, we enhance the absorption rate, revealing the extraordinary potential of nature's resources in addressing our challenges.

Recent research has revealed the remarkable potential of adsorbents (Aranda et al. 2022) in removing mercury and lead. One study explored the adsorption of lead ions from aqueous solutions onto biochars (Mahdi et al. 2018) derived from date seed biomass. Among nine different date seed-derived biochars, the one produced at 550 °C for 3 h proved to be the most effective adsorbent for Pb^{2+} ion removal. This finding led to further investigation through batch and column experiments. The study revealed the significant impact of contact time, initial Pb^{2+} concentration, and solution pH on adsorption efficiency. Notably, the pH of the solution significantly impacted the adsorption capability of the biochar (DSB550-3) for lead ions, with optimal results achieved around pH 6.0. Both experimental and replicate results demonstrated that date seed-derived biochar possesses an impressive adsorption capacity for Pb^{2+}, surpassing that of other plant-based biochars noted in the literature. Lead (Pb) (Collin et al. 2022; Khatun et al. 2022; Lead Between the Lines | Nature Chemistry, n.d.; Lead (Metal)—An Overview | ScienceDirect Topics, n.d.) It is a naturally occurring metal that forms compounds when it combines with other elements. It can also undergo chemical reactions with air and water, resulting in lead sulfate, lead carbonates, or lead oxide.

In this study, we embark on an exciting journey to explore the selective capture of lead through molecular dynamics simulations with innovative carbon nanotubes (CNTs) in the presence of aqueous TIP4PEw water molecules. Carbon nanotubes stand as remarkable candidates for the partitioning of lead salts in polar liquid mixtures. Their unique structure, comprising internal pores and outer circumferential walls, offers an opportunity to unlock new possibilities. By varying key factors, such as CNT diameter and lead molecule concentration, we will explore the adsorption and selectivity of lead molecules, paving the way for advancements in environmental protection and materials science.

2 Methods

MD simulations (Gade et al. 2018, 2019; Hansson et al. 2002; Medesety et al. 2021, 2022; Molecular Dynamics Simulations—ScienceDirect, n.d.) of lead salts (LINCS: A Linear Constraint Solver for Molecular Simulations—Hess—1997—Journal of Computational Chemistry—Wiley Online Library, n.d.; Magnetic Nanocomposite for Lead (II) Removal from Water | Scientific Reports, n.d.; Mahdi et al. 2018) for different concentrations of lead nitrate is dispersed in polar water liquid with (13,0) CNT (Siva Kumar et al. 2021) and (20,0) CNT (Single-Walled Carbon Nanotube—An Overview | ScienceDirect Topics, n.d.) that were simulated in dissimilar conditions by using GROMACS 2019.3 package (GROMACS: Fast, Flexible, and Free—Van Der Spoel—2005—Journal of Computational Chemistry—Wiley Online Library, n.d.; Hess et al. 2008; Kohnke et al. 2020). Lennard–Jones, bond-length, bond-bending, and dihedral torsional interactions for the lead nitrate molecules were modeled using the Generalized Amber Force Field (GAFF) (Development and Testing of a General Amber Force Field—Wang—2004—Journal of

Computational Chemistry—Wiley Online Library, n.d.; Wang et al. 2004) and water molecules with tip4pEw water model (Comparing Classical Water Models Using Molecular Dynamics to Find Bulk Properties | Journal of Chemical Education, n.d.; Molecular Dynamics Simulations of the Dynamic and Energetic Properties of Alkali and Halide Ions Using Water-Model-Specific Ion Parameters | The Journal of Physical Chemistry B, n.d.; Horn et al. 2004). The atomic partial charges on carbon atoms of SWCNT (Niyogi et al. 2002; Odom et al. 1998; Pramoda et al. 2015; Single-Walled Carbon Nanotube—An Overview | ScienceDirect Topics, n.d.) were kept zero. Packmol (Martínez et al. 2009) was used to combine all the lead salts, CNT, and water molecules. In each simulation, 1 and 1.5 nm diameter single-walled carbon nanotubes were solvated in an aqueous medium of tip4pEw (Comparing Classical Water Models Using Molecular Dynamics to Find Bulk Properties | Journal of Chemical Education, n.d.; Horn et al. 2004) solvent.

The solvated composition is equilibrated for five ns at the start for an isothermal-isobaric (NPT) ensemble via Berendsen thermostat and barostat (Berendsen 1991; Molecular Dynamics with Coupling to an External Bath | The Journal of Chemical Physics | AIP Publishing, n.d.). The simulation's temperature and pressure were maintained at 300 K and 1 bar, respectively. To measure the statistical averages of molecular properties, the systems were moreover simulated for 10 ns beneath NPT canonical ensemble (Molecular Dynamics Simulation in the Grand Canonical Ensemble—Eslami—2007—Journal of Computational Chemistry—Wiley Online Library, n.d.; Tobias et al. 2002) circumstances using the Nose–Hoover thermostat (Hoover 1985; Nosé 1984; Phys. Rev. A 31, 1695 (1985)—Canonical Dynamics: Equilibrium Phase-Space Distributions, n.d.) and Berendsen barostat (Berendsen 1991; Molecular Dynamics with Coupling to an External Bath | The Journal of Chemical Physics | AIP Publishing, n.d.). The two-fs simulation time step was used to combine the equations of motion. The simulation productivity configurations were saved every one ps. The cutoff distances of 1 nm were used to minimize electrostatic and van der Waals interactions (Hess et al. 2008; Scott 1991; Wang et al. 2014). The particle mesh Ewald method (Darden et al. 1993; Essmann et al. 1995) was implemented for the estimation of long-range electrostatic interactions. Similarly, long-range dispersion corrections were implemented beyond the cutoff distances. All bonds were controlled using the LINCS algorithm (Hess et al. 2008; LINCS: A Linear Constraint Solver for Molecular Simulations—Hess—1997—Journal of Computational Chemistry—Wiley Online Library, n.d.). Exceptionally, Force field parameters have been supplied to GAFF for the preparation of lead nitrate molecules in the topology file. A complex lead nitrate molecule (Lead Nitrate—An Overview | ScienceDirect Topics, n.d.) is designed to understand or analyse the adsorption behaviour of lead salts with 1 nm and 1.5 nm Zig Zag carbon nanotube (Du et al. 2008), respectively.

3 Results and Discussion

All the blue V-shaped lines in Figure represent the TIP4PEw water molecules, a vibrant symbol of the potential within. The carbon atoms of the non-polar carbon nanotube (SWCNT), with thirteen circumferential carbon atoms, are depicted in green through a van der Waals representation, highlighting the intricate nature of its design. The lead atoms, depicted in gray, alongside nitrogen atoms in blue and oxygen atoms in red, showcase the dynamic interplay of elements.

Two layers of lead nitrate molecules find their home within the (13, 0) SWCNT, forming a pseudo-second layer that envelops the carbon nanotube. Figure 1 beautifully illustrates the complete saturation of water molecules, functioning as a solvent, alongside lead nitrates in a 1 nm carbon nanotube. Approximately forty lead nitrate molecules, harmoniously combined with TIP4PEw water molecules, are expertly distributed and simulated in this model, demonstrating the wonders of scientific exploration.

In this simulation, we witness an impressive assembly of lead nitrate molecules, emerging within a polar aqueous medium of TIP4PEw water. This result stands as a testament to the significant advancements in understanding lead salt adsorption, revealing the remarkable ability of the (13, 0) zigzag CNT to act as a powerful adsorbent for removing impurities. The simulations are conducted under normal temperature and pressure conditions, specifically at 300 K and 1 bar, within a canonical ensemble, embodying the spirit of discovery and innovation (Fig. 2).

An intriguing phenomenon emerged as lead nitrate molecules began to adsorb onto the outer wall of the carbon nanotube, maintaining that captivating interaction over time in this simulation. In another remarkable simulation study, approximately ten lead nitrate molecules interacted with the tip4pEw model for a 1 nm carbon nanotube (Fig. 3).

The blue dotted spots represent the tip4pEw water molecules, while the green color illustrates the van der Waals representation of a non-polar SWCNT with

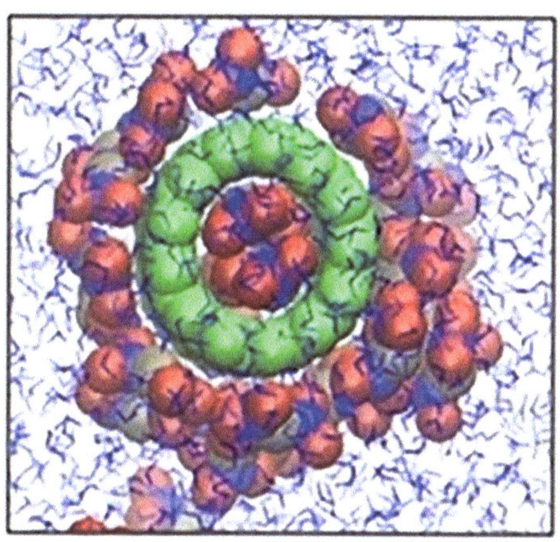

Fig. 1 The final image of the molecular dynamic simulation for lead nitrate molecules simulated with tip4pEw water molecules and (13, 0) SWCNT

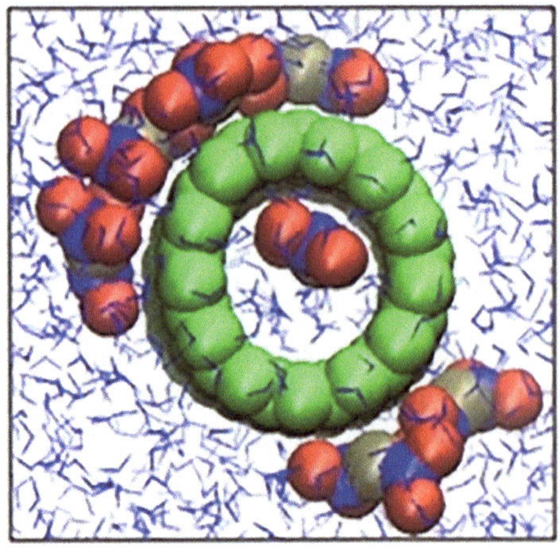

Fig. 2 The final image of the molecular dynamic simulation for a few lead nitrate molecules simulated with tip4pEw water molecules and (13, 0) SWCNT

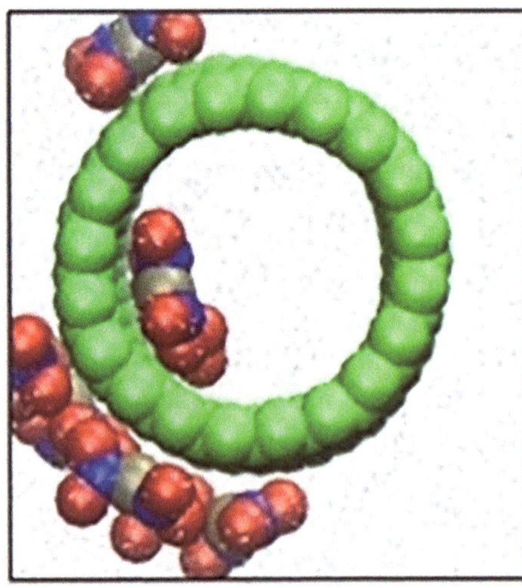

Fig. 3 The final image of the molecular dynamic simulation for a few lead nitrate molecules simulated with tip4pEw water molecules and (20, 0) SWCNT

twenty carbon atoms. Lead atoms are depicted in a grey van der Waals representation, nitrogen atoms shine in blue, and oxygen atoms are portrayed in red. In a similar simulation using a 1.5 nm diameter SWCNT, a captivating interaction between lead molecules and the carbon nanotube walls was observed. Remarkably, all lead molecules appeared to adsorb on both the outer circumference and within the walls of the carbon nanotube, showcasing the beauty of molecular interactions and the wonders of nanotechnology (Fig. 4).

Single-walled carbon nanotubes, remarkable cylindrical structures crafted through the meticulous rolling of single graphene sheets, are represented by indices (n, m), where either n or m is zero. This design leads to a unique 0° chiral angle and unlocks potential semiconducting properties. In this research study, we investigate two sizes

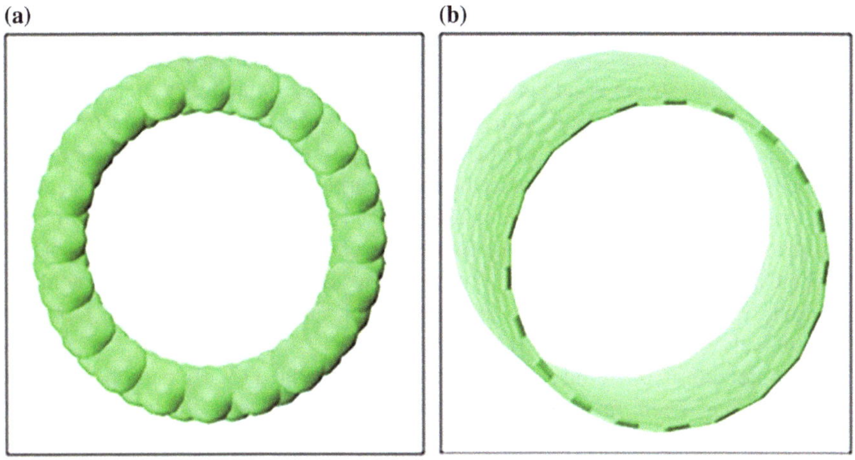

Fig. 4 The Initial images of (20, 0) SWCNT in **a** van der Waals representation and in the **b** tube model representation

of carbon nanotubes, 1 nm (13, 0) and 1.5 nm (20, 0), as principal adsorbents, showcasing the promise and innovation at the heart of nanotechnology (Fig. 5).

We explored the dynamic interactions of lead, cadmium, and mercury cations with carbon atoms confined within a 1 nm carbon nanotube, employing GAFF parameters and Tip4pEw water as our solvent. The interaction potentials of these heavy

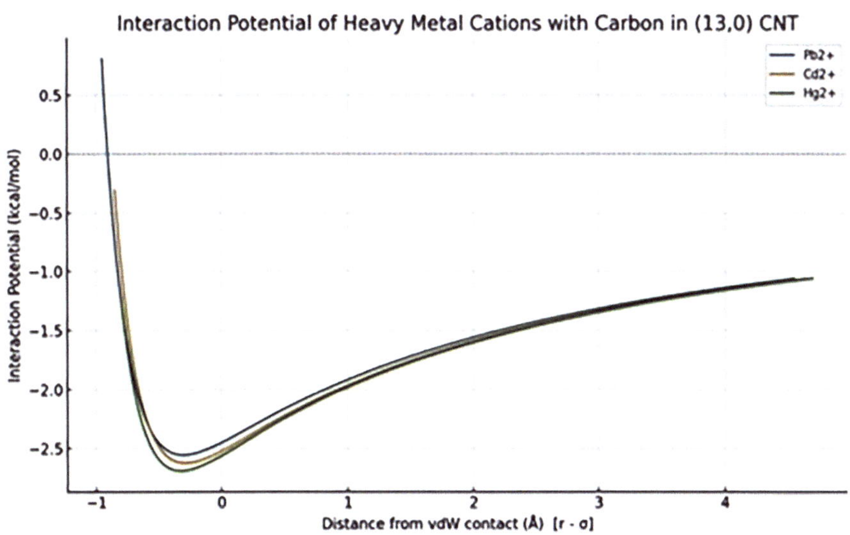

Fig. 5 The plot demonstrating the interaction potential of heavy metal cations (Pb^{2+}, Cd^{2+}, and Hg^{2+}) with carbon atoms in a (13, 0) carbon nanotube, as a function of distance offset from their vander Waals contact (i.e., $r - \sigma_{ij}$)

metals showcased impressive values ranging from – 2.5 to – 2.8 kcal/mol. The results revealed a remarkable decline in the interaction curve between – 1 and 5 nm, exemplifying the strong bond between carbon molecules and heavy metals. Beyond – 0.80 nm, the curve deepens, indicating that the optimal conditions for the potential of mean force (PMF) occur between – 0.5 nm and 0 nm. As we observed the distances for van der Waals contact between carbon and heavy metals, extending from 0 to 5 nm, the interaction potential decreased gradually from – 2.5 kcal/mol to – 1 kcal/mol, reflecting the intricate interactions between these elements (Table 1).

The Charge of the metal cations, Vander Waals diameters, interaction potentials, and ionic radius of heavy metals and the carbon atom of SWCNT are tabulated informatively with values (Fig. 6).

The interaction potential of heavy metals with the (13, 0) carbon nanotube (SWCNT) has been investigated through various solvents, including ethanol, methanol, acetonitrile, and water, as well as in a vacuum. From 2.5 nm to – 1 nm, there is a remarkable steep decrease in the curve, indicating a strong connection

Table 1 The metal ion parameters used for computing the interaction potential

Ion	Charge(e)	σ (Å)	ε (kcal/mol)	Ionic radius (Å)
Pb^{2+}	+ 2	3.5	0.012	1.19
Cd^{2+}	+ 2	3.3	0.014	0.95
Hg^{2+}	+ 2	3.2	0.010	1.02
CNT C	0	3.4	0.055	0.00

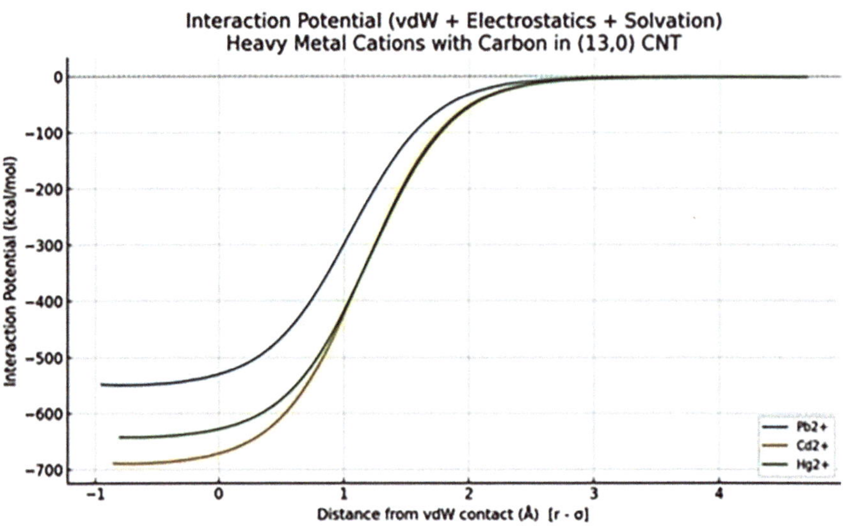

Fig. 6 The plot includes explicit solvation energy terms for the interaction of Pb^{2+}, Cd^{2+}, and Hg^{2+} with carbon atoms in a (13, 0) carbon nanotube

Table 2 Interaction strength tendency across different solvents

Solvent	Dielectric constant (ε_r)	Electrostatic screening	Solvation energy	Total interaction
Vacuum	1.0	None (strongest interaction)	No solvation	Most attractive
Ethanol	24.3	Moderate	Weak solvation	Strong interaction
Methanol	32.6	Moderate	Slightly better	Moderately strong
Acetonitrile	36.0	Moderate	Slightly better	Moderate
Water	78.5	Strongest screening	Strongest solvation	Weakest Interaction

between carbon molecules and heavy metals. After reaching 0 nm, the curve deepens, reflecting the optimal conditions for the potential of mean force (PMF) between − 1.0 and 0 nm. Furthermore, the plot reveals that as the distance increases from − 1 to 2.5 nm, for van der Waals contact between carbon atoms and heavy metals in a vacuum, the interaction potential ranges elegantly, showcasing values from − 700 to − 550 kcal/mol, respectively (Table 2).

From the table data above, we can understand that for stronger solvation in the system, there is a weaker interaction with heavy metals for SWCNT, and vice versa.

4 Conclusions

Single-walled carbon nanotubes (SWCNTs) are extraordinary nanomaterials that exhibit a remarkable ability to absorb lead salts, even at extremely low concentrations. The way their interactions with lead salts strengthen as concentration increases and tube diameter decreases reveals the exquisite design of these structures. It is fascinating how most lead salts adhere to the outer surfaces of the SWCNTs, a captivating phenomenon beautifully illustrated in our simulations. These simulations, utilizing TIP4PEw water molecules, uncover profound insights into the energetic and entropic properties of the liquid phase. As we explore innovative adsorbent materials, we edge closer to achieving exceptional selectivity and efficiency in lead capture. Our groundbreaking molecular dynamics simulations with SWCNTs in binary mixtures of $Pb(NO_3)_2$ and H_2O reveal the formation of exciting new pores and voids at varying diameters, marking a significant leap in our mission for a cleaner, brighter future. Notably, the computed interaction potential values for lead salts in a 1 nm zigzag CNT range from − 2.5 to − 2.8 kcal/mol in water and from − 700 to − 550 kcal/mol in a vacuum. The possibilities are boundless, and we are invigorated to be part of this transformative journey!

Acknowledgements Dr. Padmesh Medesety acknowledges the Chemical Engineering Departments at Visvesvaraya National Institute of Technology (VNIT), Nagpur, Maharashtra, India, as well as the Andhra University College of Engineering (A.U.), Visakhapatnam, G.M.R IT, Rajam, Andhra Pradesh, and V.S.B. Engineering College (Autonomous), Karur, Tamil Nadu-639111, India.

Conflict of Interest The authors declare that they have no conflict of interest.

References

Abass MR, El-Kenany WM, El-Masry EH (2022) High efficient removal of lead(II) and cadmium(II) ions from multi-component aqueous solutions using polyacrylic acid acrylonitrile talc nanocomposite. Environ Sci Pollut Res 29(48):72929–72945. https://doi.org/10.1007/s11356-022-21023-1

Abdel-Halim SH, Shehata AMA, El-Shahat MF (2003) Removal of lead ions from industrial waste water by different types of natural materials. Water Res 37(7):1678–1683. https://doi.org/10.1016/S0043-1354(02)00554-7

Aerobic and Anaerobic Removal of Lead and Mercury via Calcium Carbonate Precipitation Mediated by Statistically Optimized Nitrate Reductases | Scientific Reports. (n.d.). Retrieved March 8, 2025, from https://www.nature.com/articles/s41598-020-60951-1

An Overview of Heavy Metal Pollution and Control | ACS Symposium Series. (n.d.). Retrieved March 15, 2025, from https://pubs.acs.org/doi/10.1021/bk-2023-1456.ch001

Aranda FL, Rivas BL, Aranda FL, Rivas BL (2022) Removal of mercury and lead by bioadsorbents: an overview. J Chil Chem Soc 67(4):5682–5691. https://doi.org/10.4067/S0717-97072022000405682

Axon SA, Klinowski J (1990) Fourier-transform infrared, scanning electron microscopy and magic-angle-spinning nuclear magnetic resonance studies of silicalite synthesized in non-alkaline media. J Chem Soc Faraday Trans 86(21):3679–3682. https://doi.org/10.1039/FT9908603679

Berendsen HJC (1991) Transport properties computed by linear response through weak coupling to a bath. In: Meyer M, Pontikis V (eds) Computer simulation in materials science: interatomic potentials, simulation techniques and applications. Springer Netherlands, pp 139–155. https://doi.org/10.1007/978-94-011-3546-7_7

Chao EE, Cheng KL (1977) The solubility products of some slightly soluble lead salts and the potentiometric titration of molybdate, tungstate, perrhenate and fluoride with use of a lead ion-selective electrode. Talanta 24(4):247–250. https://doi.org/10.1016/0039-9140(77)80143-4

Chowdhury IR, Chowdhury S, Mazumder MAJ, Al-Ahmed A (2022) Removal of lead ions (Pb^{2+}) from water and wastewater: a review on the low-cost adsorbents. Appl Water Sci 12(8):185. https://doi.org/10.1007/s13201-022-01703-6

Collin MS, Venkatraman SK, Vijayakumar N, Kanimozhi V, Arbaaz SM, Stacey RGS, Anusha J, Choudhary R, Lvov V, Tovar GI, Senatov F, Koppala S, Swamiappan S (2022) Bioaccumulation of lead (Pb) and its effects on human: a review. J Hazard Mater Adv 7:100094. https://doi.org/10.1016/j.hazadv.2022.100094

Comparing Classical Water Models Using Molecular Dynamics to Find Bulk Properties | Journal of Chemical Education (n.d.) Retrieved July 18, 2023, from https://pubs.acs.org/doi/abs/10.1021/acs.jchemed.7b00385

Darden T, York D, Pedersen L (1993) Particle mesh Ewald: an N·log(N) method for Ewald sums in large systems. J Chem Phys 98(12):10089–10092. https://doi.org/10.1063/1.464397

Datar SD, Mane RS, Kumar N, Sawant V, Malpure S, Jha N (2023) Effective removal of heavy metal-lead and inorganic salts by microporous carbon derived from Zeolitic Imidazolate Framework-67 electrode using capacitive deionization. Desalination 558:116619. https://doi.org/10.1016/j.desal.2023.116619

Development and Testing of a General Amber Force Field—Wang—2004—Journal of Computational Chemistry—Wiley Online Library (n.d.) Retrieved July 17, 2023, from https://onlinelibrary.wiley.com/doi/abs/10.1002/jcc.20035

Du AJ, Chen Y, Lu GQ, Smith SC (2008) Half metallicty in finite-length zigzag single walled carbon nanotube: a first-principle prediction. Appl Phys Lett 93(7):073101. https://doi.org/10.1063/1.2970055

El Hammari L, Hamed R, Azzaoui K, Jodeh S, Latifi S, Saoiabi S, Boukra O, Krime A, Boukra A, Saoiabi A, Hammouti B, Khan MM, Sabbahi R, Hanbali G, Berisha A, Taleb M, Dagdag O (2023) Optimization of the adsorption of lead (II) by hydroxyapatite using a factorial design: density functional theory and molecular dynamics. Front Environ Sci 11:1112019. https://doi.org/10.3389/fenvs.2023.1112019

Essmann U, Perera L, Berkowitz ML, Darden T, Lee H, Pedersen LG (1995) A smooth particle mesh Ewald method. J Chem Phys 103(19):8577–8593. https://doi.org/10.1063/1.470117

Gade HM, Wanjari PP, Velpuri SVV (2018) Water-mediated curvature change in graphene by single-walled carbon nanotubes. Phys Chem Chem Phys 20(34):22359–22367. https://doi.org/10.1039/C8CP02394H

Gade HM, Velpuri SVV, Wanjari PP (2019) Conformational rearrangements in n-alkanes encapsulated within capsular self-assembly of capped carbon nanotubes. Chem Phys 517:198–207. https://doi.org/10.1016/j.chemphys.2018.10.014

GROMACS: Fast, Flexible, and Free—Van Der Spoel—2005—Journal of Computational Chemistry—Wiley Online Library (n.d.) Retrieved July 17, 2023, from https://onlinelibrary.wiley.com/doi/abs/10.1002/jcc.20291

Hansson T, Oostenbrink C, van Gunsteren W (2002) Molecular dynamics simulations. Curr Opin Struct Biol 12(2):190–196. https://doi.org/10.1016/S0959-440X(02)00308-1

Hess B, Kutzner C, van der Spoel D, Lindahl E (2008) GROMACS 4: algorithms for highly efficient, load-balanced, and scalable molecular simulation. J Chem Theory Comput 4(3):435–447. https://doi.org/10.1021/ct700301q

Hoover WG (1985) Canonical dynamics: equilibrium phase-space distributions. Phys Rev A 31(3):1695–1697. https://doi.org/10.1103/PhysRevA.31.1695

Horn HW, Swope WC, Pitera JW, Madura JD, Dick TJ, Hura GL, Head-Gordon T (2004) Development of an improved four-site water model for biomolecular simulations: TIP4P-Ew. J Chem Phys 120(20):9665–9678. https://doi.org/10.1063/1.1683075

Khatun J, Intekhab A, Dhak D (2022) Effect of uncontrolled fertilization and heavy metal toxicity associated with arsenic(As), lead(Pb) and cadmium (Cd), and possible remediation. Toxicology 477:153274. https://doi.org/10.1016/j.tox.2022.153274

Kohnke B, Kutzner C, Grubmüller H (2020) A GPU-accelerated fast multipole method for GROMACS: performance and accuracy. J Chem Theory Comput 16(11):6938–6949. https://doi.org/10.1021/acs.jctc.0c00744

Kumkum P, Kumar S (2020) Evaluation of lead (Pb(II)) removal potential of biochar in a fixed-bed continuous flow adsorption system. J Health Pollut 10(28):201210. https://doi.org/10.5696/2156-9614-10.28.201210

Lead (Metal)—An Overview | ScienceDirect Topics (n.d.) Retrieved March 15, 2025, from https://www.sciencedirect.com/topics/physics-and-astronomy/lead-metal

Lead Between the Lines | Nature Chemistry (n.d.) Retrieved March 15, 2025, from https://www.nature.com/articles/nchem.1761

Lead Nitrate—An Overview | ScienceDirect Topics (n.d.) Retrieved March 16, 2025, from https://www.sciencedirect.com/topics/chemistry/lead-nitrate

LINCS: A Linear Constraint Solver for Molecular Simulations—Hess—1997—Journal of Computational Chemistry—Wiley Online Library (n.d.) Retrieved July 17, 2023, from https://onlinelibrary.wiley.com/doi/abs/10.1002/(SICI)1096-987X(199709)18:12%3C1463::AID-JCC4%3E3.0.CO;2-H

Magnetic Nanocomposite for Lead (II) Removal from Water | Scientific Reports (n.d.) Retrieved March 8, 2025, from https://www.nature.com/articles/s41598-024-68491-8

Mahdi Z, Yu QJ, El Hanandeh A (2018) Removal of lead(II) from aqueous solution using date seed-derived biochar: Batch and column studies. Appl Water Sci 8(6):181. https://doi.org/10.1007/s13201-018-0829-0

Martínez L, Andrade R, Birgin EG, Martínez JM (2009) PACKMOL: a package for building initial configurations for molecular dynamics simulations. J Comput Chem 30(13):2157–2164. https://doi.org/10.1002/jcc.21224

Medesety P, Gade HM, Singh NK, Wanjari PP (2021) Highly selective carbon capture by novel graphene-carbon nanotube hybrids. Mol Simul 47(16):1326–1334. https://doi.org/10.1080/08927022.2021.1968391

Medesety P, Chaitanya K, Gade HM, Jaiswal V, Wanjari PP (2022) Carbon nanotube assisted highly selective separation of organic liquid mixtures. Chem Phys 562:111647. https://doi.org/10.1016/j.chemphys.2022.111647

Molecular Dynamics Simulation in the Grand Canonical Ensemble—Eslami—2007—Journal of Computational Chemistry—Wiley Online Library (n.d.) Retrieved July 18, 2023, from https://onlinelibrary.wiley.com/doi/abs/10.1002/jcc.20689

Molecular Dynamics Simulations of the Dynamic and Energetic Properties of Alkali and Halide Ions Using Water-Model-Specific Ion Parameters | The Journal of Physical Chemistry B (n.d.) Retrieved July 18, 2023, from https://pubs.acs.org/doi/full/10.1021/jp902584c

Molecular Dynamics Simulations—ScienceDirect (n.d.) Retrieved June 8, 2023, from https://www.sciencedirect.com/science/article/abs/pii/S0959440X02003081

Molecular Dynamics with Coupling to an External Bath | The Journal of Chemical Physics | AIP Publishing (n.d.) Retrieved July 17, 2023, from https://pubs.aip.org/aip/jcp/article-abstract/81/8/3684/565473/Molecular-dynamics-with-coupling-to-an-external

Mosconi E, Azpiroz JM, De Angelis F (2015) Ab Initio molecular dynamics simulations of methylammonium lead iodide perovskite degradation by water. Chem Mater 27(13):4885–4892. https://doi.org/10.1021/acs.chemmater.5b01991

Namkoong H, Biehler E, Namkoong G, Abdel-Fattah TM (2022) Efficient removal of lead ions from aqueous media using sustainable sources based on marine algae. ACS Omega 7(44):39931–39937. https://doi.org/10.1021/acsomega.2c04322

NHMRC Information Paper: Evidence on the Effects of Lead on Human Health (n.d.)

Niyogi S, Hamon MA, Hu H, Zhao B, Bhowmik P, Sen R, Itkis ME, Haddon RC (2002) Chemistry of single-walled carbon nanotubes. Acc Chem Res 35(12):1105–1113. https://doi.org/10.1021/ar010155r

Nosé S (1984) A molecular dynamics method for simulations in the canonical ensemble. Mol Phys 52(2):255–268. https://doi.org/10.1080/00268978400101201

Odom TW, Huang J-L, Kim P, Lieber CM (1998) Atomic structure and electronic properties of single-walled carbon nanotubes. Nature 391(6662):6662. https://doi.org/10.1038/34145

PACKMOL: A Package for Building Initial Configurations for Molecular Dynamics Simulations—Martínez—2009—Journal of Computational Chemistry—Wiley Online Library (n.d.) Retrieved July 17, 2023, from https://onlinelibrary.wiley.com/doi/full/10.1002/jcc.21224

Phys. Rev. A 31, 1695 (1985)—Canonical Dynamics: Equilibrium Phase-Space Distributions (n.d.) Retrieved July 17, 2023, from https://journals.aps.org/pra/abstract/10.1103/PhysRevA.31.1695

Pramoda K, Kumar R, Rao CNR (2015) Graphene/single-walled carbon nanotube composites generated by covalent cross-linking. Chem Asian J 10(10):2147–2152. https://doi.org/10.1002/asia.201500627

Saunders BC, Stacey GJ (1949) 197. Organo-lead compounds. Part I. Trialkyl-lead salts possessing sternutatory properties. J Chem Soc (Resumed), 919–925. https://doi.org/10.1039/JR9490000919

Scott R (1991) Review of computer simulation of liquids. Math Comput 57(195):442–444. https://doi.org/10.2307/2938686

Shimkevich AL, Shimkevich IY (2011) Molecular dynamics simulation of the clustering of minor lead additives in liquid sodium. J Metall 2011(1):890321. https://doi.org/10.1155/2011/890321

Shrestha R, Ban S, Devkota S, Sharma S, Joshi R, Tiwari AP, Kim HY, Joshi MK (2021) Technological trends in heavy metals removal from industrial wastewater: a review. J Environ Chem Eng 9(4):105688. https://doi.org/10.1016/j.jece.2021.105688

Single-Walled Carbon Nanotube—An Overview | ScienceDirect Topics (n.d.) Retrieved March 13, 2025, from https://www.sciencedirect.com/topics/engineering/single-walled-carbon-nanotube

Siva Kumar OVPR, Sundaramoorthy A, Padmapriya VS, Raman DN (2021) Preparation of free-standing films from SWCNT/PANI nanocomposites using different blending techniques and characterization of their EMI shielding effectiveness in X-band. Phosphorus Sulfur Silicon Relat Elem 197:1–9. https://doi.org/10.1080/10426507.2021.2012680

Swain CK (2024) Environmental pollution indices: a review on concentration of heavy metals in air, water, and soil near industrialization and urbanisation. Discover Environment 2(1):5. https://doi.org/10.1007/s44274-024-00030-8

Tobias DJ, Martyna GJ, Klein ML (2002) Molecular dynamics simulations of a protein in the canonical ensemble (world). ACS Publications, American Chemical Society. https://doi.org/10.1021/j100151a052

Wang J, Wolf RM, Caldwell JW, Kollman PA, Case DA (2004) Development and testing of a general amber force field. J Comput Chem 25(9):1157–1174. https://doi.org/10.1002/jcc.20035

Wang J, Huang L, Yang R, Zhang Z, Wu J, Gao Y, Wang Q, O'Hare D, Zhong Z (2014) Recent advances in solid sorbents for CO_2 capture and new development trends. Energy Environ Sci 7(11):3478–3518. https://doi.org/10.1039/C4EE01647E

Wehbe M, Abu Tarboush BJ, Shehadeh M, Ahmad M (2020) Molecular dynamics simulations of the removal of lead(II) from water using the UiO-66 metal-organic framework. Chem Eng Sci 214:115396. https://doi.org/10.1016/j.ces.2019.115396

Wise FW (2000) Lead salt quantum dots: the limit of strong quantum confinement. Acc Chem Res 33(11):773–780. https://doi.org/10.1021/ar970220q

Zeng K, Hachem K, Kuznetsova M, Chupradit S, Su C-H, Nguyen HC, El-Shafay AS (2022) Molecular dynamic simulation and artificial intelligence of lead ions removal from aqueous solution using magnetic-ash-graphene oxide nanocomposite. J Mol Liq 347:118290. https://doi.org/10.1016/j.molliq.2021.118290

Zhou Q, Xiao J, Wang W, Liu G, Shi Q, Wang J (2006) Determination of atrazine and simazine in environmental water samples using multiwalled carbon nanotubes as the adsorbents for preconcentration prior to high performance liquid chromatography with diode array detector. Talanta 68(4):1309–1315. https://doi.org/10.1016/j.talanta.2005.07.050

Zhou H-Z, Shi Z, Sun Y, Jiang H-M, Song Z, He T, Cheng M (2024) The impact of industrialization on the distribution of heavy metals in surrounding soil: evidence from the Chang-Zhu-Tan urban agglomeration, central China. Ecol Ind 166:112368. https://doi.org/10.1016/j.ecolind.2024.112368

Bio-Circular Approaches to Critical Metal Recovery: Closed-Loop Bioleaching Strategies

Prasamita Samantaray, Sailaja Priyadarsini, and Alok Prasad Das

Abstract The growing demand for key metals, fueled by rapid technological innovation and an international shift toward renewable energy sources, has underlined the critical importance of sustainable and cyclical resource recovery solutions. This chapter investigates bio-circular approaches to crucial metal recovery, with a focus on closed-loop applications bioleaching systems as an eco-friendly alternative to traditional extraction methods. Bioleaching uses microorganisms' metabolic activity to dissolve metals from ores, trash, or secondary resources, resulting in a low-energy, low-emission approach with minimal chemical input. Integrating biological leaching into sustainable development frameworks improves material reuse and recycling, linking resource recovery with stewardship and waste minimization principles. The chapter presents a complete description of closed-loop bioleaching techniques, which allow biological substances and leachates to be used continuously and repeatedly. The utilization of extremophilic and artificially produced microorganisms with improved metal sensitivity and tolerance, as well as the design of bioreactors that facilitate process optimization and scalability, are among the most significant advancements. The focus is on recovering key metals including cobalt, nickel, lithium, rare earth components, and platinum alloy metals from end-of-life devices, used batteries, and industrial leftovers. The chapter also covers technological, economic, and environmental aspects, such as life cycle evaluation along with procedure convergence with both upstream and downstream valorisation. This chapter demonstrates the potential of closed-loop applications bioleaching as an essential element of bio-circular cycle models through case studies and current breakthroughs. The synergistic confluence of biotechnology, technology development, and continuous resource management may convert waste into a lucrative resource stream, minimizing environmental impact while safeguarding essential metal supply chains for a more sustainable future.

P. Samantaray · S. Priyadarsini · A. P. Das (✉)
Department of Life Sciences, Rama Devi Women's University, Bhubaneswar, Odisha, India

A. P. Das and S. Priyadarsini (eds.), *Recovery and Recycling of Critical Metals from Electronic Waste for Environmental Sustainability*, Earth and Environmental Sciences Library, https://doi.org/10.1007/978-3-032-29064-9_9

1 Introduction

The crucial metals are essential to modern life because they are used in many high-tech military, green, and household applications, but their supplies are fundamentally unstable. The idea of criticality as it relates to the critical metals is summarised in this paper along with important concerns regarding their resources and future supply. Depending on the perspective of the organization considering criticality, certain materials are more strategically significant than others, as evidenced by the development of techniques to measure the significance of critical metals over time (Priyadarsini and Das 2025b). Nevertheless, it is still unknown how much of some essential metals are produced worldwide, and how much are their reserves and resources. There are techniques for reasonably accurately estimating the supply of critical metals, but they rely on data from the mining sector, suggesting that improved reporting procedures would advance our understanding of the worldwide assets and cycling of such vital commodities. Criticality can also be investigated in many different manners, which involves the review of known mine supply chains for facilitating profitable utilization of critical metal contaminants, determining of the critical metallic material prospectivity of mining/mineral evaluating wastes (which was given a significant amount of valuable metals currently deport regarding waste), a greater number of recovering intermediates or final use goods that contained critical metals, when the discovery of new and economic funding of the critical metals. However, in order to facilitate future discoveries and raise the possibility of critical metallic substances being extracted as primary or by-products, all of the above methods and the related policies need more data in the areas of mineral abundance finance, substance economy, material flow analysis, the processing of minerals, and economic geology. In the absence of this information, important aspects of our understanding of the supply (and security associated with this production) of the key metals will remain unknown (Jowitt et al. 2018).

Because biosurfactants may intricate, chelate, and extract pollutants from contaminated environments, they have become attractive agents for environmental remediation. With a focus on the contribution of biosurfactant-metal interaction in developing green recovery techniques and improving resource circularity, this paper assesses possibilities for recovering crucial nutrients from waste materials to assist the production of renewable energy. With recovery efficiency surpassing 75% under ideal conditions, rhamnolipids are one of the biosurfactants that exhibit the highest attraction for metals like nickel, cadmium, and copper because of their highly stable coefficients and functional molecules like carboxylates. Analytical methods such as Fourier-Transform Infrared spectroscopy (FTIR), Scanning Electron Microscopy (SEM), and Inductively Coupled Combustion Mass Spectrometry (ICP-MS) are essential for evaluating interaction mechanisms and recovery efficiency (Tripathy et al. 2026). Biosurfactant-based recovery processes were evaluated using the Green Chemistry Metrics Framework, which showed 75–90% lower environmental factors than traditional techniques. There are still a lot of unanswered questions regarding the use of biosurfactants to extract metals like cobalt and lithium from waste materials and batteries. Growing biosurfactant-based technologies have the potential to address resource shortages and environmental protection issues at the same time by facilitating effective, efficient removal of metals and resource circularity. Noble metals have grown more and more common in modern times due to their wide range of uses in industry and various facets of human life. Traditional mineral extractions must be accompanied by secondary and waste material recycling because of their finite supply and continuously rising demand. With a focus on sustainability and selectivity, this Minireview discusses the latest advancements in solvometallurgical renewal of Pd, Pt, Rh, Ru, Ag, and Au. Processing of appropriate multi-metal compounds for reusing from discarded printed circuit boards to end-of-life automotive catalysts is described, including specific antioxidant dissolution, reductive the process of precipitation solvent extraction, co-precipitation, membrane transfer, and trapping to solid media (Dey et al. 2024). A potential future course for recycling noble metals is suggested, with a focus on environmentally friendly methods. Different methods and a combination of technologies can be used to treat metals, metal-mixtures, and multi-component substrates that comprise non-metal components such as polymers and ceramics. Substrates typically need to be physically disassembled first. Particle separation using magnetic, electrostatic, or gravitational techniques typically follows techniques involving delamination, cutting, shredding, and grinding. Dust particles often evolve as a result of mechanical processing. For instance, brominated flame retardants, which are found in a variety of WEEE, are hazardous and bioaccumulative persistent organic pollutants (Deravian and Mulligan 2025; Zupanc et al. 2022).

Waste including organic compounds and heavy metals is becoming more prevalent in the environment, as evidenced by the constantly expanding industries and technologies (Priyadarsini and Das 2025a). Both the environment and people may suffer irreversible harm as a result of these toxic pollutants. One Metal elements that have a density that is at least five times that of water are generally referred to as heavy metals. Arsenic, a metalloid that is hazardous even in trace amounts, is frequently found in waste that contains hazardous metals. These metals' poisoning of the environment has sparked worries about the environment and human health in recent years. Despite the fact that heavy metals are normally found in the crust of the Earth, their concentration in soil, water, and air has increased due to the industrial, residential, and agricultural use of these compounds, as well as their recovery from mines. Due to their high conductance and chemical stability, precious metals like gold, silver, Platinum and platinum itself (PGMs) are typically discovered in a variety of waste materials (Mishra et al. 2024). Because of this, there is a greater economic need to collect these rare and valuable metallic compounds from solid waste in order to address the heavy metals' negative effects on public health. Pyrometallurgy and hydrometallurgy are two of the conventional techniques for recovering metals. Thermal techniques are used in pyrometallurgical operations to recover metals from waste. Due to the substantial use of energy and the production of harmful gases (furans and dioxin) that pollute the environment, the majority of thermal techniques are prohibitively expensive. Hydrometallurgical techniques produce a lot of byproducts that need to be further processed for disposal and use environmentally toxic chemicals to leach metals. The ability of a variety of microorganisms, including autotrophic and heterotrophic bacteria and fungi, to extract metallic ions from solid waste materials and transfer them into a solution is used in bio-hydrometallurgy or bioleaching processes. Microorganisms have two methods for removing metal ions in this process: direct bioleaching and indirect bioleaching (Dey et al. 2024). The bacteria in the intermediate mechanism only break down solid waste indirectly by producing metabolites; they do not come into direct touch with the surfaces of solid waste. It is possible for bacteria to directly engage in the contact mechanism. As a result, an electrochemical process makes metals soluble. Organic carbon is essential for the growth of autotrophic bacteria. Rather, they grow because of the carbon fixation caused by atmospheric CO_2. The carbon source for heterotrophic bacteria and fungus is organic matter. The metabolism of these microorganisms, including the organic compounds (malic acid, oxalic acid, glucose, nitric acid, and citric acid) released in the culture medium, depends critically on these organic carbon sources. These microorganisms have been shown to remove metals from a solid matrix, frequently thanks to their chelating and acidic properties. These heterotrophic microbes also form non-acidic complexes as a result of their protein catabolism, which leads to alkaline leaching systems (Gu et al. 2018).

2 Critical Metals: An Overview

Sustainable energy and emerging applications are driving long-term demand-side growth in the well-established critical metals theme. Finding and creating new sources of key metals supply is a necessary step in delivering the energy shift. The exploratory process is where new supply constraints begin, as land access and meeting stakeholder, administrative, and legal requirements have grown increasingly difficult, uncertain, expensive, and time-consuming in recent years. Because major mining corporations have steady financial flows, they are favoured over smaller explorers due to the repercussions of greater exploration access costs and longer timetables. Globally, developed mining governments with a history of settling conflicting land use issues are preferred over more frontier jurisdictions without such a history due to greater land access risks (Trench and Sykes 2025). The US National Research Council was the first to propose the idea of essential minerals (2008). Minerals that are vital to a nation's economy as a whole or to important economic sectors but susceptible to supply disruptions whether as a result of local or global supply chain issues, the concentration of supply from a small number of mines, geographic location, geopolitical concerns, trade policy, or other factors are referred to as "critical minerals" (Carr-Wilson et al. 2024; Schrijvers et al. 2020). Difficult environmental, social, and governance hazards to mine development and exploration could be additional factors. The production of a vast array of high-tech products, such as smartphones, electric cars, renewable energy systems, aerospace, and defense applications, as well as essential components to maximize performance in numerous other electronic applications, depends on critical minerals. Critical minerals are necessary for renewable energy efficiency and performance, even if they make up a very small portion of an innovative energy technology use or consumer end product by weight and value (Müller et al. 2025; Walker et al. 2025).

Valorising mine waste has two advantages: it reduces the environmental effects of legacy mining activities while also supporting resource recovery, which is consistent with the circular economy. Obtaining valuable components from tailings can enhance environmental safety and lower site rehabilitation costs (Rosario-Beltré et al. 2023). One of the biggest waste streams produced worldwide is mine tailings in particular. They are the leftover solids that remain after valuable minerals have been extracted, and they are mostly made up of calcium, iron, silicon, and aluminium oxides with trace metals. Tailings pose serious safety risks, particularly in the event of dam breaches, and frequently need a large amount of land for storage. However, they also contain significant amounts of metals that have not been retrieved, including those that are currently considered important. Because primary ores are growing lower grade and extraction technologies are improving, processing effluent is becoming more and more appealing. The mining industry can lessen its environmental impact and boost resource efficiency by utilizing these advancements. There is widespread agreement that mine tailings should be treated as secondary resources due to the growing dearth of high-grade ore deposits. By doing this, the risks connected to former mining sites are lessened, shortages in mineral supplies are lessened, and sustainable development objectives are supported. Critical raw materials, such as metals and minerals, that are necessary for contemporary technologies and environmentally friendly power systems, are at the core of the shortage of resources problem (Alajoki 2024; Mafra et al. 2022). These consist of light metals like lithium and magnesium, radioactive elements like uranium, valuable metals like platinum and rhodium, and rare earth elements (REEs) like neodymium, dysprosium, and terbium. Their distinct physical and chemical characteristics make them essential for a variety of uses, ranging from renewable energy to electronics and defense. However, because to economic, environmental, and geopolitical limitations, their supply is becoming more and more precarious. Consequently, nations have determined and ranked their own lists of essential materials. For instance, 34 materials are currently listed as critical by the European Union, while lists comparable to this have been created by the USA, Australia, Canada, India, and Taiwan based on their own national interests. A common understanding of the strategic significance of ensuring consistent supplies of vital minerals is reflected in these endeavours. In order to manage long-term environmental hazards while promoting important mineral supply, the value-adding of mine waste through thorough sampling, characterisation, and asset modelling has become a viable approach. Since the switch to renewable energy technologies is expected to result in exponential increases in demand for essential minerals, notably those used in battery power, rooftop solar cells, and wind turbines, the worldwide push for an ecological economy has only made this urgency more pressing (Nwaila et al. 2025; Sarker et al. 2023).

The spectroscopic and magnetic characteristics of rare earth elements make them especially important because they support important technologies like electronics, clean energy systems, and permanent magnets. Although secondary sources such tailings, industrial byproducts, and urban trash are becoming more and more recognized as alternate reservoirs, REEs are mostly derived from minerals like the amorphous bastnasite, and xenotime. In addition to improving resource security, recuperation from these sources encourages environmentally friendly waste management (Das et al. 2024). In a similar vein, coal gangue, a frequent byproduct of coal mining, is becoming more widely acknowledged as a low-grade substitute for traditional ore due to its substantial amounts of aluminum oxide, lithium, and rare earth elements. In addition to contributing to the crucial metal supply, taking advantage of coke gangue can ease land-use demands and lower the environmental concerns connected with them (Sarker et al. 2023). An other unusual but intriguing source of REEs is acid mine drainage (AMD). AMD is currently being reassessed for its economic possibilities after being historically viewed as an environmental concern because of its acidic nature and heavy metal concentration. According to studies, REE contents in AMD rise noticeably at lower pH values, indicating a potential for resource recycling during treatment. In particular, the trio of platinum group elements (PGEs) rhodium, palladium, and platinum—which are essential to green technologies and industrial processes—are subject to the same change in viewpoint. Secondary materials, such recovered industrial tailings and scrap, are increasingly being used to augment primary supplies. Reusing and secondary recovery provide high-efficiency alternatives to the energy-intensive and loss-prone method of recovering PGEs from primary ores, hence promoting the long-range viability of PGM supply chains (Gibson et al. 2023; Goodman et al. 2023; Zhang and Honaker 2020).

3 Fundamentals of Bioleaching

When conventional extraction techniques are not economically feasible, a process known as bioleaching uses microorganisms to recover metals from ore. This method is frequently applied to sulphide mineral deposits, which are the raw materials of many precious metals, including copper, silver, and gold. Traditional techniques for extracting metal from sulphide minerals, such as pyrometallurgy, are costly, energy-intensive, and polluting. Low-quality ores are therefore usually not treated with these methods and may be thrown away as waste. By using bacteria's metabolic processes to break down metal ores, bioleaching provides an affordable and low-input solution to this issue (Baláž 2008). Microbes oxidize sulphur and iron from sulphide rocks to provide energy during the bioleaching process. Target metals are released as a result of the oxidants that damage the sulphide minerals. The basic idea of bioleaching is straightforward, but its fundamental processes are intricate; microbial metabolisms of sulphur and iron depend on a complex network of enzymes to carry out the process. In this study, we provide an overview of the genes governing aerobic ferrous and sulphur oxidation as well as a summary of the current understanding of the general mechanisms of bioleaching. Theoretical bioleaching routes are modelled using chalcopyrite and pyrite, two prevalent bioleaching substrates. This review focuses on the bioleaching processes mechanisms of acidophilic bacteria since bioleaching usually occurs in acidic environments.

According to recent studies, bioleaching is a commercially feasible two-pronged technique for simultaneously improving the dewaterability of sludge and extracting heavy metals. The microbiological decomposition of the fundamental sulfur by the implanted sulfur bacteria is the cause of the bioleaching sludge's notable pH drop. *Acidithiobacillus ferrooxidans* and *Acidithiobacillus thiooxidans* are two acidophilic chemolithoautotrophic bacteria that are frequently involved in the bioleaching process. It has been demonstrated that, in mild conditions, the bioleaching of sludge from wastewater effectively removes heavy metals. At the same time, bioleaching frequently improves the dewaterability of sludge (for example, a 66.9% reduction in sludge filtering resistance following bioleaching) (Liu et al. 2012; Shokoohi et al. 2025). It is still unclear exactly which metabolic processes in ADS are responsible for this synergistic improvement in sewage dewaterability and contaminants removal. The dissolved organic carbon (DOC) portion in the mixture rises with any introduction of organic carbon, such as low-molecular-weight peptides that amino acids, or polysaccharides distributed during cell lysis. This is especially true for the majority of acidophilic sulphur-oxidizing bacteria, which are strictly autotrophs and only use inorganic substrates for energy. Increased DOC levels have the ability to combine with generated metal ions, changing the species they form and redox potential. They can also create a protective colloidal covering on sludge particles, which prevents the acidophilic consortia from attaching, forming biofilms, and sulfidogenesis. When combined, these impacts can lower EPS degradation and proton-driven metal solubilization rates, which will ultimately impair metal leaching efficiency and dewatering performance. For instance, Fang et al. discover that the biological leaching of heavy metals is impeded when the sludge's dissolved organic matter content exceeds 150 mg DOC/L because *Acidithiobacillus* species are less active (Fang and Zhou 2007). Pig dung extract is said to prevent iron rusting and metal solubilization because of its soluble organic components. It is discovered that the primary inhibitors of sludge bioleaching are acetic and propionic acids. The demonstrated performance of integrated heterotroph–autotroph systems over axenic cultures in mineral oxidation activities can be explained by this metabolic complementarity. The dangers of applying sludge to land have not yet been thoroughly examined. A solution that is guaranteed to improve the safety of sludge land use is sludge conditioning treatment. Investigators have removed and purified acidophilic self-sustaining strains with certain partner (heterotrophic) bacteria, including Rhodotorula mucilaginosa R30 and Mucor circinelloides ZG-3, in order to create a high-efficiency bioleaching inoculum. Using native microorganisms that have the ability to bioleach can greatly simplify this (Wang et al. 2010; Zheng et al. 2016).

4 Closed-Loop Bioleaching Concepts

The leftover debris is deemed to have too low a metal concentration for additional processing through methods such as biomining, purification of water, and land decontamination after the high-grade ore has been physically and chemically extracted from mined rock. The biological stage in every one of these methods involves using a range of organisms that retrieve the elements from a "waste" product in order in order to detoxify or increase cost-efficiency. Additional value can be extracted from this unwanted ore through the use of biological treatment. Although there is limited specificity for individual metals, this waste, which has become acidic as a result of earlier processing procedures, can be further thermally treated with solutions and cyano-compounds to dissolve the remaining metals from the ore. Bio oxidation can recover gold and silver, while bioleaching can recover copper, zinc, cobalt, and uranium, depending upon the metal composition of the ore (Panagos et al. 2013). Metals are extracted from mining waste for economic reasons, whereas water from industrial and environmental applications is treated for metal removal primarily for detoxifying rather than for commercial value. Screening and ion exchange technologies are employed on a lesser scale and with samples that are only mildly polluted. Activated sludge, a highly energetic community of microorganisms activated by aeration, is a common kind of purification on a large scale. It breaks down organic particles, absorbs metal ions, and eventually sediments and clarifies wastewater. This treatment is completely integrated and necessary to the entire water purification process, in contrast to biomining. There have been a lot of recent research on the use of this activated sludge for the selective removal of metals, such as the elimination of chromium from sewage and the recuperation of cadmium and nickel, copper, lead, and zinc (Meulepas et al. 2015). By applying engineering concepts to molecular biology and embracing the "design, build, test, and learn" mindset, synthetic biology enables iterative assembly and feedback loop analysis. By standardizing the genetic components needed to create translational units and operons, synthetic biology improves upon current methods and makes it easier to create libraries in vitro and screen for desired phenotypes later on. Biological approaches were formerly sluggish, complicated, non-standardized, and had incompatible components across many techniques. As a result, several distinct and complementary toolkits have been created for the genetic modification of numerous industrially significant organisms. From simple generation of homologous proteins in model systems to the enhancement of non-model organisms' genetic capacities through the replication of plasmids and antibiotic resistances, these toolkits cover a wide range of tasks. Since many of the toolkits are still in their infancy, there are currently few examples of their application in biological remediation and metal detection. This is not the case for the more well-known bio brick system, which has been used for identifying amounts of mercury, lead, and copper as well as for the biological elimination of gold, cobalt, nickel, and mercury (Yan et al. 2018).

Biohydrometallurgy is a novel, cleaner, and potentially environmentally beneficial technology that can be used as an alternative to hydrometallurgical and pyrometallurgical processes. The foundation of biohydrometallurgy, also known as bioleaching, is the utilization of specialized microbes whose activity produces metabolites that can remove particular metals. Compared to other recycling processes, microorganisms' ability to dissolve and recover metals has several benefits, including reduced emissions of hazardous gasses, minimal energy inputs, low operating costs, and environmental safety. Additionally, because it uses less energy, emits fewer harmful gasses, and operates at lower temperatures than traditional methods, it is more beneficial to the environment (Magoda and Mekuto 2022). Because it can lessen the negative ecological effects of disposing of e-waste and offer an environmentally friendly source of metals, bioleaching is a potential method for reclaiming metals from e-waste. Even with all of the benefits of the biologic leaching process, the technique's slow kinetics make it unable to compete with conventional approaches on an industrial scale. Thus, with more study and advancement, bioleaching could emerge as a popular technique for recovering metals and a prototype for a circular e-waste ecosystem. Through the action of certain microbes, the agent that leaches Fe^{3+} is bio produced in the first stage. A common bacterium for metal solubilization is Acidthiobacillus ferroxidase. Under aerobic conditions, it employs cellular oxygen as its final acceptor of electrons and converts ferrous ions to ferric ions to provide energy (Priya et al. 2018). Preparing the e-waste for leaching is the second step. In order to work with a specimen that has a greater concentration in the element of interest and to guarantee an ideal particle size, this step is essential. Leaching is the third stage. At this point, the pretreated waste from the second stage and the Fe^{3+} enerated in the first step come into contact. When Fe^{3+} and e-waste come into contact, spontaneous oxidation–reduction reactions take place, reducing Fe^{3+} to Fe^{2+} and oxidizing and dissolving the metals in the e-waste. The recovery of elements from the wastewater into an immovable state is the final step. Numerous metals with wildly disparate concentrations make this stage extremely complicated. As a result, it is challenging to collect all very pure metals selectively. Several techniques, including extraction by solvent, the exchange of ions, adsorption and selective precipitation, and cementation, are used to remove metals from leaching solutions with precision (Ashiq et al. 2019). In stage 1, Fe^{2+} is finally reintroduced and converted by the microbes into Fe^{3+}. Because the leaching agents are regularly reused, this completes the cycle and makes the whole procedure sustainable. By improving the effectiveness and sustainable practices of metal extraction from e-waste, this cyclical process supports the ideas of a systemic economy. Furthermore, the ability to optimize each of the four steps separately is another key benefit of closed-loop bioleaching. Thus, gathering and summarizing scientific publications that concentrate on examining each stage independently is the main goal of this review. The fact that the third stage of closed-loop bioleaching produces a solution containing different metals at varying quantities is one of its difficulties. These metals are recycled back into the bioreactor if they are not completely removed in the fourth stage. This can stop iron-oxidizing bacteria from doing their work, which lowers the rate at which Fe^{3+} is produced (Morell et al. 2025).

After the high-grade ore has been physically and chemically extracted from mined stone, the leftover waste is deemed to have too low a metal concentration to be processed further using these methods. Additional value can be extracted from this garbage of poor-quality ore through the use of biological treatment. Although there is limited specificity for individual metals, this waste, which has become acidic as a result of earlier processing procedures, can be further thermally handled by chemicals and cyano-compounds to dissolve the remaining metals from the ore. Copper, zinc, cobalt, and uranium can be extracted by bioleaching, whereas gold as well as silver are obtainable by bio-oxidation, with respect to the metal content of the ore (Kaksonen et al. 2017; Tanaka et al. 2015). Sulfuric acids are given in both situations to increase the dissolution of metal ions, which are subsequently bio absorbed or accumulated by living microbes and archaea or by single or mixed cultures of these organisms. The wasted ore heap's interior temperatures may rise to 50 °C during this treatment, and additional acid may be manually injected or created by the bacteria to keep the atmosphere acidic. In order to retrieve the metal-laden microorganisms for processing downstream, the deposits are then cleaned. To encourage the growth of cultures of microbial origin, extracellular polymers derived from bacteria can also be added. These chemical compounds form at the boundaries between the bacteria and the spent-ore deposits, stimulating biofilm development and cell-to-cell communication to increase metal recovery (Nkulu et al. 2015). Most of the microorganisms utilized in this method are acidophilic microbes or archaea that can oxidize substances that include sulphur and/or iron and have a high heat tolerance. The most common of these microorganisms in biomining cultures are Acidithiobacillus sp., either by itself or in combination with other bacteria including Acidiphilium sp., Leptospirillum sp., and Sulfobacillus sp (Gumulya et al. 2018). The acquisition of metals from unspent ore is increased by both biomining and chemical usage; however, if the wastewater containing trace levels of metals, acids, salts, and microorganisms is not contained or detoxified, it may leak into the environment and damage both land and water.

Organisms can extract and reclaim metal ions in a variety of ways. One method is to directly adsorb them onto the outermost layer, which, with the right harvesting process, enables the metallic elements to be collected alongside the organism. Metal ions bind to polysaccharides outside of cells and biofilm matrix, which is one of the main ways that metals can be extracted during the biomining process. A variety of bacteria, algae, and fungi have been shown to do this (Muthu et al. 2017). The inclusion of metallothioneins (MTs) and phyto-chelatins (PCs) may considerably improve metal adsorption, even if it can happen rather passively. These enhance the organisms' tolerance to harmful metal ions and enable the incorporation of more elements to the cell surface. Both PCs and MTs are cysteine-rich polypeptide that are present in a variety of organisms and chelate metal ions through thiolate linkages. Rich in cysteine and glutamate residues, plant-derived PCs are well-known for their ability to chelate metal ions and prevent them from interfering with vital intracellular or extracellular processes, hence promoting heavy-metal endurance to the elements lead, cadmium, and zinc. For adsorption onto the surface of copper using minimal cysteine-rich peptides, an iterative, logical synthetic design process has recently been used in conjunction with MTs. It has been demonstrated that altering the length and remaining number of PCs broadens their area of application to include a wider assortment of metal ions (Cobbett 2000). The recovery of electronic trash is the intended use, but it may be extended to waste containing platinum and rare earth elements (REEs) and can be immediately applied to the recycling procedures that are now in place. Furthermore, bacterial phages have been studied for metal recovery; the electrostatic attraction of the REE-containing molecule lanthanum phosphate has been demonstrated to be a feasible option through the surface display of short peptides with a randomized pattern. PCs and MTs have been utilized as a platform for the creation of nanotechnology for a very broad spectrum of metals employing materials from their bio recovery process, in addition to merely chelating metal ions (Choi et al. 2018; Urbina et al. 2019). Absorption permits intercellular accumulation in addition to the surface binding of beneficial metals, so long as the presence of these metals does not impair regular cellular physiology because of oxidative breakdown or mis-metalation. Since plants may accumulate significant levels of heavy metal ions and can be picked out to eliminate the offending poisons from the site, they are the primary choices for treating contaminated land in this process. Metal homeostasis, a strong oxidative stress response, and molecular chelation are necessary to accomplish this. In order to do this, both inflow and efflux mechanisms function to restrict the accumulation of metal ions within the cell. It has been demonstrated that P-type ATPases are particularly significant and play a significant role in natural soil organisms when heavy metals like copper, arsenic, and platinum are present. According to recent reports, they also play a part in plants' and Lactobacillus's cadmium and manganese responses. It has been demonstrated that a different class of metal-transporters, known as ABC permeases, which are present in Gram-positive bacteria, facilitate the uptake of minerals from host cells or the surrounding media, such as soil or water. In Staphylococcus species, these metal ions include the element copper, the metal cobalt zinc, and nickel. This capability is immediately applicable to the entire metal bio recovery format. Similarly, gold can be scavenged and reabsorbed by bacterial cells for metal buildup by the usage of delftibactin, a non-ribosomal peptide that resembles a siderophore. Since its identification, it has been suggested that it may be used in the procedure for recycling and to recover low-concentration platinum from mineral ore and electronic waste (Johnston et al. 2013; Song et al. 2018).

5 Future Perspectives

Critical metals like the element lithium, cobalt, nickel, and other common and rare earth elements (REEs) are in high demand as the world economy shifts toward electrification, digitalization, and renewable energy. These metals are essential to green technology such as solar cells, wind turbines, and batteries for electric cars. Conventional mining techniques, however, are insufficient to meet foreseeable expectations and are energy-intensive and environmentally harmful. As a result, bio-circular methods—in particular, closed-loop bioleaching techniques—are showing promise as substitutes that complement sustainability and the circular economy. Microbial leaching, another name for bioleaching, is the process of removing minerals from ores, streams of trash, or recycled materials by means of microbes. It has historically been applied to low-grade ores, especially copper and gold, but new developments in microbial ecology and biotechnology are expanding its use to include important metals. To produce oxidizing sulfide of metal and solubilize target elements, the procedure uses bacteria and archaea such Acidithiobacillus ferrooxidans and Leptospirillum ferrooxidans.

Low energy input, less usage of hazardous chemicals, and the possibility of on-site, decentralized operation are characteristics of contemporary bioleaching. By allowing metal recovery from mine tailings, industrial residues, and end-of-life products, bioleaching's inclusion into closed-loop systems improves its sustainability credentials while lowering waste and reliance on virgin ores. The circular economy, which turns trash into feedstock for novel manufacturing cycles, is the foundation of closed-loop bioleaching techniques. Recovering vital metals from secondhand goods, purifying them for future use, and reintegrating them into their manufacturing supply chain are the goals of these systems. Examples include fly ash from burning coal, slag from making steel, and the bioleaching of waste from lithium-ion batteries.

Bioreactors built for particular waste streams, in conjunction with post-leaching treatment technologies including extracting solvents, anion exchange, and membrane filtering, will be progressively incorporated into future closed-loop systems. Self-sustaining recovery units for dispersed or distant facilities could be created through incorporation with renewable energy sources, such as solar-powered bioreactors.

1. Expansion to Complex Waste Streams: The ability of closed-loop bioleaching to handle increasingly complex and heterogeneous waste sources, such as e-waste, industrial slags, and hybrid batteries that contain multiple metals in varying oxidation states, as well as the development of microbial consortia and genetically engineered strains to target specific metals or degrade synthetic polymers that encapsulate valuable elements, is what will determine its future.
2. Biotechnological Innovations: Synthetic biology is poised to revolutionize bioleaching; engineered microbes with improved metal-binding affinity, enhanced tolerance to toxic metals, or the ability to secrete customized organic acids could greatly speed up recovery rates. CRISPR and other gene-editing tools provide precise control over microbial metabolism, allowing the creation of bio-catalysts for particular recovery tasks.
3. The integration of closed-loop bioleaching strategies into product life cycles is expected to increase: Future factories may be designed with end-of-life recuperation in mind, ensuring products more conducive to microbial treatment. This systemic examine will allow for continuous surveillance and sorting of materials, ensuring effectiveness extraction and disposal at the end of product life.

4. Policy and Economic Incentives: Encouragement policies, carbon pricing, and advantages for secondary consumption of resources are necessary for the scalability and promotional adoption of closed-loop bioleaching. Regulatory frameworks must change to recognize biological technology processes in metal recovery as viable industrial practices, and green accreditation, subsidies, and partnerships between the private and public sectors can spur innovation and infrastructure development.
5. Decentralization Worldwide: Additionally, closed-loop bioleaching has potential for small-scale, decentralized operations. This is especially advantageous for poorer nations where there is a surplus of mining waste but little access to advanced recycling facilities. For local communities, portable bioleaching systems can be an affordable way to recover important metals, improve resource security, and generate income.

Closed-loop bioleaching-based bio-circular techniques provide a revolutionary route to long-term crucial metal recovery. These strategies bring environmental stewardship and technological innovation together as society transitions to a low-carbon future. Closed-loop bioleaching has the potential to develop from a specialized biotechnology into a key component of the international circular economy with sustained research, funding, and governmental support.

6 Conclusion

While many engineering efforts have focused on the metal chelation process, other considerations including temperature, pH, osmolarity, and competition with the current microbial community in the waste streams must also be made when creating strains for metal recovery. These must be taken into account for an efficient bio recovery system, which may include the use of model or non-model organisms for the metal recovery task. Many elements that address the constraints of present methods as well as the first issues with the adoption of novel innovations and the challenges they bring with them must be taken into consideration whenever invention is to supersede existing technologies. Although this will depend on the circumstances, it is crucial to innovate responsibly, particularly when genetically modified organisms and the environment are both involved, as this will be the limiting factor for many. In order to demonstrate proof of principle profits, synthetic biological activities will initially be implemented on a modest scale because they will undoubtedly be a costly endeavor requiring new infrastructure and training. The shift to circular and sustainable resource use is now a global necessity rather than an option. Bio-circular methods, especially closed-loop bioleaching techniques, are noteworthy in this regard as novel and environmentally responsible means to recover important metals from primary as well as secondary sources. In a society that is becoming more and more reliant on high-tech, green, and internet infrastructure, these tactics are not only technically possible but also strategically essential for attaining efficiency of resources, reliability in supply chains, and sustainable environmental practices. End-of-life electronics, used batteries, pharmaceutical slags, and mine tailings are just a few of the waste products from which valuable metals can be extracted and recovered using the closed-loop bioleaching paradigm, which combines various chemical, biological, and systems engineering methodologies. It differs from conventional metallurgical methods in that it uses microbial digestion to mobilize metals under moderate circumstances, uses less energy, has a smaller environmental impact, and can handle complicated and low-grade waste matrices. The scientific foundation, technological development, and application areas of closed-loop bioleaching techniques have all been covered in this chapter. We have demonstrated the use of synthetic or naturally occurring microorganisms in controlled settings to generate acid or oxidize metals biochemically. We investigated the effects of process variables on extraction efficiency and metal selectivity, including pH, oxidation potential, temperature, and the makeup of the microbial population. What becomes evident is that biological leaching is not just a substitute; rather, it is a progressive addition to current recovery technologies that can be used to a wide range of sectors, scales, and geographical locations. Given the growing amount of e-waste, the growing need for lithium-ion batteries and parts, and the global push to cut greenhouse gas emissions, its use is especially pertinent. By converting liabilities into assets and promoting waste valorisation and key raw material recovery, bio-circular methods help achieve the objectives of sustainable development and the circular economy. There are still a number of obstacles in the way of its potential. Further investigation and regulatory oversight are needed in the areas of feedstock material diversity and complexity, potential biosecurity issues with modified microorganisms, and the requirement for uniformity in bioleaching procedures. Additionally, economies of scale and technological innovation must be used to continuously increase economic viability with conventional quarrying and smelting. In summary, microbiological science, ecologic engineering, and environmentally friendly science come together in closed-loop bioleaching. It offers a practical, adaptable, and forward-thinking strategy for recovering key metals in a circular economy. This approach demonstrates how nature-inspired inventions can be at the core of the next corporate revolution—one that is regenerating, inclusive, and ecologically grounded—by utilizing the power of microbial existence to address one of the most urgent material concerns of our time.

References

Alajoki J (2024) Reprocessing and leaching of valuable metals from weathered flotation tailings. https://aaltodoc.aalto.fi/handle/123456789/126313

Ashiq A, Kulkarni J, Vithanage M (2019) Hydrometallurgical recovery of metals from e-waste. Elsevier, pp 225–246. https://doi.org/10.1016/B978-0-12-816190-6.00010-8

Baláž P (2008) Mechanochemistry in minerals engineering. In: Mechanochemistry in nanoscience and minerals engineering, pp 257–296. https://doi.org/10.1007/978-3-540-74855-7_5

Carr-Wilson S, Pattanayak SK, Weinthal E (2024) Critical mineral mining in the energy transition: a systematic review of environmental, social, and governance risks and opportunities. Energy Res Soc Sci 116:103672. https://www.sciencedirect.com/science/article/pii/S2214629624002639

Choi Y, Park TJ, Lee DC, Lee SY (2018) Recombinant Escherichia coli as a biofactory for various single- and multi-element nanomaterials. Proc Natl Acad Sci 115(23):5944–5949. https://doi.org/10.1073/PNAS.1804543115

Cobbett C (2000) Phytochelatin biosynthesis and function in heavy-metal detoxification. Curr Opin Plant Biol 3(3):211–216. https://doi.org/10.1016/S1369-5266(00)00066-2

Das A, Dey S, Das AP (2024) Ecotoxicological impacts of synthetic microfiber pollutants and development of sustainable mitigation strategies. Environ Chem Ecotoxicol. https://doi.org/10.1016/J.ENCECO.2024.12.003

Deravian B, Mulligan CN (2025) Sustainable recovery of critical minerals from wastes by green biosurfactants: a review. Molecules 30(11):112461. https://doi.org/10.3390/MOLECULES30112461

Dey, S., Dash, A., & Das, A. P. (2024). Bioremediation as a Sustainable and Circular Solution for Mining Waste Management and Bioenergy Generation. Sustainable Management of Mining Waste and Tailings: A Circular Economy Approach, 107–119. https://doi.org/10.1201/9781003442455-6

Fang D, Zhou LX (2007) Enhanced Cr bioleaching efficiency from tannery sludge with coinoculation of Acidithiobacillus thiooxidans TS6 and Brettanomyces B65 in an air-lift reactor. Chemosphere 69(2):303–310. https://doi.org/10.1016/J.CHEMOSPHERE.2007.03.059

Gibson B, Nwaila G, Manzi M et al (2023) The valorisation of platinum group metals from flotation tailings: a review of challenges and opportunities. J Petersen Miner Eng. https://www.sciencedirect.com/science/article/pii/S0892687523002303

Goodman A, Bednar A, Geochemistry JR-A (2023) Rare earth element recovery in hard-rock acid mine drainage and mine waste: a case study in Idaho Springs, Colorado. Elsevier. https://www.sciencedirect.com/science/article/pii/S0883292723000288

Gu T, Rastegar SO, Mousavi SM, Li M, Zhou M (2018) Advances in bioleaching for recovery of metals and bioremediation of fuel ash and sewage sludge. Biores Technol 261:428–440. https://doi.org/10.1016/J.BIORTECH.2018.04.033

Gumulya Y, Boxall NJ, Khaleque HN, Santala V, Carlson RP, Kaksonen AH (2018) In a quest for engineering acidophiles for biomining applications: challenges and opportunities. Genes 9(2):116. https://doi.org/10.3390/GENES9020116

Johnston CW, Wyatt MA, Li X, Ibrahim A, Shuster J, Southam G, Magarvey NA (2013) Gold biomineralization by a metallophore from a gold-associated microbe. Nat Chem Biol 9(4):241–243. https://doi.org/10.1038/NCHEMBIO.1179

Jowitt SM, Mudd GM, Werner TT, Weng Z, Barkoff DW, McCaffrey D (2018) The critical metals: an overview and opportunities and concerns for the future. In: Metals, minerals, and society, pp 25–38. https://doi.org/10.5382/SP.21.02

Kaksonen A, Boxall N, Usher K et al (2017) Biosolubilisation of metals and metalloids. In: Sustainable heavy metal remediation: volume 1: principles and processes, pp 233–283. https://doi.org/10.1007/978-3-319-58622-9_8

Liu F, Zhou L, Zhou J, Song X, Wang D (2012) Improvement of sludge dewaterability and removal of sludge-borne metals by bioleaching at optimum pH. J Hazard Mater 221–222:170–177. https://doi.org/10.1016/J.JHAZMAT.2012.04.028

Mafra C, Bouzahzah H, Stamenov L, Gaydardzhiev S (2022) An integrated management strategy for acid mine drainage control of sulfidic tailings. Miner Eng 185:107709. https://doi.org/10.1016/j.mineng.2022.107709

Magoda K, Mekuto L (2022) Biohydrometallurgical recovery of metals from waste electronic equipment: current status and proposed process. https://www.mdpi.com/2313-4321/7/5/67

Meulepas RJW, Gonzalez-Gil G, Teshager FM, Witharana A, Saikaly PE, Lens PNL (2015) Anaerobic bioleaching of metals from waste activated sludge. Sci Total Environ 514:60–67. https://doi.org/10.1016/J.SCITOTENV.2014.12.073

Mishra S, Dash D, Das AP (2024) Isolation and molecular characterization of polyester-degrading bacterial strains from municipal wastewater and their application in microplastic degradation in contaminated wastewater. Geomicrobiol J. https://doi.org/10.1080/01490451.2024.2376101

Morell J, Lao-Luque C, Sole-Sardans M, Ribas D, Guimerà X, Dorado AD (2025) Closed-loop bioleaching for mobile phones from the bioregeneration of the leaching agent to the recovery of metal: a review. J Ind Eng Chem. https://doi.org/10.1016/J.JIEC.2025.04.020

Müller D, Groves D, Santosh M (2025) Critical metals: their applications with emphasis on the clean energy transition. Elsevier. https://www.sciencedirect.com/science/article/pii/S2772883824000608

Muthu M, Wu HF, Gopal J, Sivanesan I, Chun S (2017) Exploiting microbial polysaccharides for biosorption of trace elements in aqueous environments—scope for expansion via nanomaterial intervention. Polymers 9(12):120721. https://doi.org/10.3390/POLYM9120721

Nkulu G, Gaydardzhiev S, Mwema E, Compere P (2015) SEM and EDS observations of carrollite bioleaching with a mixed culture of acidophilic bacteria. Miner Eng 75:70–76. https://doi.org/10.1016/J.MINENG.2014.12.005

Nwaila G, Ghorbani Y, Zhang S et al (2025) Valorisation of mine waste-Part II: resource evaluation for consolidated and mineralised mine waste using the Central African Copperbelt as an example. J Environ Manag. https://www.sciencedirect.com/science/article/pii/S0301479721016157

Panagos P, Van Liedekerke M, Yigini Y, Montanarella L (2013) Contaminated sites in Europe: review of the current situation based on data collected through a European network. J Environ Public Health. https://doi.org/10.1155/2013/158764

Priya A, Hydrometallurgy SH et al (2018) Extraction of metals from high grade waste printed circuit board by conventional and hybrid bioleaching using Acidithiobacillus ferrooxidans. https://www.sciencedirect.com/science/article/pii/S0304386X17309465

Priyadarsini S, Das AP (2025a) Isolation and identification of native lithium-resistant bacteria from electronic waste contaminated sites: a focused investigation on tolerance assessment with potential lithium bio-recovery application. Geomicrobiol J. https://doi.org/10.1080/01490451.2025.2539509

Priyadarsini S, Das AP (2025b) Lithium bioleaching: a review on microbial-assisted sustainable technology for lithium bio-circularity. J Water Process Eng 69:106744. https://doi.org/10.1016/j.jwpe.2024.106744

Rosario-Beltré A et al (2023) Critical raw materials recovery potential from Spanish mine wastes: a national-scale preliminary assessment. J Clean Prod. https://www.sciencedirect.com/science/article/pii/S0959652623013215

Sarker S, Bruckard W, Haque N et al (2023) Characterization of a carbonatite-derived mining tailing for the assessment of rare earth potential. https://www.sciencedirect.com/science/article/pii/S0957582023002239

Schrijvers D, Hool A, Blengini G, Resources WC (2020) A review of methods and data to determine raw material criticality. In: Resources, conservation and recycling. https://www.sciencedirect.com/science/article/pii/S0921344919305233

Shokoohi R, Najafi-Vosough R, Shabanloo A, Torkshavand Z, Mahmoudi MM, Abasi MA (2025) Removing heavy metals and improving the dewaterability of sewage sludge with the bioleaching process by Thiobacillus Ferrooxidans bacteria. Water Sci 15(2):34. https://doi.org/10.1007/S13201-025-02365-W

Song L, Zhang Y, Chen W, Gu T, Zhang SY, Ji Q (2018) Mechanistic insights into staphylopine-mediated metal acquisition. Proc Natl Acad Sci 115(15):3942–3947. https://doi.org/10.1073/PNAS.1718382115

Tanaka M, Yamaji Y, Fukano Y, Shimada K, Ishibashi JI, Hirajima T, Sasaki K, Sawada M, Okibe N (2015) Biooxidation of gold-, silver, and antimony-bearing highly refractory polymetallic sulfide concentrates, and its comparison with abiotic pretreatment techniques. Geomicrobiol J 32(6):538–548. https://doi.org/10.1080/01490451.2014.981645

Trench A, Sykes J (2025) Critical metals exploration and energy transition—a perspective. Geosystems Geoenvironment 4(1):100353. https://doi.org/10.1016/J.GEOGEO.2025.100353

Tripathy B, Singh S, Behera ID, Mishra S, Das AP (2026) Metagenomic profiling of diversified marine microbiome across microplastic-contaminated niches of Bay of Bengal, India. Mar Pollut Bull 222:118872. https://doi.org/10.1016/J.MARPOLBUL.2025.118872

Urbina J, Patil A, Fujishima K, Paulino-Lima G, Saltikov C, Rothschild LJ (2019) A new approach to biomining: bioengineering surfaces for metal recovery from aqueous solutions. Sci Rep 9(1):16422. https://doi.org/10.1038/s41598-019-52778-2

Walker B, Belton E, Tess A (2025) Energy: transition trade, climate, and net zero pathways: scenarios and implications for developing countries and climate-resilient development. https://repository.graduateinstitute.ch/record/320221/files/TESS%20Briefing%20Note%20-%20Energy%20Transition%20-%20Net%20Zero.pdf

Wang S, Zheng G, Zhou L (2010) Heterotrophic microorganism Rhodotorula mucilaginosa R30 improves tannery sludge bioleaching through elevating dissolved CO_2 and extracellular polymeric substances levels in bioleach solution as well as scavenging toxic DOM to Acidithiobacillus species. Water Res 44(18):5423–5431. https://doi.org/10.1016/J.WATRES.2010.06.055

Yan L, Sun P, Xu Y, Zhang S, Wei W, Zhao J (2018) Integration of a gold-specific whole E. coli cell sensing and adsorption based on BioBrick. Int J Mol Sci 19(12):3741. https://doi.org/10.3390/IJMS19123741

Zhang W, Honaker R (2020) Process development for the recovery of rare earth elements and critical metals from an acid mine leachate. Miner Eng 153:106382. https://www.sciencedirect.com/science/article/pii/S0892687520302028

Zheng G, Wang Z, Wang D, Zhou L (2016) Enhancement of sludge dewaterability by sequential inoculation of filamentous fungus Mucor circinelloides ZG-3 and Acidithiobacillus ferrooxidans LX5. Chem Eng J 284:216–223. https://doi.org/10.1016/J.CEJ.2015.08.119

Zupanc A, Install J, Jereb M, Repo T (2022) Sustainable and selective modern methods of noble metal recycling. Angewandte Chemie (Int Ed. in English) 62(5):e202214453. https://doi.org/10.1002/ANIE.202214453

Innovative Approaches and Recent Progress in Critical Metal Recovery from E-Waste

Nageswara Rao Lakkimsetty, Nourhan Hilal El Mustapha El Mohamad, Yahya Ali Hamadi, and Rahma Juma

Abstract The rapid proliferation of electronic devices has led to a significant increase in electronic waste (e-waste), which poses both environmental challenges and opportunities for resource recovery. E-waste contains valuable critical metals such as lithium, cobalt, rare earth elements, and precious metals, which are essential for modern technologies including renewable energy systems, electric vehicles, and advanced electronics. This chapter provides a comprehensive overview of recent innovations and emerging technologies for the recovery of critical metals from e-waste. It explores advancements in hydrometallurgical and bio-hydrometallurgical processes, selective leaching, solvent extraction, and electrochemical recovery methods. Emphasis is also placed on sustainable approaches such as green solvents, low-energy processes, and the integration of artificial intelligence (AI) and machine learning (ML) for process optimization and real-time monitoring. The chapter highlights the role of circular economic principles, policy frameworks, and industrial scalability in enabling efficient and environmentally sound metal recovery. By showcasing current research and industrial practices, this work underscores the need for interdisciplinary collaboration to transform e-waste into a sustainable source of critical raw materials.

Keywords E-waste recovery · Critical metals · Hydrometallurgy · Sustainable technologies · Circular economy

---

N. R. Lakkimsetty (✉) · N. H. E. M. El Mohamad · Y. A. Hamadi · R. Juma
Department of Chemical and Petroleum Engineering, School of Engineering and Computing, American University of Ras Al Khaimah, Ras Al Khaimah, United Arab Emirates
e-mail: lakkimsetty.rao@aurak.ac.ae

A. P. Das and S. Priyadarsini (eds.), *Recovery and Recycling of Critical Metals from Electronic Waste for Environmental Sustainability*, Earth and Environmental Sciences Library, https://doi.org/10.1007/978-3-032-29064-9_10

1 Introduction

The rapid growth of the electronics industry has revolutionized modern society, enabling advancements in communication, healthcare, automation, and renewable energy systems. However, this surge in the production and consumption of electronic devices such as smartphones, computers, batteries, and electric vehicles has led to a parallel rise in electronic waste (e-waste) generation. Globally, e-waste is the fastest-growing waste stream, projected to exceed 75 million tons by 2030. This trend is driven by rapid technological innovation, shorter product lifespans, and consumer demand for constant upgrades. E-waste contains a complex mixture of materials, including plastics, glass, and metals such as copper, gold, silver, palladium, lithium, cobalt, and rare earth elements. These critical and precious metals are essential for the functioning of modern electronic devices but are finite and unevenly distributed across the globe. Improper disposal of e-waste through landfilling, open burning, or informal recycling poses severe environmental and health hazards, releasing toxic substances such as lead, mercury, and brominated flame retardants into soil and water systems. Therefore, efficient recovery and recycling of valuable metals from discarded electronics are both an environmental necessity and an economic opportunity (Adetunji et al. 2023).

Recovering critical metals from e-waste supports a circular economy by reducing the need for virgin mining, lowering greenhouse gas emissions, and conserving natural resources. Environmentally sustainable recovery methods such as hydrometallurgical leaching, bioleaching, and solvent extraction are being explored to replace traditional energy-intensive smelting processes. Economically, urban e-waste mining can be far more profitable than extracting the same metals from natural ores; for example, one ton of mobile phone waste may contain more gold than a ton of mined gold ore. Furthermore, the recovery of critical materials strengthens supply chain resilience and supports the development of renewable energy and high-tech industries. Nations that adopt effective e-waste management policies and invest in metal recovery technologies can achieve significant progress toward circular economy objectives and the United Nations Sustainable Development Goals (SDGs). In summary, the escalating e-waste challenge presents both an environmental threat and a strategic resource opportunity. Developing efficient, green, and economically viable recovery technologies is essential to ensure sustainable resource management and environmental protection in the digital age (Balasubramanian 2020).

2 Critical Metals in E-Waste: Composition and Importance

E-waste has become one of the fastest growing types of waste in the world with its growth projected to reach over 74 million tons by 2030. Besides posing serious environmental threats, e-waste represents a large stock of extremely important metals needed by clean energy technologies, sophisticated electronics, and the manufacture of the world. These metals include base, precious, rare-earth and technology-critical metals, which are classified as critical metals because of their high economic significance, lack of substitutes, and extreme risks in supply. In this section, the compositional properties of e-waste are defined, the important groups of metals are identified, and their strategic worth as secondary resources are assessed (Ilankoon et al. 2024; Kamran et al. 2023).

2.1 Types of Critical Metals in E-Waste

E-waste is one of the richest sources of metals and more often has higher elemental concentrations than those found in natural ores. The largest fraction of mass is made up of metals which are concentrated in circuit boards, wiring, batteries, displays, and magnets, as shown in Fig. 1.

The most common recoverable fraction is base metals. Copper is the prevailing material due to its widespread use in circuits and connectors, while iron, aluminum, nickel, and tin are found in casings, solders and structural parts. Their comparatively high purity and concentration make their recovery economically worthy (Van Yken et al. 2021).

Even the trace amounts of precious metals are overvalued in monetary value. Gold, silver, palladium, and platinum are valued due to their conductivity and resistance to corrosion, and they are used in connectors, semiconductor chips, and soldered joints, as shown in Table 1. Although in small proportions, they contribute to a large share of the economic potential of e-waste, especially in the mobile phone and high-grade circuit boards (Gómez Soto et al. 2023).

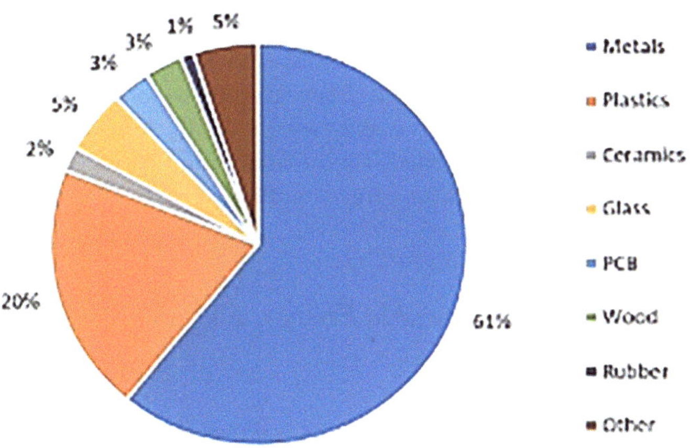

Fig. 1 Average percentage of materials in a clean e-waste stream. Adapted from Kaya (2018) and Van Yken et al. (2021)

Table 1 Typical composition of printed circuit boards in e-waste (wt%)

Component	Range (wt%)	Types and Uses
Overall, fractions		
Metals	30–35	Main recoverable fractions
Resins	24–30	Polymer, epoxy
Ceramics	32–35	Fiberglass, fillers
Key METALS		
Copper (Cu)	10–27	Dominant conductor
Nickel (Ni)	0.3–2	Alloys, contacts
Aluminum (Al)	2–19	Heat sinks, casings
Tin (Sn)	0.08–0.9	Solders
Iron (Fe)	8–38	Structural parts

Data adapted from Van Yken et al. (2021) and Oke and Potgieter (2024a)

High-strength permanent magnets, needed in speakers, motors, and hard drives, cannot be made without rare-earth elements (REEs), including neodymium, dysprosium, and praseodymium. Europium, terbium, and yttrium are essential in phosphors that are used in lighting and display technologies. These are also the elements that have limited alternatives and are concentrated geographical areas and thus recovery of e-waste becomes strategically essential (Graedel et al. 2015; Liang 2023). Metals that are technology-intensive, like indium, gallium and tantalum, allow advanced uses including thin-film displays and high-capacitance electronics. Rechargeable batteries are dominated by lithium and cobalt, and they form the basis of portable devices and electric vehicles. Antimony, selenium, and tellurium are other components that are found useful in semiconductors and photovoltaic systems. Together, they are of great strategic value even though they have small mass fractions (Ilankoon et al. 2024; Kamran et al. 2023). Lastly, e-waste has lead, cadmium, mercury, and chromium, which are hazardous metals. They are compounds that are generated by solders, batteries, and older displays, and they are deadly substances to health and the environment when not controlled. Their existence highlights the importance of recovery and safe disposal practices being controlled (Oke and Potgieter 2024a).

2.2 Applications in Renewable Energy, EVs, and Advanced Electronics

The technologies that drive the global energy transition are intrinsically connected to the demand of critical metals. Renewable energy systems, electric mobility, and high-performance electronics heavily rely on a limited group of strategic factors, which makes e-waste a more useful secondary source. Lithium, cobalt, and nickel are battery metals that are essential in lithium-ion batteries that are used to charge portable devices and electric cars (Priyadarsini and Das 2024, 2026). Their increasing use has increased concerns about the security of supply, thus making end-of-life batteries a priority material to be recovered to help in saving finite resources and stabilizing market reliance (Kamran et al. 2023). Rare earths such as neodymium, praseodymium and dysprosium are essential to permanent magnets in electric motors, generators and other electronic parts. The recovery of these components in e-waste is a substitute to primary mining, which is frequently dispersed geographically and pollutes the environment (Graedel et al. 2015).

Thin-film solar cells, light emitting diodes, and touch screens are based on technology-critical metals, especially indium, gallium, and tellurium. Even though all these components are found in minimal amounts, when they are combined in a global e-waste, they become strategically important due to their scarcity in nature. Metals like gold, silver, and palladium are precious metals that are highly conductive, resistant to corrosion, and allow miniaturization of electronic circuits and sensors (Gómez Soto et al. 2023).

They are the most important element of the financial feasibility of recycling operations due to their relatively high concentrations in discarded printed circuit boards and mobile devices (Gómez Soto et al. 2023). In addition to energy and consumer technologies, cross-sectoral applications support the strategic significance of the former. Tantalum capacitors are imperative in aerospace and defense systems because they are stable in dire circumstances, whereas platinum-group metals are used in catalytic converters and hydrogen fuel cells. Together, these applications demonstrate how a small number of critical components support a wide range of modern technologies, which explains the need to ensure efficient recovery and closed-loop systems of recycling (Ilankoon et al. 2024). Rare earth elements are most found in electronic components, including magnets, phosphors, catalysts, and glass, as shown in Fig. 2.

2.3 *Resource Scarcity and Strategic Importance*

Even though e-waste is a good source of secondhand metals, the quantity of many rare elements is limited by geological, geopolitical, and economic factors. Therefore, extracting these materials out of e-waste is not only an environmental intervention measure, but a necessary step to maintaining technological sustainability and energy security (Ilankoon et al. 2024; Kamran et al. 2023). Concentration in geopolitical supply creates a huge supply risk. Most of the world's productions of cobalt are in the Democratic Republic of Congo, while most of the rare-earth production and processing is done by China. This excessive dependence on a few regions presents global industries with inconsistency, trade controls and policy limitations, thus validating the case in favor of diversification through urban mining and recycling (Supply Chain Risks Report 2022).

Critical evaluations of rare elements are made based on the supply risk, economic value and the environmental impact of the metals. Cobalt, indium, gallium,

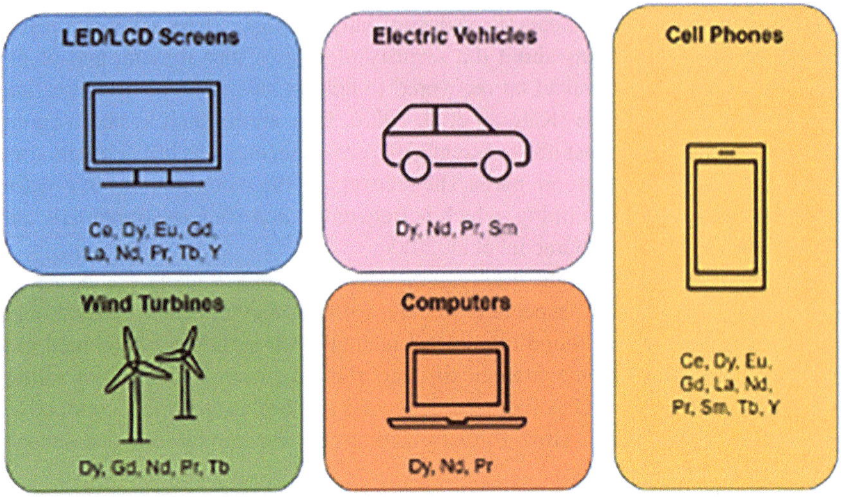

Fig. 2 Rare earth elements are most found in electronic components, including magnets, phosphors, catalysts, and glass. Adapted from USEA (2023)

neodymium and platinum-group metals are always rated as high-risk, hence inadequate supply might have a negative impact on renewable energy, electronics and defense industries (Graedel et al. 2015). These pressures are aggravated by increased demand (Mishra et al. 2022; Ghosh and Das 2018). It is estimated that the shift to electric mobility and grid-scale storage will increase the demand of lithium, cobalt, and nickel by many times, and the development of wind and motor technologies is expected to increase the demand of neodymium and dysprosium even higher. The trajectories are unlikely to be met solely through primary mining, thus making efficient recovery of e-waste an essential complementary measure (Ilankoon et al. 2024). Recycling is also urgently needed for environmental and ethical reasons. The extraction of resources in resource-endowed areas like Central Africa has been linked severally with the destruction of ecosystems, labor hazards, and human rights infractions. Similarly, rare-earth processing is often accompanied by toxic and radioactive waste. Secondly, recovery compensates for negative effects and contributes to acquiring responsibly sourced materials (Wall and Pell 2017).

Lastly, critical-metal recycling is now on the national policy agenda due to strategic energy security. E-waste recovery is seen by many jurisdictions as a method of decreasing reliance on imports, domestic level supply chain stabilization, and industrial competitiveness. Metals like dysprosium, neodymium, gallium and cobalt are now categorized as short-term critical and nickel, platinum, and magnesium are likely to become similar with the speed of the transition of energy (Center for Strategic Studies [CSS] 2024). U.S. DOE Materials Criticality Matrix (medium term, 2025–2035), showing supply risk versus importance to the energy sector, is shown in Fig. 3.

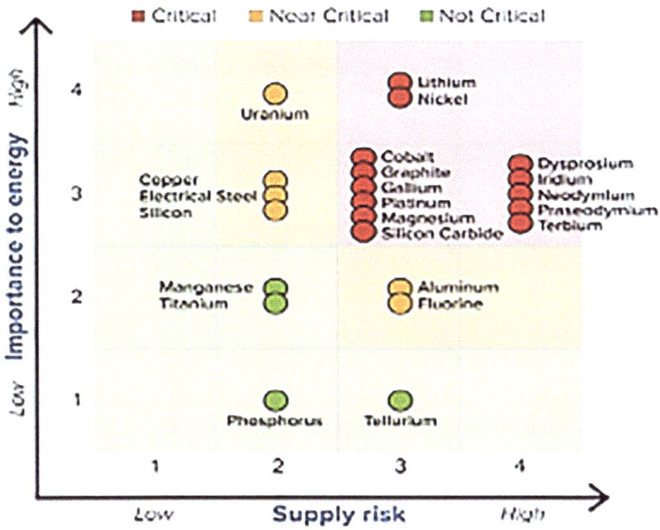

Fig. 3 U.S. DOE Materials Criticality Matrix (medium term, 2025–2035), showing supply risk versus importance to the energy sector. Adapted from Center for Strategic Studies (CSS) (2024)

3 Conventional and Emerging Recovery Techniques

The extraction of critical metals in e-waste involves the treatment of highly heterogeneous and chemically complex materials, which can only be recovered through specialized processes. E-waste is a significant secondary resource because modern devices also have valuable base, precious, and rare-earth elements which are often present in higher concentrations than in primary ores. Recycling processes are based on traditional metallurgical techniques that were developed out of the mining process. Although pyrometallurgy was initially utilized, its excessive requirement of energy and the lack of selectivity have led to the shift to hydrometallurgy and its biological and electrochemical improvements. This section describes conventional methods and emerging alternatives that strive to enhance selectivity, reduce the consumption of reagents, and reduce the impact on the environment (Gulliani et al. 2023). Overview of metal recovery techniques from e-waste, including conventional and emerging approaches, is shown in Fig. 4.

3.1 Hydrometallurgical Processes

Hydrometallurgy is the foundation of metal recovery in e-waste due to its moderate conditions of operation, high selectivity, and pure metals obtained. The treatment usually entails leaching, purification, and recovery of metals. In leaching, minerals are dissolved in solutions with the help of mineral acids like sulfuric, hydrochloric

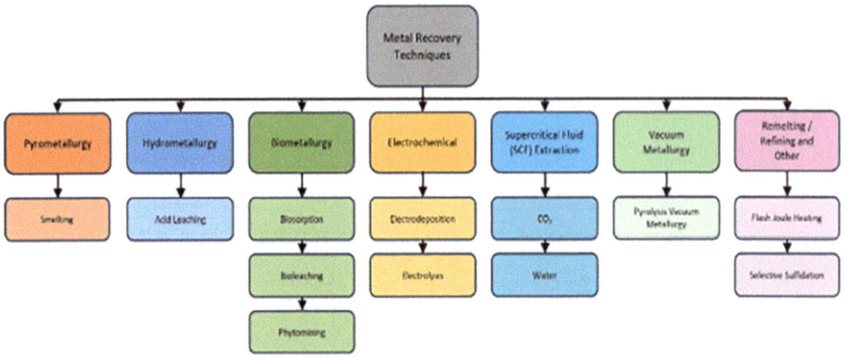

Fig. 4 Overview of metal recovery techniques from e-waste, including conventional and emerging approaches. Adapted from Gulliani et al. (2023)

or nitric acid or a combination of these like aqua regia. These reagents are effective in dissolving metals of printed-circuit boards and batteries to release copper, nickel, gold, and others. The ammonia or organic acid-based milder systems can also be used to enhance selectivity and minimize undesirable dissolution (Gulliani et al. 2023; Rodriguez et al. 2023).

The pregnant leach solution has both desirable and pollutant ions. The purification of the target metals is done by precipitation, solvent extraction, or ion-exchange: the target metals are precipitated by adjusting the pH or using a particular reagent, or by solvent extraction or ion-exchange to fine-resolution closely related elements such as cobalt and nickel (Enemona and Balogun 2023). Lastly, the recovery of metal transforms the dissolved ions into utility products. Electrowinning gives high-purity copper and nickel; crystallization gives lithium salt and cementation gives precious metals such as gold and silver by replacement with a more reactive metal (Gulliani et al. 2023). The method used depends on the type of metals, concentration, and the economics of the processes.

Printed-circuit boards and lithium-ion batteries are some of the typical applications of hydrometallurgical routes (Botelho Junior et al. 2021). An example is that sulfuric acid used in conjunction with hydrogen peroxide is effective in leaching cobalt, nickel and lithium-ion batteries, and aqua regia is effective in extracting gold and copper in circuitry. Nevertheless, traditional hydrometallurgy has a few environmental challenges. The acid concentration and oxidants are costly and produce acidic effluents which need to be neutralized (Rodriguez et al. 2023). Variability in feedstock complicates selectivity and, in most cases, co-dissolution of toxic metals will occur thus increasing the difficulty of purification (Oke and Potgieter 2024b). Corrosive conditions also increase the wear and tear of equipment and reduce scale-up. These deficiencies can be countered in recent developments through use of greener leaching systems, such as citric or oxalic acid in combination with gentle oxidants. These systems achieve similar efficiencies that are reached by using mineral acids and produce fewer secondary pollutants. Moreover, the control of pH, redox potential,

Table 2 Examples of hydrometallurgical metal recovery from e-waste

E-Waste type	Leaching system	Target metals	Recovery efficiency (%)	References
PCBs	HNO_3; Aqua Regia	Cu, Ag, Au	Cu: 88.5–99.9 Ag: 68 Au: 97	Gulliani et al. (2023)
PCBs	HCl + Cyanex 272 SX	Cu, Fe	Cu: 96.8 Fe: 97.1	Enemona and Balogun (2023)
LIBs	$H_3PO_4 + H_2O_2$ $H_2SO_4 + H_2O_2$	Li, Co	Li: 93 Co: 98.5	Botelho Junior et al. (2021)
LIBs	Citric acid + H_2O_2	Co, Li	Co: 95 Li: 93	Botelho Junior et al. (2021)

Compiled from Gulliani et al. (2023), Enemona and Balogun (2023), Botelho Junior et al. (2021)

and the use of complexing agents are further improving the selectivity of lithium, cobalt, and other important metals, improving yield and environmental outcomes (Manikandan et al. 2023).

To conclude, hydrometallurgy offers a flexible and developed basis for e-waste recycling. Continuous improvement of the reagent choice, process control, as well as waste management are increasingly enhancing not only sustainability but also recovery efficiency, thus creating a framework on which the enhanced recovery technologies are to be further developed in the following sections. Hydrometallurgical metal recovery from e-waste is shown in Table 2.

3.2 Bio-hydrometallurgy

Bio-hydrometallurgy, also known as bioleaching, is a biodegradation process that involves the usage of microorganisms to dissolve e-waste metals under mild environmental requirements. The mechanism takes place through two major pathways, (i) direct bioleaching, whereby the microorganisms attach themselves to the solid matter, and mediate the surface oxidation reactions; and (ii) indirect bioleaching, where the microorganisms secrete leaching agents like the ferric ions or organic acids, which in turn dissolve the metals. Common examples of microbial species used are *Acidithiobacillus ferrooxidans* and *Leptospirillum ferrooxidans* which oxidize metal and sulfur minerals, and fungi like Aspergillus niger which produce organic acids that can dissolve metal bearing matrices. Such systems have been demonstrated to work with printed-circuit boards, lithium-ion batteries, and phosphor powders with high recovery of metal base materials (copper, nickel, cobalt, zinc) recovered. Other precious metals such as gold and palladium can also be retrieved, but the quantities are usually lower (Jaiswal and Lee 2024; Mohsin et al. 2024).

The major benefit of bio-hydrometallurgy is the fact that it is environmentally friendly. It eliminates the use of concentrated acids, and it can be used at comparatively low temperature and pressure, thus minimizing chemical use, energy use, and secondary waste. Besides, the biological flexibility of such systems makes it easy to treat complex and heterogeneous e-waste fractions, which are challenging to treat by purely chemical processes. However, limitations persist. The rate of dissolution in metals is lower than that of chemical leaching and limits throughput (Jaiswal and Lee 2024). E-waste contains toxic additives and heavy metals that may inhibit the activity of microbes, which require careful regulation of pH, aeration, and nutrient supply (Mohsin et al. 2024). Also, there is an increased complexity of the process of scaling to the industrial level, including the design of the reactor, stability of the process, and management of contamination (Tunsu et al. 2019). Altogether, bio-hydrometallurgy is a promising addition to traditional hydrometallurgy. Further streamlining of microbial resistance, kinetic performance, and bioreactor design can make it an acceptable large-scale route to retrieve valuable metals with reduced environmental consequences.

3.3 Selective Leaching Approaches

Selective leaching aims at dissolving valuable metals in e-waste and reducing dissolution of unwanted materials (Gulliani et al. 2023). This contrasts with traditional hydrometallurgy, which normally produces a wide range of elements simultaneously, instead focusing on high-value or supply-essential metals, including lithium, cobalt, and rare-earth elements. Increased selectivity lowers the need of purification, chemical use, and the general cost of the process. The control of the chemical and operational conditions is used to achieve selectivity. Adjustment of pH and redox potential allows selective recovery of target metals and leaves others in a solid state. The desired ions in the solution can be stabilized using complexing agents like ammonia, citric acid or oxalate and prevent the leaching of lower-value species (Rodriguez et al. 2023). Similarly, the cautious addition or removal of oxidizing or reducing agents can tune the dissolution environment, for example, sulfuric acid in combination with hydrogen peroxide can increase the recycling of lithium and cobalt in spent lithium-ion batteries. Sequential or staged leaching also increases efficiency. It is a method that divides the process into several steps that are performed under differentiated conditions to attack specific groups of metals. As an example, rare-earth elements can be recovered in mildly acidic conditions, then base metals in more oxidizing leaching and precious metals using special reagents. This kind of organized development allows maximum recovery with the lowest cross-contamination of metal fractions (Manikandan et al. 2023).

Selective leaching has technical problems despite its benefits. Exercising tight control over the parameters of leaching and reagent concentrations makes the process more complex and expensive, especially on an industrial scale. Three-stage selective leaching process for printed circuit assembly powders. Rare earth elements (REEs) were first leached with sulfuric acid diluted, followed by base metal leaching using sulfuric acid with hydrogen peroxide, and finally selective leaching of precious metals with sulfuric acid, thiourea, and ferric sulfate, leaving a palladium-rich residue, as shown in Fig. 5.

The other issue is incomplete recovery whereby the conditions that are optimum when extracting one metal can reduce the extraction of the other metals. Multi-step systems will tend to enhance yields but also increase the time of operation and increase the number of reagents. The recyclability and toxicity of the chemical agents chosen in the process determines sustainability. In conclusion, selective leaching is the refinement of conventional hydrometallurgy, which incorporates specific recovery with higher efficiency and cleaner products. As it advances, it will provide a feasible pathway towards high-purity metals recovery of complex e-waste

Fig. 5 Three-stage selective leaching process for printed circuit assembly powders. Rare earth elements (REEs) were first leached with sulfuric acid diluted, followed by base metal leaching using sulfuric acid with hydrogen peroxide, and finally selective leaching of precious metals with sulfuric acid, thiourea, and ferric sulfate, leaving a palladium-rich residue. Adapted from Diallo et al. (2025)

streams and will be a crucial bridge between the innovation of small-scale laboratories and massive sustainable recycling (Gulliani et al. 2023).

3.4 Solvent Extraction Techniques

Solvent extraction (SX) is an already proven method used to extract and purify metals in hydrometallurgical leach solutions. The procedure is based on the transfer of metal ions in an aqueous solution to the organic solution where a given extractant is dissolved in a carrier solvent, like kerosene (Rodriguez et al. 2023; Gkika et al. 2024). The separation is achieved by a difference in behavior of the complexity and chemical affinity, hence allowing the recovery of the target metals and leaving the rest of the metals in the aqueous raffinate. Organophosphorus reagents D2EHPA and Cyanex 272 are typical extractants, as well as amine reagents. Such agents create stable complexes of coordination with certain metal ions, which allows control of selectivity with fine precision. In the example, Cyanex 272 selectively removes cobalt against nickel in a sulfate media and has extensive use in the purification of cobalt (Rodriguez et al. 2023).

E-waste recycling involves the wide use of solvent extraction. In printed-circuit boards, it is an efficient copper-iron separator in leachates and is a cobalt–nickel separator in lithium-ion battery processes. In addition, it is the most advanced process of separating rare earths, where there are several extraction and scrubbing processes used to fractionate closely related lanthanides (Gkika et al. 2024). The main strengths of the technique are high selectivity, scalability, and ability to make high-purity products. Restrictions include the loss of solvents to the environment, volatility, and processing losses that may create environmental and safety issues (Rodriguez et al. 2023). Several extraction processes are often required to obtain complete separation, which results in increased cost and waste. Therefore, extractants should be regenerated and recycled to operate sustainably. In general, solvent extraction is a very important downstream process in hydrometallurgy and a powerful intermediate between leaching and recovery of the final metal. Further optimization of alternative greens, such as ionic liquids and deep eutectic solvents, will ensure that it will remain efficient and be more environmentally friendly, as well as less operationally complex (Gulliani et al. 2023).

3.5 Electrochemical Recovery Methods

In the recycling of e-waste, electrochemical techniques are widely used to extract the metals using leachates. The two most common methods, electrowinning and electrodeposition, involve cathodic reduction of metal ions in an electrolytic cell, thus producing metallic deposits. The electrolyte used in electrowinning is typically the same as that used in the electrodeposition process. Hence, large-scale recovery of base metals, including copper and nickel, is by electrowinning, but selective or small-scale recovery of precious metals, including gold and silver, is by electrodeposition. Both models are capable of a high purity of the products and a high process control through fine-tuning of the potential, current density, and electrolyte composition.

Recent developments have brought in emergent techniques in electrochemical methods aimed at improving selectivity and energy efficiency. Electro sorption is based on the adsorption of ions onto charged electrode surfaces, and can be recovered with ease, and has potential in lithium and cobalt extraction of dilute leachates. Electrodialysis is an ion exchange separation that uses ion exchange membranes in conjunction with electric fields to separate the concentrate metals without the organic solvents so that it can run continuously. Electrocoagulation, which involves the production of coagulants using sacrificial electrodes, can eliminate contaminants, and the metals produced as hydroxides, which are used as a pre-treatment step prior to other recovery processes (Tunsu et al. 2019). Table 3 shows the summary of electrochemical methods applied to e-waste recycling.

The key advantages of the electrochemical processes lie in the fact that they produce high-purity metals and produce very little secondary waste. Electrowinning copper and nickel cathodes regularly yields purities of over 99% and eliminates further purification. Furthermore, strict regulation of operating conditions also allows selective recovery of metals with different redox potentials, thus making multimetallic solutions more efficient. However, there are still major shortcomings. The problem of high energy consumption is a chronic one, especially the

Table 3 Summary of electrochemical methods applied to e-waste recycling

Method	Typical applications in e-waste	Strengths	Limitations
Electrowinning	Large-scale Cu, Ni recovery from PCB and LIB leachates	Produces high-purity cathodes; industrially proven	Energy intensive; less efficient at low ion concentrations
Electrodeposition	Selective Au, Ag, Pd recovery; Co/Ni from LIBs	High selectivity by tuning potential; > 90% recovery for precious metals	Requires careful potential control; side reactions (H_2 evolution)
Electrosorption	Li and Co capture from dilute LIB leachates	Low energy use; reversible adsorption–desorption cycles	Limited scalability; electrode fouling
Electrodialysis	Li, rare metals separation using ion-exchange membranes	Solvent-free; continuous operation possible	Membrane fouling; high capital costs
Electrocoagulation	Removal of toxic or non-valuable metals from effluents	Simple setup; in situ generation of coagulants	Produces sludge; less selective for valuable metals

Adapted from Tunsu et al. (2019)

problems with dilute solutions which require high current densities. Side reactions, including hydrogen evolution, may reduce efficiency and shorten electrode life, and cell leachates are highly corrosive and can limit cell components to durability limits (Tunsu et al. 2019). Also, the upscaling of such systems requires careful attention to electrode fouling and membrane stability to maintain performance. Altogether, electrochemical recovery is a clean and versatile solution to recover useful metals in e-waste. Although its energy requirements and complexity of maintenance hinder wider use in industry, its demonstrated ability to yield high-purity metals has guaranteed a key place in future hybrid and sustainable recycling processes (Phogat et al. 2025). A schematic representation of electrochemical methods applied in e-waste recycling is shown in Fig. 6.

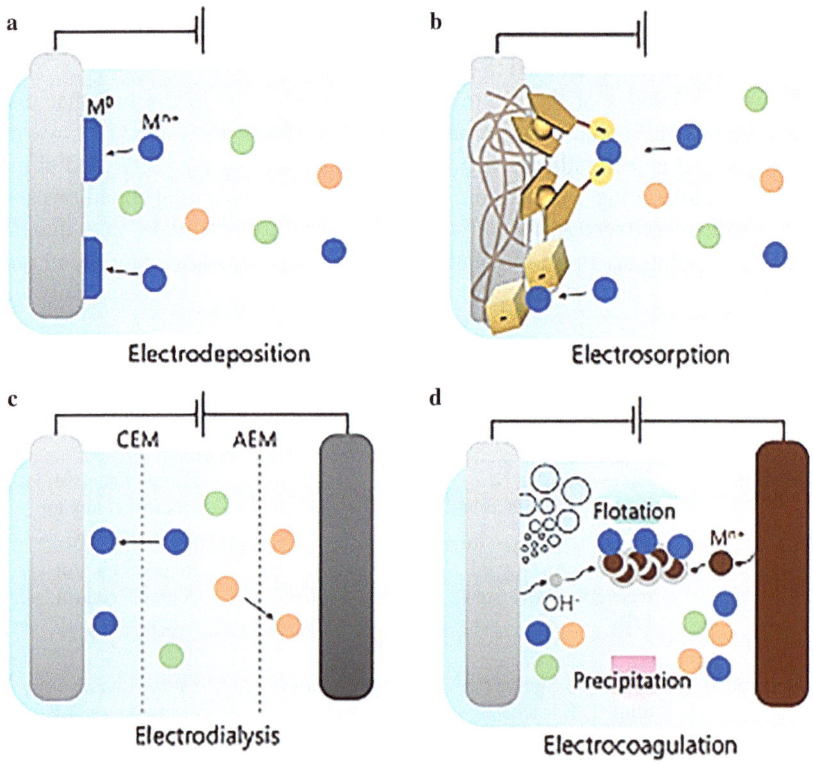

Fig. 6 Schematic representation of electrochemical methods applied in e-waste recycling: **a** electrodeposition, **b** electrosorption, **c** electrodialysis, and **d** electrocoagulation.

3.6 Comparative Insights

Recycling the critical metals in e-waste requires a collection of processes each having unique benefits and constraints. Hydrometallurgy is the most common method, due to its high dissolution levels, scalability, and good recovery of both base and precious metal (Gulliani et al. 2023). Nonetheless, it is dependent on strong acids and oxidants, which facilitates the formation of reagents consumption, corrosion, and secondary waste, which increase the long-term sustainability issue. Bio-hydrometallurgy is also a greener approach, whereby microorganisms are used to dissolve metals under mild conditions with reduced chemical usage (Jaiswal and Lee 2024). However, slow kinetics, low tolerance to toxic compounds, and complicated scale-up demands limit its application in industry. The selective leaching is efficient and aims at retrieving high-valued materials like lithium, cobalt, and rare earths, hence less purification is required, yet at the cost of increasing the complexity of the processes. Solvent extraction is an industrially proven method of metal separation of such metals as cobalt, nickel, and rare earth elements but faces some problems such as solvent degradation and waste. Electrochemical recovery is very effective in the extraction of high-purity metals and selective deposition, but it is energy-intensive, and electrode replacement is often required, hindering its use.

Hybrid systems are also being utilized more often to ensure that the best of both these approaches is merged, and thus better efficiency is achieved. Bioleaching combined with solvent extraction or electrowinning may produce high-purity copper and cobalt, but selective leaching in combination with solvent extraction enhances the separation of rare-earth (Gulliani et al. 2023). These combined flowsheets depict the increasing pressure to support the flexible, low-impact recovery processes that can support diverse e-waste compositions (Rodriguez et al. 2023). To conclude, the problem of e-waste recycling is too complicated to implement a single approach to cover the whole problem. Traditional hydrometallurgy provides a strong base, but new and hybrid processes are more selective, less damaging to the environment and drive the industry to circular and sustainable recovery models.

4 Innovations in Sustainable Metal Recovery

The increasing demand for metals driven by industrial growth, renewable energy systems, and electronics manufacturing has intensified the need for sustainable and energy-efficient recovery methods. Conventional metallurgical processes such as pyrometallurgy and hydrometallurgy are often energy-intensive and environmentally detrimental due to the generation of toxic emissions, large volumes of waste, and chemical hazards. In response, researchers and industries are focusing on innovative approaches that promote green chemistry, resource efficiency, and circular economy principles.

4.1 Use of Green Solvents and Ionic Liquids

Green solvents and ionic liquids have emerged as promising alternatives to traditional chemical reagents in metal recovery. Unlike conventional acids and organic solvents, ionic liquids possess tunable physicochemical properties, negligible vapor pressure, and high thermal stability, making them effective and environmentally safer for dissolving and separating metals. They can selectively extract critical metals such as lithium, cobalt, and rare earth elements from complex waste matrices, including spent batteries and e-waste. Deep eutectic solvents (DESs), a subclass of ionic liquids, further enhance sustainability due to their biodegradability, low cost, and ease of preparation. These green solvents reduce the need for hazardous chemicals while improving metal selectivity and recovery efficiency.

4.2 Low-Energy Recovery Processes

Reducing energy consumption is central to sustainable metal recovery. Novel low-temperature and low-pressure extraction methods, such as bioleaching and electrochemical recovery, are gaining prominence. Bioleaching utilizes microorganisms to mobilize metals from solid waste through natural biochemical reactions, eliminating the need for high heat or strong acids. Similarly, electrochemical techniques can recover metals with high purity under mild conditions using renewable electricity sources. Process intensification strategies, including microwave-assisted extraction and supercritical fluid systems, are also being explored to minimize energy use while maintaining high recovery yields.

4.3 Waste Minimization and Eco-friendly Approaches

A key objective in sustainable metal recovery is the reduction of secondary waste generation. Closed-loop systems are being designed to recover reagents and solvents for reuse, minimizing environmental impact. Solid residues from recovery processes are being repurposed into construction materials, catalysts, or adsorbents, further promoting resource circularity. Integrating green solvents, low-energy operations, and waste reuse establishes a holistic eco-friendly framework that aligns with global sustainability goals. In summary, innovations such as ionic liquids, bioleaching, and circular process design represent transformative steps toward sustainable metal recovery. These approaches not only reduce environmental burdens but also offer economically viable solutions for resource conservation and cleaner industrial production.

5 Integration of Artificial Intelligence (AI) and Machine Learning (ML)

The rapid exponential development of electronic goods has been followed by a parallel rise in e-waste, which poses environmental and resource challenges. Precious metals such as gold, silver, palladium, and rare earth elements are contained in e-waste that play a crucial role in modern technologies. Traditional recycling methods are lacking in terms of efficiency, scalability, and precision. Artificial intelligence (AI) and machine learning (ML) have in recent times emerged as revolutionary technology in optimizing the extraction of these strategic metals, offering evidence-based solutions for process control, sorting, and extraction. By enabling data-driven optimization of a complex process, AI and ML facilitate improved efficiency, higher yields, and reduced environmental impact. These technologies are particularly valuable given the heterogeneous nature of e-waste, the variability of its composition, and the multi-stage processing routes required for extracting critical raw materials such as lithium, cobalt, and rare earth elements.

5.1 *AI for Process Optimization and Predictive Modeling*

ML algorithms can be trained on experimental and operational datasets to predict metal recovery under varying leaching, solvent extraction, or electrochemical conditions. Srivastava et al. (2025) used AI and ML techniques to predict the recovery of copper from printed circuit boards. The model was developed using an Artificial Neural Network (ANN) and a Boosting Algorithm (BA) to provide a comprehensive understanding of copper recovery achieving over 94% yield under optimized conditions. Such predictive capabilities reduce experimental workload, accelerate scale-up, and minimize reagent and energy consumption. Other studies used machine learning (ML) powered optimization framework for sustainable recovery of Cu, Ag and Au from the WPCBs. The process uses NH_4Cl and low-temperature roasting for the metal recovery for the designed experimental conditions in the lab and confirms 99%, 90% and 80% respectively recovery of Cu, Ag and Au from the WPCBs (Srivastava et al. 2025).

5.2 *Real-Time Monitoring and Process Control*

Artificial Intelligence (AI) and Machine Learning (ML) have revolutionized process control in the important recovery of metals from e-waste through real-time monitoring and adaptive optimization. Although most of the applications are in the research stage, there is potential for AI/ML to enable real-time monitoring, adaptive control, and digital-twin applications in metal recovery plants. For instance, the scientometrics analysis recognizes machine-learning algorithms being applied to optimize process parameters in hydrometallurgical and bio-hydrometallurgical recovery for yield improvement and reduction of chemical waste (Phogat et al. 2025).

5.3 Intelligent Pre-processing and Sorting

AI-based computer vision and classification systems have been increasingly applied to improve dismantling, sorting, and separation of e-waste components. Mohsin et al. (2024) describes the measurement of recyclability of waste electronic components (WECs) from waste printed circuit boards (WCPBs) by integrating AI models for improved disassembly and sorting. This assessment of recyclability of individual electronic components present on WPCBs provides an insight into the recovery potential of valuable materials. Artificial Intelligence based models also have the potential to recognize the different WECs on PCBs boards and sort them in similar categories using robotic manipulators. This approach enhances the purity of downstream metal streams, lowers contamination, and reduces reliance on manual labor, which is often hazardous and inefficient. Madhav et al. (2022) also applied Deep Learning techniques such as Convolutional Neural Networks (CNNs) for real-time classification of metallic versus non-metallic fractions and hence, facilitating more efficient segregation at collection and recycling facilities. The use of CNNs in categorizing the e-wastes resulted in 96% accuracy.

6 Circular Economy and Policy Perspectives

With the increased generation of e-wastes in the world, the issue of reclaiming the metals of critical importance requires a solution that goes beyond technology to include systemic change in resource management. The circular economy (CE) offers a framework of this type by reducing reliance on primary mining, lengthening the lifecycle of products, and making the most out of material use through reuse, repair, and recycling. Such a transition is, however, possible only through the help of regulations, market incentives, and cooperation between industry, academia, and governments. This part focuses on the influence of CE principles on e-waste management, the policy that facilitates the application of these principles, and the models of cooperation that relate innovation to governance (Srivastav et al. 2023; Elgarahy et al. 2023).

6.1 The Role of the Circular Economy in Metal Recovery

The circular economy redefines waste as a resource and encourages closed-loop material flows that keep products and components in circulation for as long as possible. Applied to e-waste, CE transforms discarded electronics into secondary raw-material reservoirs rich in critical metals such as copper, cobalt, and rare earth elements. By replacing the linear "take-make-dispose" model with reuse, refurbishment, and recycling, CE reduces pressure on mining, mitigates pollution, and strengthens material security for renewable energy and digital industries (Srivastav et al. 2023).

Urban mining is an example of this change; whereby outdated devices are processed as high-quality ores. Due to the high levels of metal in e-waste that are usually higher than those found in natural deposits, recovery may be profitable as well as conducive to the environment. But its success lies in the efficient product take-back mechanisms, open markets in secondary materials and accountable producers. In the absence of these enabling conditions, urban mining will only be technically feasible but economically limited (Rizki et al. 2025).

Circular economic practices are based on technological innovations. The processes of hydrometallurgical, bio-hydrometallurgical, solvent-extraction and electrochemical are increasingly included in circular recovery chains instead of being considered an individual recycling step. Of these, the area of bio-hydrometallurgy best fits the aims of the CE with the goals of achieving high yields of metals with low energy requirements. Artificial intelligence, machine learning, and blockchain-based traceability are additional digital tools that can be used to enhance efficiency by enhancing sorting, monitoring flows, and verifying compliance within the global value chains (Srivastav et al. 2023). There are also socio-economic gains that come because of the circular approach. When e-waste is treated as a domestic resource, the countries can lower their dependence on imports, promote domestic industries, and create some jobs in collection and refurbishment. However, these benefits need institutional support and government control; otherwise, informal recycling will continue, and workers will be exposed to risks and poor-quality materials. As a result, informal actors should be incorporated in the process of developing inclusive circular systems by training, certifying, and motivating them (Maphosa and Maphosa 2020). Global e-waste generation in 2014, 2019, and projected levels for 2030 are shown in Fig. 7.

The effectiveness of the CE strategies is shown to be different in the regions. Countries that are well regulated and have well-developed collection networks will have a better recovery rate and safer processing, as opposed to those that do not enforce regulations and depend on informal sectors as their primary sources. The creation of public awareness, convenient collection sites, and consumer incentives are also necessary towards increasing participation (Elgarahy et al. 2023). Finally, the circular economy is not only a framework of environmental functioning but also an industrial and policy approach of ensuring access to strategic materials. Categorizing e-waste as a national resource pool, CE connects technological innovation to economic resilience and social responsibility which provide the theoretical basis

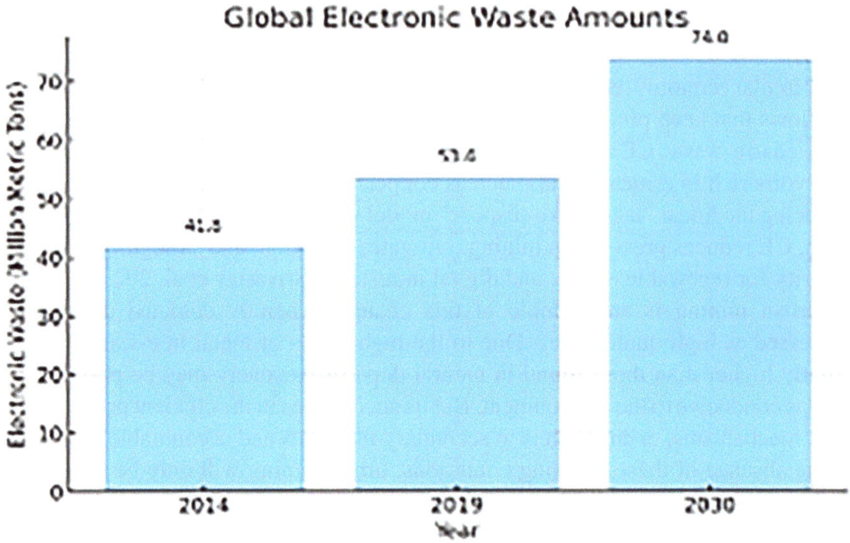

Fig. 7 Global e-waste generation in 2014, 2019, and projected levels for 2030 (million metric tons). Adapted from Global E-waste Monitor (2020)

of the policy frameworks and collaborative models that are covered in subsequent subsections.

6.2 International Regulations and Policy Frameworks

E-waste policies that are based on the circular economy will only be effective when they are supplemented by consistent legal and policy frameworks. Such rules specify the producer's requirements, promote recycling that is environmentally friendly, and encourage investment in recovery technologies. Since e-waste is a commodity that is traded globally, its governance needs to be carried out at various levels, starting with international agreements and down to the national systems of enforcement (Srivastav et al. 2023; Elgarahy et al. 2023).

6.2.1 Global Landscape

E-waste is also often transnational, often between industrialized and developing countries where the cost of processing is low, but the control is weak. Global structures set the standards of safe trade and recycling to avoid dumping into the environment (Elgarahy et al. 2023). The foundation is still held by Basel Convention (1989), which demands the exporting states to ensure that the importers consent and that

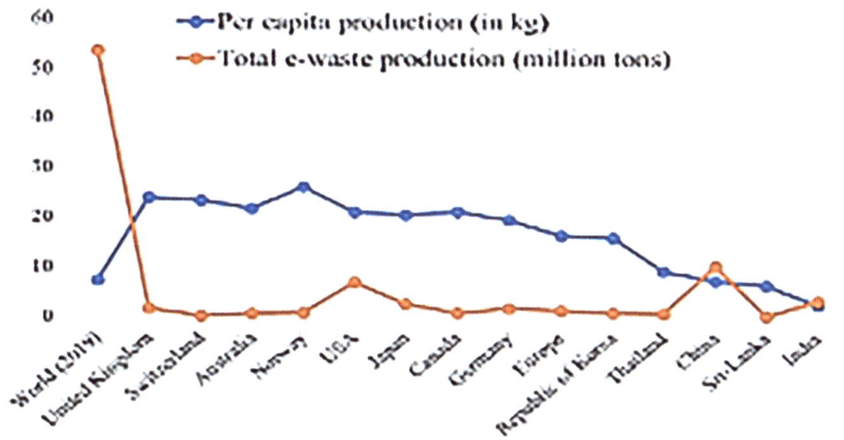

Fig. 8 Per capita and total e-waste production by selected countries and regions in 2019 (Forti et al. 2020)

the facilities can handle the hazardous materials in a responsible manner. However, enforcement is not even, and the difference between reusable and waste electronics is often abused to evade regulations. The Organization for Economic Co-operation and Development (OECD) has supplemented the Basel Convention by promoting the concept of Extended Producer Responsibility (EPR), which entails that producers should handle end-of-life products. In most states, implementation is not mandatory despite its influence, which inhibits uniformity. The Global E-waste Monitor by the United Nations provides vital information, as it shows that of the 53 million tons of e-waste produced in the world in the year 2019, less than one-fifth of the total was officially recycled; a figure that is likely to increase unless there is more co-ordination of policies. Per capita and total e-waste production by selected countries and regions in 2019 are shown in Fig. 8. The United States is an example of this conflict between dedication and action. It is not a Basel signatory, and even with national plans of stewardship and procurement reforms, used devices labeled as "reusable" are still being exported. Illegitimate flows are still prevalent all over the world due to demand for secondary materials and the inability to enforce the regulations. Collectively, these frameworks offer a platform that the global community can be held responsible: Basel gives the legal framework, the OECD formalizes the responsibility of producers, and the UN gives transparent information. Their success will remain in their country in translation into enforceable laws and infrastructure of safe collection and recovery (Srivastav et al. 2023).

6.2.2 Regional Policies

On regional and national levels, global principles are manifested in terms of collection schemes, design standards, and financing schemes. The methods vary depending on the management abilities and economic interests. The most inclusive model is the one developed by the European Union which is a combination of the Waste from Electrical and Electronic Equipment (WEEE) Directive, Restriction of Hazardous Substances (RoHS) Directive, and Circular Economy Action Plan. These tools establish recoverable binding goals, limit hazardous materials at the design phase, and encourage durability and recyclability. Despite the differences in performance between the Member States, the EU system can be used as the world model of how to incorporate electronics into a more comprehensive circular-economy approach (European Commission 2020).

A combination of state legislation and federal programs is used by the United States. The National Strategy of Electronics Stewardship has objectives of sustainable design and domestic recycling, and lack of a federal requirement has resulted in a patchwork of state programs with different targets and implementation. Japan has a well-structured and early system. Its Home Appliance Recycling Law and Small Home Appliance Recycling Law give consumers, retailers, and manufacturers clear roles and high recovery rates and less dangerous treatment. Traceability and mandatory recycling fees guarantee transparency, yet apparent expenses can deter action. In addition, China incorporates e-waste management in the industrial policy through the Circular Economy Promotion Law. An extensive formal network has been formed on the basis of national EPR scheme financed by producer fees and recycler subsidies, but informal recycling and uneven enforcement still exists. In developing economies like India, Brazil and some African and Southeast Asian states, the law more refers to the principles of the circular-economy but does not know how to enforce and provide infrastructure. There is still dominance of informal recyclers, but pilot projects raise some gradual steps towards formalization. Overall, the practice at the regional level proves that robust regulation, good infrastructure, and consumer involvement are the key to successful e-waste governance (Elgarahy et al. 2023).

6.2.3 Policy Instruments and Implementation Tools

The strength of e-waste governance depends on the policy instruments that turn broad principles into practice. Extended Producer Responsibility (EPR) shifts end-of-life obligations from municipalities to producers, incentivizing better product design and financing for collection and recycling. Where enforcement is robust, EPR has measurably increased recovery rates; elsewhere, weak oversight allows informal sectors to persist (Grandhi et al. 2020).

Eco-design and substance constraints are supplements of EPR that exclude dangerous inputs, making recycling safer. Curtaining lead, mercury, and similar toxins enhances the safety of workers and environmental performance as well as promoting modular and repairable product design. Public procurement policies make the markets sustainable by compelling the government agencies to buy certified electronics. Extended Producer Responsibility (EPR) policy instruments and their focus areas are shown in Table 4.

Table 4 Examples of extended producer responsibility (EPR) policy instruments and their focus areas

Instrument	Stage in product chain	Main effect
Deposit-refund	Disposal, signals to design	Encourages reuse and design improvements
Take-back schemes	Disposal and resource extraction	Promotes reuse, recycling, and design
Materials tax	Resource extraction and design	Reduces use of targeted raw materials
Advance disposal fee	Disposal	Supports recycling and recovery
Upstream tax/subsidy	Design and disposal	Reduces material input and encourages recycling
Recycled content	Design to disposal	Promotes use of secondary raw materials

Adapted from OECD (2016)

These programs leverage the purchasing power of the state to incentivize durable and recyclable products and encourage wider industry compliance. Traceability and monitoring systems, such as Japan's manifest tracking and EU producer reporting, give visibility to the data and give the consumer confidence that the e-waste is handled in accredited facilities.

Financial mechanisms maintain collection and formal recycling. The model of co-payment used in Japan and the producer-fund subsidy system in China illustrate two opposite ways of recovering the costs. Each of them balances the cost versus the necessity to finance responsible processing. Lastly, cooperative governance platforms among producers, recyclers, regulators, and communities increase compliance and responsiveness of policy instruments to local realities. Such participation systems prove particularly useful in areas where informal recycling is still quite important (Maphosa and Maphosa 2020).

6.3 Industry-Academia-Government Collaboration

The shift to circular economy of e-waste cannot be achieved only through technology or legislation; rather, it should be based on inter-sectoral collaboration. Complex value chains, such as product design for recycling, need to be coordinated between universities, industry, governments and more between civil society. These actors bring different skills on board: universities produce knowledge and skilled professionals; industry translates research into solutions in the market, and governments create policy frameworks and incentives. The functionality of these spheres is disjointed when they act separately and speed up innovation when they interact and make sustainability goals a possibility (Srivastav et al. 2023).

6.3.1 The Triple Helix Model

The conceptualization of the Triple Helix model is that of the interaction between academia, industry, and government in terms of innovation (Etzkowitz 2008). It focuses on collaboration by way of research networks and technology parks, and policy partnerships that can convert knowledge into practicable solutions.

In e-waste management, universities are leading cleaner recovery techniques including hydrometallurgy and bioleaching; industry is offering infrastructure and commercialization, and governments are guaranteeing regulation and funding. These collaborations can be illustrated by joint projects like European pilot projects in Horizon programs or Asian public–private research centers that can overcome the gap between laboratory research and industrial-scale recycling. A clash of interests is inevitable: industry is interested in cost-efficiency, and policymakers are concerned with safety and compliance with the environment's requirements. Academia usually helps bridge the gap between the two through evidence-based advice. In that regard, the Triple Helix does not only promote technological advancement but also promotes even-handed governance in which both innovation and regulation proceed concurrently (Etzkowitz 2010).

6.3.2 Beyond the Triple Helix: Quadruple and Quintuple Extensions

Although the Triple Helix is concerned with collaboration at the institutional level, the Quadruple and Quintuple Helices extend the model to include social and environmental aspects that are required to be truly circular (Carayannis and Campbell 2010). The Quadruple Helix also has a fourth factor in the form of civil society, as it acknowledges that the behavior of the people is what dictates whether the devices follow formal recycling or informal streams. Consumer participation should be encouraged by creating awareness campaigns, having convenient collection points, and involving the media. The Quintuple Helix also includes the natural environment, in that the recovery systems should minimize environmental damage. In this model, the concept of innovation is also directly connected with the concept of sustainability to make sure that the process of recycling does not pollute and does not harm the ecosystems. Collectively, these frameworks note that technological and institutional collaboration should exist with the community and environmental stewardship (Carayannis and Campbell 2012).

6.3.3 Practical Applications in E-Waste Management

There is a growing manifestation of collaborative models in real-world projects. Universities in Brazil have played the role of arbitrating between regulators, recyclers, and producers based on data-driven evaluations of the need to balance economic and environmental concerns. Pilot projects on e-waste recovery and e-waste tracking have been developed in Europe through partnerships between academia, industry and government. The examples of the subsidy-based collection programs in China demonstrate that the integration between government funding and logistics of the industry and the involvement of citizens can institutionalize recycling systems.

The support of the international donors in the projects in developing economies like Malaysia and some African countries is an example of potential and vulnerability. Incorporating informal recyclers into formal systems will need a continuous working relationship, education, and building of trust to achieve inclusive and sustained success. Across these contexts, cooperation transforms policy into practice. By pooling expertise and sharing risk, collaborative networks convert innovation into sustainable operations and scale local pilots into viable national systems (CORDIS 2020).

6.3.4 Challenges and Opportunities

Regardless of their potential, collaborative practices are characterized by continual obstacles. The lack of coordination is usually caused by divergent institutional priorities and limited resources. Poor governance and divided roles in most nations slow down the implementation. Another challenge has been the prevalence of informal recycling, which offers livelihoods based on unsafe and polluting methods (Maphosa and Maphosa 2020). The informal workers can be moved to the formal sector by replacing the exclusionary policies with training, certification, and financial rewards. There is also a lack of continuity due to financial insecurity; a lot of partnerships are short-term. To have a long-lasting partnership, it is necessary to have stable financing, accountable governance, and fair cost sharing. Nevertheless, there are significant possibilities. Digital applications like AI and blockchain increase the level of traceability and regulatory oversight, whereas international collaboration promotes knowledge exchange and technical education. These mechanisms can hasten the worldwide transition to safe, inclusive, and circular e-waste management when these mechanisms are aligned with local conditions.

In conclusion, successful partnership between academia, industry, government, and society is what would make the principles of the circular economy theory a reality. Breaking the cycle of barriers to coordination and informality, as well as funding, can open the door to innovation and resilience, such that e-waste management can be used to achieve sustainability and equitable economic growth (Srivastav et al. 2023).

7 Challenges and Opportunities in Industrial Scalability

The transition from laboratory-scale innovations to industrial-scale application in the recovery of critical metals from e-waste is a dual landscape of opportunity and challenge. While significant progress has been made in the discovery of effective hydrometallurgical, bio-hydrometallurgical, and electrochemical technologies, translating these technologies to mass industrial application requires careful balancing of economic, technical, and regulatory factors.

7.1 Complexity of E-Waste Streams

E-waste is very heterogeneous and includes plastics, ceramics, glass, and metals with diverse composition. The heterogeneous and complex composition of e-wastes varies depending on type of electronic device and its components. In general, e-waste contains 40% metal, 30% plastic polymers, and 30% oxides of different materials (Hossain and Sahajwalla 2024). However, these quantities can vary due to uneven mix of electronic and electrical equipment in e-waste across different countries. Because of this complexity, it is challenging to create a metal recovery strategy that works for all types of e-waste, requiring the creation of specialized procedures.

7.2 Process Economics and Cost Competitiveness

One major obstacle is the substantial upfront cost needed to build extensive sustainable recovery facilities (Phogat et al. 2025). Scaling up recovery technologies generally include high CAPEX and OPEX. Competition with the low market price of primary metals can make secondary recovery economically less favorable in the absence of subsidies, policy incentives, or successful value chain integration. Techno-economic analyses highlight that reactor design, solvent regeneration, and energy integration are key drivers of feasibility, and in the absence of intensification or integration of processes, the environmental benefits seen at the bench scale can be offset by prohibitive treatment and energy costs at the plant scale (Van Yken et al. 2021).

7.3 Future Directions for Large-Scale Adoption

While innovative green technologies such as ionic liquids, deep eutectic solvents, and bioleaching have shown strong potential for sustainable metal recovery at laboratory scale, their transition to industrial implementation requires strategic advancements in technology integration, policy support, and economic feasibility. Future research should focus on scaling up these processes while ensuring environmental safety, cost-effectiveness, and operational reliability.

7.4 Process Optimization and Scale-Up

Large-scale adoption depends on improving process efficiency and stability under industrial conditions. Pilot-scale demonstrations are essential to address challenges related to solvent recovery, process kinetics, and system corrosion (Tunsu et al. 2019). Optimization of solvent composition, microbial performance, and electrochemical parameters can improve recovery yields while reducing energy consumption. Integration of hybrid technologies, combining green solvents, electrochemical recovery, and bio-based methods offers opportunities for process intensification and continuous operation (Zhang et al. 2022).

7.5 Economic Viability and Supply Chain Integration

For sustainable metal recovery technologies to compete with conventional methods, cost reduction through solvent recycling, renewable energy use, and waste valorization is crucial. The development of localized collection and processing networks for e-waste and spent batteries can reduce transportation costs and ensure a steady supply of feedstock (Li et al. 2023). Public–private partnerships and government incentives can further accelerate investment in green recovery infrastructure and technology transfer.

7.6 Policy, Regulation, and Circular Economy Frameworks

Strong policy frameworks are vital to promote the circular use of critical metals. Governments should establish extended producer responsibility (EPR) schemes, enforce e-waste collection targets, and encourage the adoption of eco-friendly recycling practices (European Commission 2020). International collaboration can harmonize standards, ensuring the safe use of ionic liquids and other novel solvents at industrial scale. Incorporating sustainability metrics such as life cycle assessment (LCA) and techno-economic analysis (TEA) will guide decision-making for industrial deployment.

7.7 Digitalization and Smart Process Control

The integration of artificial intelligence (AI), digital twins, and data analytics into recovery systems represents a promising frontier. These tools can optimize real-time process parameters, predict solvent degradation, and enhance metal separation efficiency (Wang et al. 2022). Automated process monitoring also supports compliance with environmental standards and improves scalability. In summary, the large-scale adoption of sustainable metal recovery technologies will depend on advancing pilot-scale demonstrations, integrating digital and hybrid approaches, and establishing robust policy and economic frameworks. With coordinated global efforts, these innovations can redefine metal recovery as a circular, low-carbon, and resource-efficient industry.

8 Next-Generation Technologies for E-Waste Recycling

The future generation of e-waste recycling technologies is moving beyond incremental optimization to system transformations that combine efficiency with environmental friendliness. The concept of innovation flows through all of the phases of the product life cycle, starting with design and proceeding to recovery and automation.

At the design stage, the researchers are designing recyclable materials to facilitate end-of-life processing. The vitrimer-based printed circuit board is one such example, a new type of dynamic polymer substrates that can be dissolved in mild solvents, as opposed to severe chemical or thermal solvents. Vitrimers can recover embedded metals and polymers with minimum waste production unlike traditional thermoset boards, which are not easily split apart because of the cross-linked structures. Life-cycle assessments indicate significant emissions and energy savings, which makes such materials a feasible basis of circular electronic development (Kaya 2018).

Similar advancements in the field of biotechnology recovery are redefining the process of extracting metal into discarded devices. Passive biomass like biochar, which is a by-product of agricultural residues, has the potential to act as a renewable catalyst and sorbent, which can leach precious metals under mild conditions, and therefore does not require concentrated acids. At the same time, there is the development of engineered microbial systems that use bacteria and fungi with greater resistance to the toxicity of e-waste. These organisms dissolve copper, nickel, and gold faster with the help of enzymatic and redox reactions. The research of synthetic biology further increases this possibility by allowing the introduction of specific genetic modifications to increase selectivity, resilience, and stability of processes under industrial conditions (Kamran et al. 2023).

Nanotechnology provides an additional aspect to this development. It has been shown that metallic nanoparticles including gold, silver, and copper oxides can be produced directly from e-waste through environmentally friendly chemical and biological processes. Such nanoparticles are of high commercial importance not only as refined raw materials but also as catalysts, conductors, and antimicrobial agents. Their production provides new value chains, which connect recycling with sophisticated uses in electronics, energy, and medicine, and expand the economic sphere of e-waste use (Kumar et al. 2025).

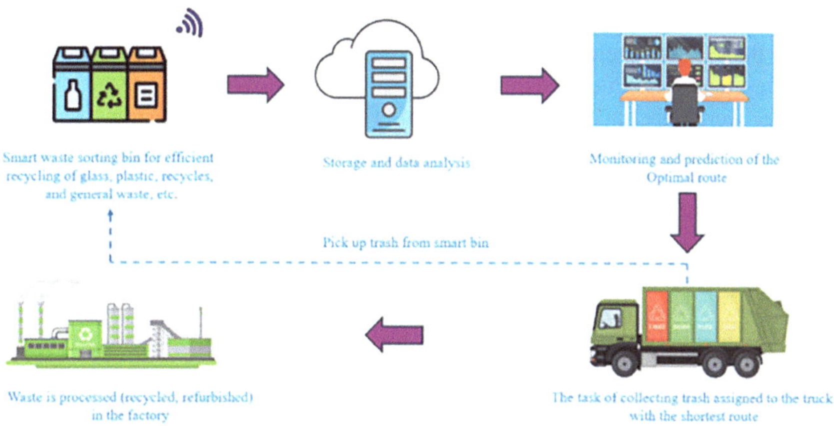

Fig. 9 Example of an AI-enabled waste management system integrating smart bins, data analysis, and optimized collection logistics. Adapted from Dao et al. (2025)

Recycling operations are changing their front end through automation and artificial intelligence. Precision rates in the identification, sorting, and disassembling of components are now enabled by vision-guided robotics and machine-learning algorithms with precision rates of over 90% in pilot studies. These systems improve the safety of workers by minimizing manual handling, boosting throughput, and providing uniform feedstock to subsequent hydrometallurgical or biological processing. Combination of AI and data analytics also assists predictive maintenance and optimization of the process and prepares fully digital recycling ecosystems (Madhav et al. 2022).

Taken together, these changes indicate the transition to smart, low-impact, and circular recovery mechanisms. Recyclable material convergence, bio-based extraction, nanotechnological innovation, and automation promise a future where e-waste recycling is not just an activity of reactive waste-management but also an active part of sustainable production and resource security. An example of an AI-enabled waste management system integrating smart bins, data analysis, and optimized collection logistics is shown in Fig. 9.

8.1 Integration with Renewable Energy and Sustainable Manufacturing

The combination of e-waste recycling and renewable energy systems is becoming one of the most important directions toward the achievement of low-carbon and circular manufacturing processes (Hossain and Sahajwalla 2024). Traditional recovery methods, such as smelting and electrochemical refining, often utilize fossil-generated power, thus undermining the environmental benefits of metal recovery. Researchers are in turn coming up with protocols that run on solar, wind, or hydrogen-powered energy, thus allowing electrochemical leaching and electrowinning to run on non-polluting electricity. Pilot tests have recorded that photovoltaic-based electrolysis could recover copper and cobalt with equal efficiency as the conventional systems, and hydrogen-based buffers in storage can provide stability during fluctuating production requirements. Such a direct combination of recycling infrastructure with renewable resources will thus result in significant improvements in the number of greenhouse-gas emissions and in the costs of operations, especially in those territories, where a rich potential of clean energy is provided (Manikandan et al. 2023).

Environmentally friendly manufacturing supplements these energy innovations by ensuring recovered metals are reintroduced into the supply chains in an efficient way. A design-for-recycling approach, including modular circuit boards or vitrimer based circuit boards, is an example of how recyclability can be captured at the product level, allowing the components to be disassembled and re-processed with minimal environmental impact. Through the coordination of product design and recovery technology, material streams can be kept in closed loops, thus making the process of recycling a proactive process as opposed to a reactive one, a part of production systems (AL-Reyamia et al. 2020; Shaik et al. 2026; Zhang et al. 2023).

The advantages of this type of integration are always supported by life-cycle assessments. There is empirical evidence that recycling, which is powered by renewable energy, could cut carbon emissions by over fifty per cent compared to traditional metallurgical recycling, particularly in energy-intensive processes like the recovery of rare-earth magnets. The combination of green electricity and solvent innovations, such as deep eutectic solvents or ionic liquids, increases selectivity and sustainability. These solvent systems achieve greater recovery efficiencies and with a low amount of secondary waste when run on renewable energy (Jaiswal and Lee 2024).

On the systems level, the division between recycling and manufacturing is becoming more permeable. The metals recovered in e-waste are directly recycled into battery, catalyst, and additively manufactured part production of clean energy technologies. As an example, the salvaged cobalt and nickel in used electronics can be used as cathodes to lithium-ion batteries, but recovered platinum-group metals are used in fuel-cell catalysts. This mutual support between recovery technologies and renewable-powered industries leads to a sustainable cycle: clean energy stimulates cleaner recycling, and recycling provides the materials that are necessary to maintain clean energy systems. Essentially, e-waste recycling coupled with renewable energy and sustainable production redefines the purpose of recycling in industrial ecology. It puts recovery in the middle of the consumption process, but as the key to a regenerative economy, where energy, materials, and design have been brought together to reduce waste and ensure long-term resource security (Wang et al. 2022).

8.2 Interdisciplinary Collaboration for Resource Security

The long-term security of the critical e-waste-derived metals is not only based on technological innovation but also on the concerted communication of various fields and disciplines. Chemistry, materials science, microbiology, data analytics, environmental engineering, and policy should work together to transform laboratory discoveries into industrial use. This multi-faceted strategy ensures that the technologies used in recycling are economically viable, technically feasible, and environmentally acceptable (Wibowo et al. 2021).

Such cooperation has proven to be valuable with recent efforts. There has been the development of collaborative pilot plants where universities, industries and technology companies are able to test new solvent systems, microbial recovery pathways, and automated disassembly procedures in realistic operations. Such projects fill the old divide between research and commercialization so that scientists can optimize their techniques on industrial feedback, and recyclers can more readily implement innovations at lower risk. The government and the funding agencies have a very strong supporting role, as they create consortia which would bridge academic research, industrial infrastructure, and regulatory oversight under the same circular-economy agenda (Zhang et al. 2022).

International alliances also expand this pattern. Asia, Europe and Africa are linking networks across continents and exchanging information, aligning safety standards and recycling technologies to developing nations and regions that have very little infrastructure. This kind of collaboration does not only amplify the recovery capacity of the world but also resilience in the economy through the diversification of the sources of important metals. The attempts to create international databases on e-waste composition and trade flows also contribute to transparency, allowing policymakers to plan the resourcing based on actual and timely information. Sustainable collaboration is based on education and workforce development. Universities are increasingly incorporating the ideas of circular economy, green chemistry, and digital recycling systems into engineering programs. Such programs equip future professionals to handle complex waste streams, as well as integrate sustainability principles in design and manufacturing. In addition to technical competence, this education develops systems of thinking and policy awareness and provides graduates with the capacity to bridge the gap between scientific research and societal demands.

Finally, interdisciplinary and international cooperation will help to make e-waste management not a bifurcated technical issue but a coordinated international business. Through the alignment of research, industry, governance and education, societies can transform e-waste that is increasingly becoming a liability to a renewable and secure source of vital materials, which will enhance both environmental sustainability and economic stability (Diallo et al. 2025).

9 Conclusions

The increasing generation of electronic waste (e-waste) presents both a global environmental challenge and a valuable opportunity for sustainable resource recovery. As electronic devices proliferate, the demand for critical and precious metals such as gold, silver, copper, and rare earth elements continues to escalate. Traditional recovery methods, though effective, are often energy-intensive and environmentally damaging. Recent innovations have focused on integrating sustainability into recovery processes, utilizing green solvents, ionic liquids, and bio-based extractants to reduce chemical hazards and minimize waste generation. Emerging low-energy recovery processes, such as bioleaching, electrochemical extraction, and solvent-based recycling, offer promising pathways to enhance metal recovery efficiency while lowering carbon footprints. These approaches align with circular economic principles by converting e-waste into a secondary resource stream rather than landfill material. Additionally, the integration of digital monitoring, process optimization, and machine learning tools can further improve recovery yields and economic viability. Future progress depends on scaling up laboratory successes into industrial applications through policy support, investment in research infrastructure, and collaboration between academia and industry. By adopting greener and more efficient recovery technologies, nations can reduce dependence on primary mining, conserve resources, and contribute to global sustainability goals. Ultimately, sustainable metal recovery represents not just an environmental necessity but a strategic economic opportunity for advancing a more resilient and circular materials economy.

References

Adetunji AI, Adeniyi OD, Ajayi OO (2023) Microbial-assisted recovery of metals from electronic waste: current progress and future prospects. J Environ Chem Eng 11:109876

AL-Reyamia NS, Shaik F, Lakkimsetty NR (2020) Environmental impact on usage of treated effluents from ISTP for irrigation. J Water Process Eng 36:101363. https://doi.org/10.1016/j.jwpe.2020.101363

Balasubramanian A (2020) Rare earth elements: occurrences, extraction, and applications. Earth Sci India 13:1–21

Botelho Junior AB, Oliveira VD, Afonso JC (2021) Leaching and recovery of critical metals from lithium-ion batteries: advances and prospects. Waste Manag 120:741–753

Carayannis EG, Campbell DFJ (2010) Triple helix, quadruple helix and quintuple helix and how do knowledge, innovation and environment relate to each other? Int J Soc Ecol Sustain Dev 1:41–69

Carayannis EG, Campbell DFJ (2012) Mode 3 knowledge production in quadruple helix innovation systems. Springer

CORDIS (2020) Horizon 2020 projects database. European Commission

CSS (2024) Critical materials strategy 2024. Center for Strategic Studies, US Department of Energy

Dao SVT, Dao HVT, Dao HH (2025) Integrating artificial intelligence for sustainable waste management. Waste Sustain Energy 3:12–28

Diallo S, Mbaye M, Ndiaye B (2025) Selective leaching strategies for rare earths and precious metals from complex e-waste streams. Hydrometallurgy 220:105741

Elgarahy AM et al (2023) E-waste management: global challenges and sustainable solutions. J Clean Prod 396:136–149

Enemona AA, Balogun TO (2023) Application of solvent extraction and electrodeposition in printed circuit board recycling. Sep Purif Technol 308:123092

Etzkowitz H (2008) The triple helix: university–industry–government innovation in action. Routledge

Etzkowitz H (2010) The triple helix: a new mission for the university. Springer

European Commission (2020) Circular economy action plan. EU Publications Office

Forti V et al (2020) The global e-waste monitor 2020. United Nations University

Gkika DA, Papadopoulos T, Balomenos E (2024) Solvent extraction of rare earths: advances and perspectives for recycling applications. Miner Eng 198:107405

Gómez Soto J, van der Voet E, Kleijn R (2023) Metal value recovery from e-waste: prospects for critical materials. Resour Conserv Recycl 191:106930

Graedel TE, Harper EM, Nassar NT, Reck BK (2015) On the materials basis of modern society. Proc Natl Acad Sci USA 112:6295–6300

Grandhi R, Patwa N, Saleem M (2020) The role of EPR in India's e-waste management: a policy review. Waste Manag Res 38:879–891

Gulliani J, Chen Y, Zhao H (2023) Innovative hydrometallurgical routes for metal recovery from waste electrical and electronic equipment. Resour Conserv Recycl 190:106799

Ghosh S, Das AP (2018) Metagenomic insights into the microbial diversity in manganese-contaminated mine tailings and their role in biogeochemical cycling of manganese. Sci Rep 2018 8:1, 8(1), 1–12. https://doi.org/10.1038/s41598-018-26311-w

Hossain R, Sahajwalla V (2024) Current recycling innovations to utilize e-waste in sustainable green metal manufacturing. Philos Trans A Math Phys Eng Sci 382:20230239

Ilankoon IMSK, Ghorbani Y, Chong MN, Herath G, Moyo T, Petersen J (2024) Critical metal supply challenges and recycling opportunities. J Clean Prod 427:139530

Jaiswal R, Lee JC (2024) Bioleaching of metals from electronic waste using bacterial and fungal systems: opportunities and challenges. J Hazard Mater 459:132028

Kamran M, Fazal MR, Bilal M, Rashid MI (2023) Metal recovery from spent batteries and e-waste: a sustainable resource perspective. Renew Sustain Energy Rev 174:113120

Kaya M (2018) Recovery of metals and nonmetals from electronic waste by physical and chemical recycling processes. Waste Manag 57:64–90

Kumar S et al (2025) Extraction of metallic nanoparticles from electronic waste for biomedical applications: pioneering sustainable technological progress. Sustainability 17:2100

Li Y, Sun Z, Chen M (2023) Scaling up green metal recovery technologies: barriers and opportunities in industrial implementation. J Clean Prod 407:136928

Liang Y (2023) Phosphors and rare earth utilization in lighting and display technologies. J Lumin 255:119557

Madhav AVS, Yadav A, Tiwari A, Kumar R (2022) Application of artificial intelligence in sustainable e-waste management: challenges and future directions. J Environ Manag 320:115843

Manikandan S, Kumar A, Babu P (2023) Leaching and separation processes for lithium-ion battery recycling: emerging trends and challenges. J Power Sources 567:232312

Maphosa V, Maphosa M (2020) E-waste management in Sub-Saharan Africa: challenges and opportunities. Cogent Environ Sci 6:1783174

Mishra S, Dash D, Al-Tawaha ARMS, Das AP (2022) A review on heavy metal ion adsorption on synthetic microfiber surface in aquatic environments. Appl Biochem Biotechnol 194(10), 4639–4654. https://doi.org/10.1007/S12010-022-04029-W/METRICS

Mohsin A et al (2024) Green recovery of critical metals from electronic waste using eco-friendly hydrometallurgical processes. J Clean Prod 420:138921

OECD (2016) Extended producer responsibility: updated guidance for efficient waste management

Oke SA, Potgieter JH (2024a) A review of printed circuit board recycling: metal recovery and sustainability perspectives. J Environ Chem Eng 12:110251

Oke SA, Potgieter JH (2024b) Comparative evaluation of acid and ammonia-based leaching systems for e-waste recycling. J Sustain Metall 10:567–582
Phogat V, Singh R, Kumar S (2025) Sustainable extraction of rare earth elements from e-waste: green hydrometallurgical approaches. J Sustain Metall 11:345–359
Priyadarsini S, Das AP (2024) Analytical and structural characterization of waste lithium-ion batteries for their effective recycling strategy. Environ Chem Ecotoxicol. https://doi.org/10.1016/j.enceco.2024.12.004
Priyadarsini S, Das AP (2026) Lithium in waste water: impacts on pollution and human health. Decarbonization of Wastewater Pollutants as a Sustainable Solution, 353–364. https://doi.org/10.1016/B978-0-443-33062-9.00010-2
Rizki A et al (2025) Urban mining and circular economy in Southeast Asia: a socio-economic perspective. Resour Conserv Recycl 196:106–120
Rodriguez C, Martínez P, Herrera D (2023) Hydrometallurgical recovery of base and precious metals from e-waste: opportunities and constraints. Hydrometallurgy 217:105632
Shaik F, Mohammed N, Lakkimsetty NR (2026) Sustainable treatment technique for the reuse of wastewater generated from the automobile cleaning process. Interactions 247:16. https://doi.org/10.1007/s10751-026-02357-5
Srivastav R et al (2023) Circular economy and e-waste: policy, technology, and social perspectives. J Clean Prod 383:135–149
Srivastava RK, Jha MK, Kumar V (2025) Critical metals in e-waste: resource recovery and environmental implications. J Hazard Mater 470:134229
Supply Chain Risks Report (2022) Critical minerals and supply chain vulnerabilities. International Energy Agency
Tunsu C, Petranikova M, Ekberg C (2019) Scale-up challenges in hydrometallurgical recycling of rare earths from electronic waste. Resour Conserv Recycl 141:388–398
USEA (2023) Critical minerals and materials for energy transition. United States Energy Association
Van Yken J, Van Acker K, Dewulf J (2021) Resource potential of urban mines: a case study on e-waste. Waste Manag 121:357–366
Wall F, Pell RS (2017) Critical raw materials and the circular economy. Mineral Mag 81:397–415
Wang R, Yang L, Zhang J (2022) Digital transformation in sustainable metal recovery: opportunities and challenges. J Environ Manag 320:115879
Wibowo N, Suharyono S, Pradana G (2021) Strategies for improving the e-waste management supply chain sustainability in Jakarta, Indonesia. Sustainability 13:13955
Zhang X, Liu Y, Huang Q (2022) Hybrid green technologies for metal recovery: toward sustainable and circular processing. Sep Purif Technol 296:121338
Zhang Z, Wang L, Liu X (2023) Recyclable vitrimer-based printed circuit board for circular electronics. J Mater Chem A 11:8234–8245

Biotechnological Approaches for Metal Recovery and Recycling Beyond Bioleaching

Suraj Phukan, Diganta Pegu, Tina Kalita, and Manabendra Mandal

Abstract Metal recovery and recycling play a crucial role in achieving overall sustainability and meeting the growing demands of today's world. Although conventional approaches can recover metals to some extent, they often pose significant environmental risks. Biotechnology bridges this gap by offering eco-friendly and sustainable alternatives. In the biotechnological approach, *metal–microbe interactions* facilitate metal recovery and recycling through two major processes: metal mobilization (solubilization of metals) and metal immobilization (metals do not solubilize). Among the biotechnological methods, bioleaching has been extensively studied and well-documented in the literature. However, other microbial processes beyond bioleaching remain underexplored and represent an important research gap. This chapter focuses on these alternative biotechnological approaches and highlights how emerging tools such as artificial intelligence (AI), machine learning (ML) and nanotechnology could further enhance the efficiency and innovation in metal recovery and recycling systems.

Keywords Biotechnology · Metal recovery and recycling · Artificial intelligence · Machine learning

S. Phukan · D. Pegu · M. Mandal (✉)
Department of Molecular Biology and Biotechnology, Tezpur University, NapaamAssam, Tezpur, India
e-mail: mandal@tezu.ernet.in

S. Phukan
e-mail: mbp24103@tezu.ac.in

D. Pegu
e-mail: dpegu62@gmail.com

T. Kalita
Department of Botany, Gauhati University, Gopinath Bordoloi Nagar, JalukbariAssam, Guwahati, India
e-mail: tkalita968@gmail.com

A. P. Das and S. Priyadarsini (eds.), *Recovery and Recycling of Critical Metals from Electronic Waste for Environmental Sustainability*, Earth and Environmental Sciences Library, https://doi.org/10.1007/978-3-032-29064-9_11

1 Introduction

Industries around the world keep developing by using natural resources from the environment or by reusing and recycling materials from waste produced by themselves or others. Among the raw materials used in industry, metals are key elements in manufacturing a large number of products, whose fabrication depends on metals availability. Therefore, the global demand for metals has increased in recent decades, while the main resources are declining, and the content of valuable metals in ores is diminishing (Dominguez-Benetton et al. 2018; van der Ent et al. 2021). There are several conventional strategies for metal recovery, such as pyrometallurgy, which uses high temperatures to extract metals; hydrometallurgy, which employs chemical solutions for metal dissolution and recovery; and electrometallurgy, which relies on electrical energy to separate metals. However, these methods are often expensive due to high energy consumption and the use of chemicals that can be harmful to the environment (Sethuranjan et al. 2019; He and Kappler 2017). Efforts are always directed toward developing sustainable, eco-friendly, and cost-effective processes for metal recovery, which has led to advanced research on the application of biotechnology (Das et al. 2011). Therefore, biotechnological approaches play an important role in addressing these challenges by promoting more sustainable, cleaner, and cost-efficient technologies for metal recovery and recycling (Gavrilescu 2022). Bioleaching is a well studied biotechnological process that utilizes both autotrophic and heterotrophic microorganisms for the extraction and removal of metals from ores or waste materials. However, in this chapter, we will primarily focus on other aspects of biotechnological approaches for metal recovery and recycling beyond bioleaching. Biotechnological approaches generally employ microorganisms that interact with metals, leading to their mobilization or immobilization in the surrounding solution, thereby facilitating the recovery and recycling of metals in the form of precipitation. A basic schematic representation is shown in Fig. 1.

2 Biomineralization

Biomineralization is a new branch of environmental and geotechnical engineering that has attracted widespread attention in the past two decades (Yu et al. 2023). Here, the free metal are deposited as precipitates through the metabolism of microorganisms. E.g. are Ca, Mg, Cd, Pb, Zn and Ni (Lai et al. 2021; Zhang et al. 2021; Fang et al. 2021; Kang et al. 2016; Jalilvand et al. 2020; Zhu et al. 2016). There are two ways that the microorganisms can recover and recycle the metals. They are: (a) Phosphate precipitation and (b) carbonate precipitation. The mechanisms of both the processes are explained below.

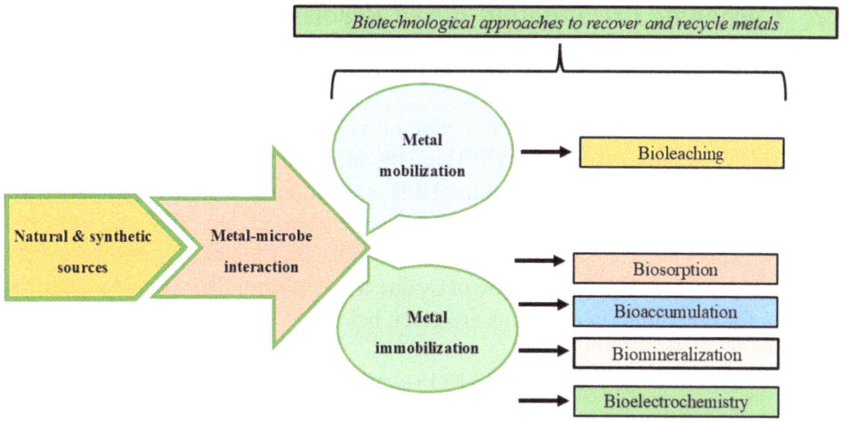

Fig. 1 Schematic diagram of the biotechnological approaches to recover and recycle metals from the natural and synthetic sources

2.1 Phosphate Precipitation Based Recovery and Recycling of Metals

It dissolves the organic phosphorus and inorganic phosphorus into PO_4^{3-} by phosphate solubilizing microorganisms (PSM). The PO_4^{3-} then reacts with the surrounding availability of metal ions and get precipitated. The chemical reactions are given below:

$$2PO_4^{3-} + 3M^{2+} \rightarrow M_3(PO_4)$$

$$3PO_4^{3-} + 5M^{2+} + X^{X-} \rightarrow M_5(PO_4)X$$

where M represents metals, and X^{X-} represents halogen ions or hydroxide groups.

Phosphate solubilizing bacteria (PSB) is the most studied phosphate precipitation based recovery and recycling of metals. PSB can solubilized both organic phosphate eg. glycerol 2-phosphate, myo-inositol 1,2,3,4,5,6 hexakisphosphate and glycerophosphate; and inorganic phosphate, e.g. phosphate rock and $Ca_3(PO_4)$.. Therefore, due to this solubility nature of PSB they can be divided into organic and inorganic PSB and hence differ in mechanism point of view. In case of organic PSB, enzymes (such as phosphatase nuclease and phytase) can catalyze the hydrolysis of phosphate ester bonds and thus decompose PO_4^{3-} from organic phosphorus. Whereas in case of inorganic PSB, they secrete organic acids instead of enzymes (such as oxalic acid, citric acid, and acetic acid) that's lower the pH level which results I chelation of phosphate bound cations through their hydroxyl and carboxyl groups or competes with the adsorption sites of phosphates. In both cases of PSB, enzymes and organic acids are secreted extracellularly. Therefore, the phospholysis process occurs extracellularly.

2.2 Carbonate Precipitation Based Recovery and Recycling of Metals

Carbonate precipitation based bioremediation is to produce CO_3^{2-} by microorganisms. Urease producing bacteria (UPB) are the most studied microorganism and we will discuss the mechanism showcasing UPB as the model. Here, UPB catalyze urea into NH_3 and H_2CO_3. The release NH_3 will get converted into NH_4^+ and OH^- that leads into rise of pH in the solution which results into the conversion of H_2CO_3 into HCO_3^- and H^+. It decreases the pH value thereby neutralizes the solution. The equation for precipitation of metals are given below:

$$CO(NH_2) + H_2O \rightarrow NH_3 + H_2CO_3$$

$$NH_3 + H_2O \leftrightarrow NH_4^+ + OH^-$$

$$H_2CO_3 \leftrightarrow HCO_3^- + H^+ \leftrightarrow CO_3^{2-} + 2H^+$$

$$M^{2+} + CO_3^{2-} \rightarrow MCO_3 \downarrow$$

where M represents metal.

Note that, PSB secretes their enzymes as well as organic acid extracellularly and UPB secretes intracellular urease. Therefore, in Fig. 1 equations are outside the cell and in Fig. 2, the equations are inside the cell (Fig. 3).

3 Bioelectrochemistry

Although the biological use of microorganisms, including both autotrophs and heterotrophs, can contribute to sustainable metal recovery and recycling, it is challenging to rely solely on these methods for large-scale applications. There is also a limitation while using the biological process is that in the disposable process of remediated contaminants or wastes. Therefore, the integration of biological, physical as well as chemical method known as Bio-electro-chemical system (BES) is used to mitigate the problem. Here, microorganisms perform the oxidation and reduction reactions around the anode and cathode respectively. By doing so they can convert chemical energy that is stored in organic wastes to electrical energy and other valuable products. The process is simplified in the Fig. 4. There are two cells in the BTS system. They are: a) Microbial Fuel Cells (MFC); and b) Microbial electrode cells (MEC). In MFCs, the anode oxidation is coupled with a cathode reduction having a higher redox potential, generating electricity spontaneously, whereas in MECs, the cathode reaction has a lower redox potential, requiring an external voltage input.

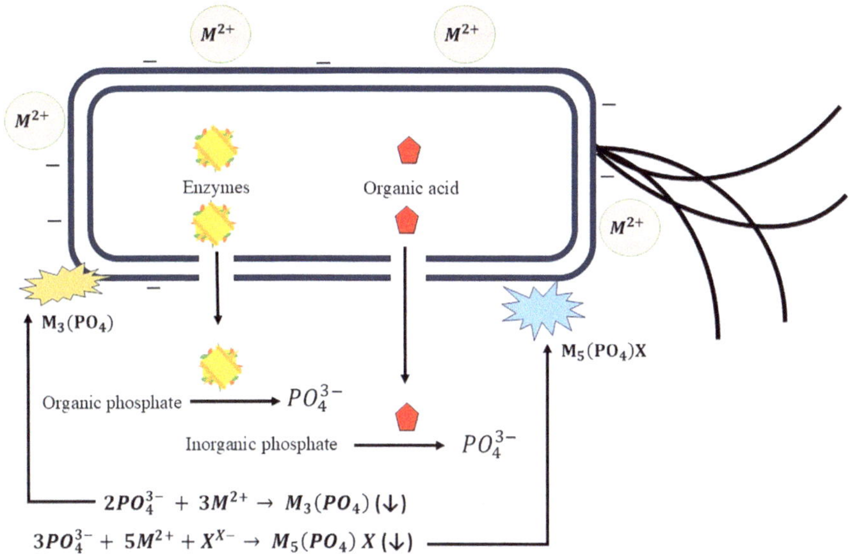

Fig. 2 Biomineralization through phosphate precipitation by PSB

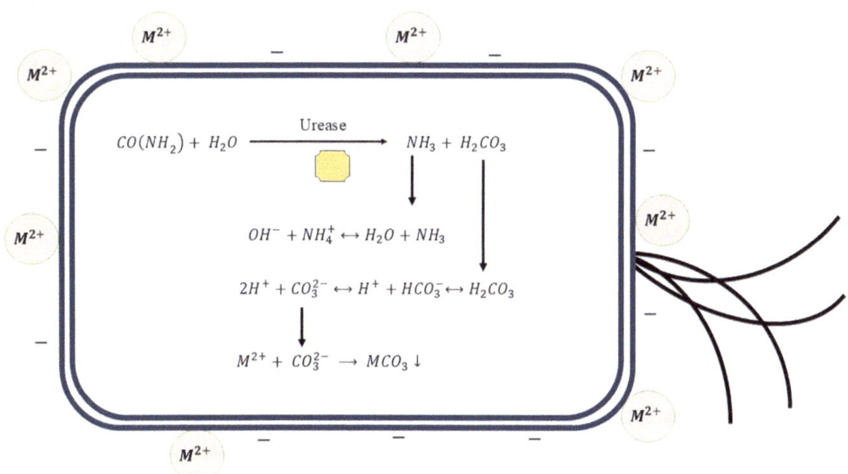

Fig. 3 Biomineralization of metals through carbonate precipitation

Therefore, BES metal recovery and removal methods are promising for practical application in the metal removal and rccovery fields (Wang et al. 2022).

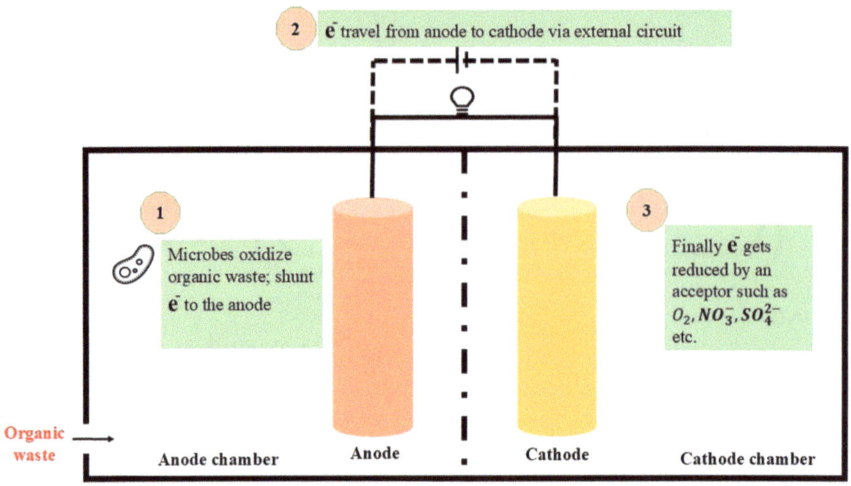

Fig. 4 Schematic diagram of BES in metal removal and recovery mechanism

4 Microbial Biosorption and Bioaccumulation

Biosorption is a category of sorption technique. Due to its biological origin, it is considered as an attractive alternative for removing pollutants to decontaminate different ecosystem. It is also a sustainable practice to adopt because it is simple, ecofriendly, economical in its approach. The biosorption mechanism is adopted in both aqueous and soil decontamination (Torres 2020). The predominant mechanism for metal biosorption includes ion exchange (most common), surface complexation, microprecipitation (Giese 2020), for effective biosorption charges on the functional groups are important determined by the pH of the process. The recovery process of biosorbed substances and its modifiable property has increased the attraction for adaption of biosorption in the industrial and commercially process (Ojima et al. 2019; Benis et al. 2020). The mechanism of biosorption is heavily dependent on microbial metabolism and metal removal site (extracellular precipitation/accumulation, cell surface precipitation/sorption). The type of biomass is an important criterion that decides the pollutant removal to be metabolic dependent or independent (Priya et al. 2022). The recovery and recycling of metals from aqueous solutions by biomass (biosorption) is a complex, multi-step process primarily driven by physicochemical interactions on the cell surface. The transport of ions from the liquid to the solid biosorbent phase begins with the movement of ions through the boundary layer and then internal diffusion within the cell wall. The core mechanisms responsible for binding these metal ions are predominantly metabolically passive (not requiring living cell energy) and includes;

(i) physical adsorption (electrostatic attraction) is rapid process involves non-covalent bonding between metal ions in the liquid and functional groups (like deprotonated carboxylates, sulfates, and phosphates) on the cell walls of inactive biomass (algae, fungi, bacteria). This attraction is highly dependent on the pH of the solution, which affects the charge of the surface groups,

(ii) ion exchange of metal ions from the solution displaces counter-ions (such as H^+, K^+ or Ca^{2+}) already bound to the cell wall's functional groups, particularly the exopolysaccharides. The simultaneous release of protons (H^+) and the uptake of metal ions often confirm this as a dominant mechanism, demonstrated by the biosorption of Pb^{2+} and Cd^{2+}.

(iii) Complexation of metal ions interact with various functional groups (carboxy, amino, thiol, hydroxyl) on the cell surface to form stable complex compounds.[1] Microorganisms can also excrete organic acids (like citric or oxalic acid) that aid in the complexation and subsequent solubilization or removal of hazardous metals.

(iv) Precipitation is through immobilization of either on the cell surface or within the cell walls, particularly at higher pH levels. It can be a chemical interaction dependent or independent from the cell metabolism, which is a defense mechanism of the microorganisms (Xu et al. 2020). While some metal uptake involves intracellular accumulation (a slower, metabolism-dependent process also known as bioaccumulation) where metals are transported across the cell membrane, the most prominent and commercially applicable forms of biosorption utilize the rapid, surface-based, and non-metabolic mechanisms listed above (Medfu et al. 2020) (Fig. 5).

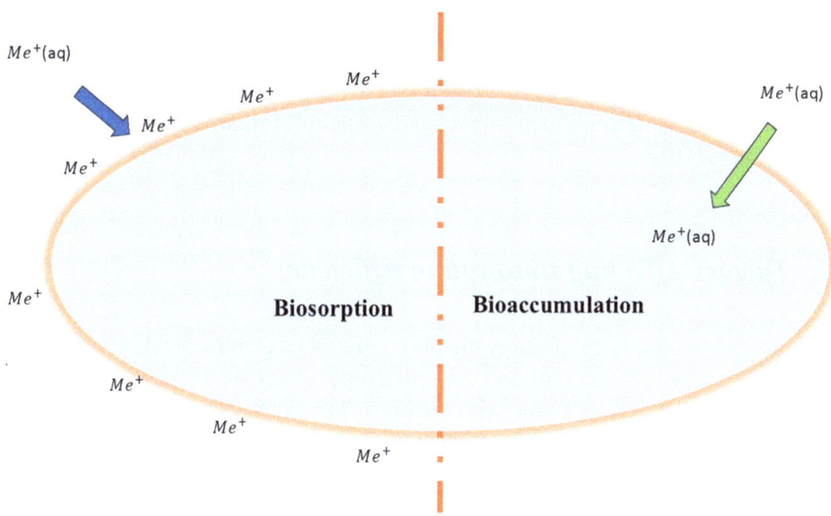

Fig. 5 Schematic diagram of Biosorption and Bioaccumulation in bacteria cells

4.1 Role of Dead Biomass Versus Live Cells

Biomass (live or dead) is the main element of biosorption technique. The advantages over, the biomass used dead or alive depends upon toxicity resistance (Rogowska et al. 2019; Peng et al. 2019; Reddy et al. 2020), biotransformation potential (biodegradation and bioaccumulation), requirement of special environment for effective biosorption, resistant strains, properties of the biomass, binding ability. The factors like the resistance of live microbes used (*Pseudomonas* sp.) more effectively removed cadmium from the environment when live microbes were used compared to dead biomass because of its high resistance (Santaeufemia et al. 2016), toxicity resistance of biomass, manipulation capacity of the biomass type, types of pollutants (Rogswska et al. 2019). Thus, characterizing materials by their morphology and composition is essential in biosorption studies to better understand sorption mechanisms FTIR, SEM, and EDS are commonly used techniques for this purpose (Torres 2020). Thermogravimetric analysis, elemental composition, zeta potential, and pH measurements are additional parameters used to assess the physicochemical properties of biosorbents. Recently, the electrokinetic technique known as dielectrophoresis (DEP) has been utilized for biosorption characterization (Wang et al. 2019a, b; Adekanmbi et al. 2020). However, the kinetic and isotherm models serve as mathematical tools to describe biosorption processes some common kinetic models include the pseudo-first order, pseudo-second order, and intraparticle diffusion models, which help determine process rate, equilibrium time, and limiting steps. Recently, alternative models like the Brouers-Sotolongo family have gained attention, while the Langmuir and Freundlich isotherms remain the most widely applied (Giese 2020). Besides its application in pollutants remover, it can be used in recovering the biosorbed materials commonly used in the extractive metallurgy (rare-earth recovery) (Dinh et al. 2021). The fungi are very efficient in removal of noxious metals and possess metal recovery capabilities. The fungal dead biomass of *Rhizopus oryzae*, *Saccharomyces cerevisiae*, *Aspergillus niger* and *Penicillium chrysogenum* was able to convert poisonous Cr (VI) to less toxic Cr (III).

4.2 Factors Affecting Biosorption Efficiency

The Biosorption efficiency of heavy metals by microbes is influenced by wide range of parameters like biomass type and dosage, temperatures, heavy metals, pH, initial metal concentration, contact time with the biosorbent and status (Priya et al. 2022). The metal sorption pH is a key a factor that influences the effectives of the biosorption by biomass, most metals are exists in the cationic form in aqueous solution in this case the higher pH (7.0–8.0) of the biosorbent facilitates its effectiveness of functional groups to find with metal ions. However, some metals behave predominately as anionic such as chromium, arsenic, molybdenum. In this case the acidic pH (2.0–4.0) favors the mechanism of biosorption because of higher positive charges on the functional groups allow attraction of anions. Otherwise, if the pH is not suitable it may either lead to precipitation of metal ions, reducing the amount of biosorbed materials. The effect of temperature is more pronounced in living biomass than dead cells. As the temperature increase the live cells are able to biosorbed more metal ions than dead cells, because the live cells metabolic process is active, however the effectiveness of biosorption mechanism depends upon the ionic strength and complex concentration of different cations (Libatique et al. 2020), as it increases the competition for binding sites on the functional groups (Zang et al. 2020). Thus, this becomes a drawback for its application in real effluents (Torres 2020). Nevertheless there are some exception to this depending upon the amount of metals biosorbed which is influenced by the exothermic and endothermic process with respect to rise in temperature that takes place, during the former process the increased temperature the surface damage of the biosorbent while in the latter case the structural changes and breaking of bonds between sorbent materials allow maximum absorption (Ojima et al. 2019).The biomass offering diverse heterogenous functional groups to interact with metals like (carboxyl, hydroxyl, sulfate, phosphoryl and amino groups) (Benis et al. 2020) Successful modification, of biomass such as immobilization, sophisticated manipulations provide wide range of versatility and great efficiency.

4.3 Overexpression of Metal-Binding Proteins

Various biomass is used for the biosorption process, but use of microorganism confers an additional advantage as they can be genetically manipulated for the interest of biosorption this field is receiving a lot of attention. The surface expression of the EC20 protein, a synthetic phyto-chelating, in different bacterial species has been applied for the biosorption of metals such as Pb, Zn, Cu, Cd, Mn, Ni, and more recently Pt. Similarly, introducing a non-metallothionein cadmium-binding protein from *Lentinula edodes* significantly improved the cadmium uptake capacity of genetically modified *Escherichia coli.* Furthermore, transforming wild-type *Saccharomyces cerevisiae* with two variants of the *Populus trichocarpa* metallothionein gene (PtMT2b) resulted in higher intracellular cadmium accumulation compared to the native strain (De Oliveira et al. 2020).

5 Integration Along with Nanotechnology

5.1 Biosynthesis of Metallic Nanoparticles

Picture a microbial cell factory: even simple bacteria, fungi, or algae could reduce metal ions, nucleate small particles, and wrap them with biomolecules. These biosynthetic pathways could be very enticing alternatives to the physical or chemical methods of synthesizing nanoparticles. Microbes work in mild environments, use liquid water mediums and frequently create biosynthetically coated particles, which add to their stability and adaptability (Ghosh et al. 2021a, b).

Mechanistically what we see is that the process often reports of microbial enzyme systems (for example nitrate reductases) or secreted reducing metabolites (peptides, polyphenols, sugars) play a role. They reduce metal ions (of Au^{3+}, Ag^{+}, Pd^{2+} for instance) to zero valent which in turn initiates nucleation, also extracellular polymeric substances or proteins act as scaffolds and caps which in turn control size, shape and aggregation. For example microbial extracellular reduction followed by biomolecule mediated stabilization is a very well-studied model (Ghosh et al. 2021a).

Two main strategies exist: intracellular reduction, where ions are taken up by cells and reduced within, and extracellular reduction, where molecules secreted into the culture supernatant do the reduction. The extracellular option is frequently favoured for the subsequent separation of nanoparticles (Ghosh et al, 2021b). The various microbes employed, from Bacillus and Pseudomonas bacteria to Aspergillus fungi and algae, yeast, and even algae, demonstrate the broad potential of this approach (Koul et al. 2021).

5.2 Recycling of Metals in Catalysis and Electronics

In the biological realm microorganisms play a growing role in metal recovery which includes capture, concentration and transformation of metals from waste streams, tailings and used electronics (e-waste). Bioadsorption, bioaccumulation and biomineralization technologies enable the select uptake of precious metals (Au, Ag, Pd, Cu, even rare earths) which in turn is then turned into nanoparticles or catalytic supports thus facilitating closed loop economies (Pati et al. 2016).

Once the metals are recovered via biological means they may be put to use in catalytic processes (e.g. hydrogenation, C–C coupling) or returned to the e–waste stream. The bio based route presents a chance for energy savings and reduction in chemical use as compared to high temperature pyrometallurgy or aggressive hydrometallurgical methods. That said large scale implementation into electronic manufacturing or catalyst supply is still in its infancy.

5.3 Role of Microorganisms in the Production of Nanoparticles

A critical component of the microbial NP synthesis is what we put under the microscope. Enzymes (for instance nitrate reductase, sulphate reductase) play a role of catalysts in the electron transfer from microbial metabolism to metal ions which in turn cause reduction. Also we see that other more general reducing agents (peptides, phenolic compounds, reducing sugars) play a support role in this. At the same time cell surface proteins, EPS and secreted polysaccharides act as nucleating templates which in turn control particle morphology also they play a role as capping and stabilizing agents.

Recent research in the area has brought to the fore how we may manipulate culture conditions, enzyme gene expression or metabolite profiles to in turn tune nanoparticle size, shape and functionality (Ghosh et al. 2021a). Also, in the early stages of development we see the use of synthetic biology to engineer microbes with better performance in nanoparticle production which is a sign of this field's growth.

6 Environmental and Economic Considerations

6.1 *Life Cycle Analysis of Biotechnological Methods*

Avoiding Environmental and Economic Risks Assessing the life cycle impacts of biotechnological interventions It is easy to think that "microbial = green," but life cycle assessment (LCA) reveals important qualifications. A comprehensive review of 71 LCA studies showed that many "green" nano-processes still overlook end-of-life stages, uncertainty analysis, and full system boundaries (Nizam et al. 2021). An additional thorough assessment identified the lack of recyclability of eco-efficient technologies for gold nanoparticle production as a significant environmental drawback, whereas recovery and reuse of gold nano-particles substantially lessened that environmental impact (Pati et al. 2016).

There are hotspots: nutrient and media production for microbial growth, energy for separation and purification of biomass and particles, and the purification/drying steps. Thus, only when the microbial route is fully integrated (feedstock recycling, continuous operation, and low downstream processing) does it often outperform traditional methods in terms of environmental burden.

6.2 *Cost–Benefit Comparison with Conventional Techniques*

When compared to traditional approaches, biological methods have several economic benefits, such as lower temperature and pressure requirements, lower aggressiveness of chemicals used, possibilities of waste-feedstock (organic residues) integration and coupling disposal biomass adsorbent coproducts. On the other side of the coin, biological methods have to deal with slow reaction kinetics on some strains, expensive growth media, complex downstream separation, and scale-up challenges to achieve consistent nanoparticle characteristics. Techno-economic assessments are still quite limited, with the main drivers being low-cost feedstocks, high productivity strains, and streamlined purification processes.

6.3 Regulatory and Scalability Challenges

Regulatory and scalability challenges From the point of view of regulation and scale-up possibilities, two primary areas stand out: (1) biocontainment/biosafety of microbial operations (especially with genetically modified organisms) and (2) the safety, movement, and transport (toxicology and environmental release) of the nanoparticles. Compared to biotechnological processing, regulations on nanomaterials are more stringent. This, along with variations in regulations in different parts of the world, can slow down the process of translating and commercializing the process. From an engineering standpoint, there are also considerable challenges in scaling from a flask to large bioreactor and uniformity of strains with respect to monitoring, mixing mass transfer, and dependable downstream separation. Achieving narrow size distribution, reproducible particle characteristics, and robust processing are also considerable challenges.

7 Future Prospects and Research Directions

7.1 AI and Machine Learning for Process Optimization

Artificial intelligence (AI) and machine learning (ML) for process optimization The integration of microbial nanoparticle production with data-driven technologies opens new horizons. Bioprocessing applications (strain optimization, media formulation, process control) have seen increasingly successful uses of ML and AI techniques. In the context of microbial NP synthesis, the use of surrogate models that connect culture and metabolite conditions with size and distribution of the nanoparticle, Bayesian optimization for experimental burden reduction, and real-time process control using sensors and ML algorithms exhibit growing interest. While the literature is still emerging, the potential is development speed, yield, and reproducibility. Beyond optimizing bioprocess yields, AI/ML is also transforming detection of heavy metals closes the loop with our bio-nano pipeline: instead of relying solely on slow, reagent-intensive assays, recent work shows that fast electrochemical or physicochemical readouts coupled to AI/ML can classify and quantify mixed burdens (e.g., square-wave voltammetry signals resolved into As, Pb, and Cd by machine-learning classifiers), generate spatial risk maps from low-cost field data, and even "soft-sense" metals from proxy process variables like BOD/COD with high accuracy (R^2 up to ~ 0.98 to 0.99); importantly, the same review frames these tools across prediction, source detection, WWTP performance monitoring, and in-silico burden modelling while flagging methodological and ethical considerations making them a natural front-end for real-time control and targeting in microbe-enabled recovery and nanoparticle production, and a health-centric back-end for communicating risk for Pb/Hg/Cd/As exposure (Fig. 6).

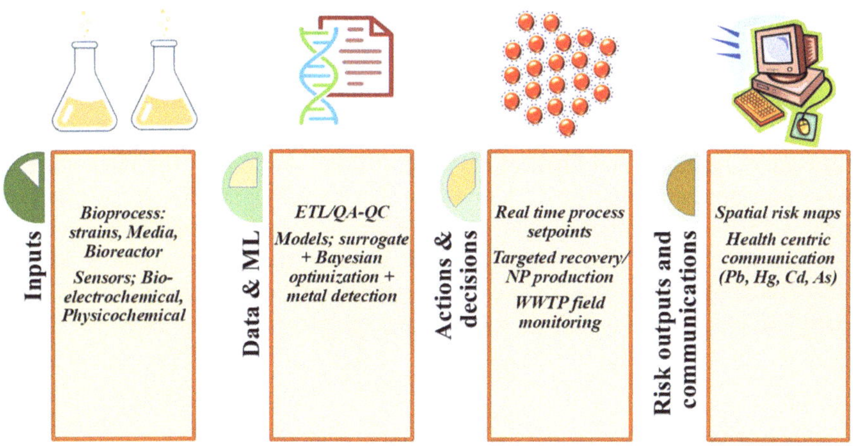

Fig. 6 Schematic diagram of AI/ML- enabled closed loop for microbial NP production

7.2 Bioreactor Design and Automation

Designing and automating bioreactors will be important to efficiently scale the production of microbial nanoparticles. Introducing continuous-flow bioreactors with separate growth and reduction phase subprocesses, integrated immobilized cell/ enzyme bioreactors for easier separations, and in-line monitoring systems (e.g., UV–Vis, nanoparticle tracking, metal ion sensors) will all be important and in-line pivotal. Furthermore, automation and the use of digital-twin modelling for scale-up will be essential. So far, pilot studies have shown potential, yet controlling scale-formation nucleation and aggregation will continue to be a key bottleneck.

7.3 Exploration of Extremophiles and Metagenomics

The potential of the microbial world that lies beyond the known laboratory strains is promising. Extremophiles (thermophiles, halophiles, acidophiles, and alkaliphiles) possess unique and attractive enzyme systems, considerable toughness, and metal tolerance which makes them useful for NP synthesis or metal recovery. Metagenomics and functional screening of environmental samples can discover novel reductases, metal-binding peptides, and scaffold proteins. In the field of nanoparticle synthesis, this is an underdeveloped frontier that offers immense potential.

7.4 Policy and Industry Adoption Pathways

In (a) we see that which goes in through niche markets which have a very special advantage from biogenic nanoparticles (for instance in the areas of biocompatibility for diagnostics and medical devices); (b) also we look at the integration of these with present metal recovery or electronics recycling supply chains (bio pretreatment before hydrometallurgy); (c) use of procurement and regulatory incentives (for instance recycled content requirements, nanomaterial safety standards) to push for adoption. We will see that harmonized standards for biogenic nanomaterials, clear safety and lifecycle disclosure protocols, and demo plants will play a key role in this.

8 Conclusion

As much as advancement of day to day life of humans will grow, the volume of waste will continue to increase. Therefore, Biotechnology will play an important role for the recovery and recycling of metals towards an eco-friendly and sustainable environment. Microbes that transform metal ions, produce functional nanoparticles, and also recover precious metals from waste. While the path to large scale implementation still includes which we term as key challenges: scale up of processes and achieving robustness, which we have also seen to include definitive techno-economic and lifecycle analysis, data frameworks for AI/ML, discovery of next generation microbial catalysts (which we see to include from the extremophiles), and regulatory harmonization. As these pieces fall into place, we see microbial nanotech to become a base element of regenerative materials infrastructure in catalysis, electronics and remediation.

Acknowledgements I would like to express my sincere gratitude to my supervisor, Prof. Manabendra Mandal, for his invaluable guidance, support, and encouragement throughout the preparation of this chapter. I also extend my heartfelt thanks to my co-authors for their contributions and collaborative efforts that helped in the successful completion of this work. Additionally, I am deeply thankful to the editorial members of this book for providing me with the opportunity to contribute a chapter to their esteemed publication.

References

Adekanmbi EO, Giduthuri AT, Carv BA, Counts J, Moberly JG, Srivastava SK (2020) Application of dielectrophoresis towards characterization of rare earth elements biosorption by *Cupriavidus necator*. Anal Chim Acta 1129:150–157. https://doi.org/10.1016/j.aca.2020.07.042

Benis KZ, Damuchali AM, McPhedran KN, Soltan J (2020) Treatment of aqueous arsenic–a review of biosorbent preparation methods. J Environ Manage 273:111126. https://doi.org/10.1016/j.jenvman.2020.111126

Chang J, Deng S, Liang Y, Chen J (2019) Cr (VI) removal performance from aqueous solution by *Pseudomonas* sp. strain DC-B3 isolated from mine soil: characterization of both Cr (VI) bioreduction and total Cr biosorption processes. Environ Sci Pollut Res 26(27):28135–28145. https://doi.org/10.1007/s11356-019-06017-w

Das AP, Sukla LB, Pradhan N, Nayak S (2011) Manganese biomining: a review. Biores Technol 102(16):7381–7387. https://doi.org/10.1016/J.BIORTECH.2011.05.018

De Oliveira VH, Ullah I, Dunwell JM, Tibbett M (2020) Bioremediation potential of Cd by transgenic yeast expressing a metallothionein gene from *Populus trichocarpa*. Ecotoxicol Environ Saf 202:110917. https://doi.org/10.1016/j.ecoenv.2020.110917

Dinh VP, Xuan TD, Hung NQ, Luu TT, Do TTT, Nguyen TD, Nguyen VD, Anh TT, Tran NQ (2021) Primary biosorption mechanism of lead (II) and cadmium (II) cations from aqueous solution by pomelo (Citrus maxima) fruit peels. Environ Sci Pollut Res 28(45):63504–63515. https://doi.org/10.1007/s11356-020-10176-6

Dominguez-Benetton X, Varia JC, Pozo G, Modin O, Ter Heijne A, Fransaer J, Rabaey K (2018) Metal recovery by microbial electro-metallurgy. Prog Mater Sci 94:435–461. https://doi.org/10.1016/j.pmatsci.2018.01.007

Fang L, Niu Q, Cheng L, Jiang J, Yu Y-Y, Chu J, Achal V, You T (2021) Ca-mediated alleviation of Cd^{2+} induced toxicity and improved Cd^{2+} biomineralization by *Sporosarcina pasteurii*. Sci Total Environ 787:147627. https://doi.org/10.1016/j.scitotenv.2021.147627

Gavrilescu M (2022) Microbial recovery of critical metals from secondary sources. Biores Technol 344:126208. https://doi.org/10.1016/j.biortech.2021.126208

Ghosh S, Ahmad R, Banerjee K, AlAjmi MF, Rahman S (2021a) Mechanistic aspects of microbe-mediated nanoparticle synthesis. Front Microbiol 12:638068. https://doi.org/10.3389/fmicb.2021.638068.

Ghosh S, Ahmad R, Zeyaullah M, Khare SK (2021b) Microbial nano-factories: synthesis and biomedical applications. Front Chem 9:626834. https://doi.org/10.3389/fchem.2021.626834]

Giese EC (2020) Biosorption as green technology for the recovery and separation of rare earth elements. World J Microbiol Biotechnol 36(4):52. https://doi.org/10.1007/s11274-020-02821-6

He J, Kappler A (2017) Recovery of precious metals from waste streams. Microb Biotechnol 10(5):1194–1198. https://doi.org/10.1111/1751-7915.12759

Jalilvand N, Akhgar A, Alikhani HA, Rahmani HA, Rejali F (2020) Removal of heavy metals zinc, lead, and cadmium by biomineralization of urease-producing bacteria isolated from Iranian mine calcareous soils. J Soil Sci Plant Nutr 20(1):206–219. https://doi.org/10.1007/s42729-019-00121-z

Kang C-H, Kwon Y-J, So J-S (2016) Bioremediation of heavy metals by using bacterial mixtures. Ecol Eng 89:64–69. https://doi.org/10.1016/j.ecoleng.2016.01.023

Koul B, Poonia AK, Yadav D, Jin JO (2021) Microbe-mediated biosynthesis of nanoparticles: applications and future prospects. Biomolecules 11(6):886. https://doi.org/10.3390/biom11060886

Lai HJ, Cui MJ, Wu SF, Yang Y, Chu J (2021) Retarding effect of concentration of cementation solution on biocementation of soil. Acta Geotech 16(5):1457–1472. https://doi.org/10.1007/s11440-021-01149-1

Libatique MJH, Lee MC, Yeh HY, Jhang FJ (2020) Total and inorganic arsenic biosorption by *Sarcodia suiae* (Rhodophyta), as affected by controlled environmental conditions. Chemosphere 248:126084. https://doi.org/10.1016/j.chemosphere.2020.126084

Medfu Tarekegn M, Zewdu Salilih F, Ishetu AI (2020) Microbes used as a tool for bioremediation of heavy metal from the environment. Cogent Food Agric 6(1):1783174. https://doi.org/10.1080/23311932.2020.1783174

Nizam NUM, Hanafiah MM, Woon KS (2021) A content review of life cycle assessment of nanomaterials: current practices, challenges, and future prospects. Nanomaterials 11(12):3324. https://doi.org/10.3390/nano11123324

Ojima Y, Kosako S, Kihara M, Miyoshi N, Igarashi K, Azuma M (2019) Recovering metals from aqueous solutions by biosorption onto phosphorylated dry baker's yeast. Sci Rep 9(1):225. https://doi.org/10.1038/s41598-018-36306-2
Pati P, McGinnis S, Vikesland PJ (2016) Waste not want not: life cycle implications of gold recovery and recycling from nanowaste. Environ Sci: Nano 3(5):1133–1143. https://doi.org/10.1039/C6EN00181E
Peng H, Li D, Ye J, Xu H, Xie W, Zhang Y, Wu M, Xu L, Liang Y, Liu W (2019) Biosorption behavior of the Ochrobactrum MT180101 on ionic copper and chelate copper. J Environ Manag 235:224–230.https://doi.org/10.1016/j.jenvman.2019.01.060
Priya AK, Gnanasekaran L, Dutta K, Rajendran S, Balakrishnan D, Soto-Moscoso M (2022) Biosorption of heavy metals by microorganisms: evaluation of different underlying mechanisms. Chemosphere 307:135957. https://doi.org/10.1016/j.chemosphere.2022.135957
Reddy S, Osborne JW (2020) Biodegradation and biosorption of reactive red 120 dye by immobilized Pseudomonas guariconensis: kinetic and toxicity study. Water Environ Res 92(8):1230–1241. https://doi.org/10.1002/wer.1319
Rogowska A, Pomastowski P, Rafińska K, Railean-Plugaru V, Złoch M, Walczak J, Buszewski B (2019) A study of zearalenone biosorption and metabolisation by prokaryotic and eukaryotic cells. Toxicon 169:81–90. https://doi.org/10.1016/j.toxicon.2019.09.008
Santaeufemia S, Torres E, Mera R, Abalde J (2016) Bioremediation of oxytetracycline in seawater by living and dead biomass of the microalga *Phaeodactylum tricornutum*. J Hazard Mater 320:315–325. https://doi.org/10.3390/pr8121584
Sethurajan M, van Hullebusch ED, Fontana D, Akcil A, Deveci H, Batinic B, Leal JP, Gasche TA, Ali Kucuker M, Kuchta K, Chmielarz, A. (2019). Recent advances on hydrometallurgical recovery of critical and precious elements from end of life electronic wastes—a review. Critical Rev Environ Sci Technol 49(3):212–275.https://doi.org/10.1080/10643389.2018.1540760
Torres E (2020) Biosorption: a review of the latest advances. Processes 8(12):1584
van der Ent A, Parbhakar-Fox A, Erskine PD (2021) Treasure from trash: mining critical metals from waste and unconventional sources. Sci Total Environ 758:143673. https://doi.org/10.1016/j.scitotenv.2020.143673
Wang L, Xiao H, He N, Sun D, Duan S (2019a) Biosorption and biodegradation of the environmental hormone nonylphenol by four marine microalgae. Sci Rep 9(1):5277. https://doi.org/10.1038/s41598-019-41808-8
Wang N, Qiu Y, Xiao T, Wang J, Chen Y, Xu X, Kang Z, Fan L, Yu H (2019) Comparative studies on Pb (II) biosorption with three spongy microbe-based biosorbents: high performance, selectivity and application. J Hazard Mater 373:39–49.https://doi.org/10.1016/j.jhazmat.2019.03.056
Wang S, Adekunle A, Raghavan V (2022) Bioelectrochemical systems-based metal removal and recovery from wastewater and polluted soil: key factors, development, and perspective. J Environ Manage 317:115333. https://doi.org/10.1016/j.jenvman.2022.115333
Xu S, Xing Y, Liu S, Hao X, Chen W, Huang Q (2020) Characterization of Cd^{2+} biosorption by Pseudomonas sp. strain 375, a novel biosorbent isolated from soil polluted with heavy metals in Southern China. Chemosphere 240:124893. https://doi.org/10.1016/j.chemosphere.2019.124893
Yu X, Jiang N, Yang Y, Liu H, Gao X, Cheng L (2023) Heavy metals remediation through bio-solidification: potential application in environmental geotechnics. Ecotoxicol Environ Saf 263:115305. https://doi.org/10.1016/j.ecoenv.2023.115305
Zhang J, Chen X, Zhou J, Luo X (2020) Uranium biosorption mechanism model of protonated *Saccharomyces cerevisiae*. J Hazard Mater 385:121588. https://doi.org/10.1016/j.jhazmat.2019.121588
Zhang M, Zhao L, Li GK, Zhu C, Dong S, Li Z, Chen J (2021) Microbially induced magnesium carbonate precipitation and its potential application in combating desertification. Geomicrobiol J 38(6):549–560. https://doi.org/10.1080/01490451.2021.1900461

Zhu X, Li W, Zhan L, Huang M, Zhang Q, Achal V (2016) The large-scale process of microbial carbonate precipitation for nickel remediation from an industrial soil. Environ Pollut 219:149–155. https://doi.org/10.1016/j.envpol.2016.10.047

Fuzzy Logic and Traditional Ecological Knowledge Based Decision Model for Sustainable Electronic Waste Recycling

Nandini Gupta and Ajoy Kanti Das

Abstract The exponential growth of electronic waste poses critical challenges for sustainable resource recovery, particularly of valuable metals. This chapter presents a decision-making framework that integrates *Fuzzy Logic* with *Traditional Ecological Knowledge (TEK)* to address these challenges. Fuzzy Logic enables systematic treatment of uncertainty, while TEK contributes context-specific, culturally embedded practices for ecological balance. The proposed model evaluates recycling options not only on cost and efficiency, but also on cultural acceptability, social equity, and long-term environmental sustainability. Application of the model to methods such as bioleaching demonstrates its capacity to identify recycling strategies that are both technologically viable and socio-ecologically responsible.

Keywords Electronic waste · Recycling · Biotechnology · Sustainability · Microorganism

1 Introduction

The rapid advancement of modern technology has significantly enhanced human convenience and quality of life. However, this progress is accompanied by an ever-increasing turnover of electronic products. Each year, millions of devices—including mobile phones, computers, batteries, and household appliances—are manufactured, utilized, and subsequently discarded. Once disposed of, these products are collectively categorized as electronic waste (e-waste). Recent estimates indicate that global e-waste generation exceeds 60 million tonnes annually, with projections showing a continuous upward trajectory.

N. Gupta (✉)
Department of Environmental Science, Bir Bikram Memorial College, Agartala, Tripura, India
e-mail: nandini1836@gmail.com

A. K. Das
Department of Mathematics, Tripura University, Agartala, Tripura, India

A. P. Das and S. Priyadarsini (eds.), *Recovery and Recycling of Critical Metals from Electronic Waste for Environmental Sustainability*, Earth and Environmental Sciences Library, https://doi.org/10.1007/978-3-032-29064-9_12

Electronic waste (e-waste) represents a significant secondary resource, containing critical metals such as lithium, cobalt, nickel, copper, gold, palladium, and various rare earth elements. These metals are indispensable for the development of advanced technologies, including electric vehicles, photovoltaic systems, and energy storage devices. Despite their strategic importance, efficient recovery of these elements from e-waste remains a considerable challenge (Mishra et al. 2023). The primary obstacles stem from the highly heterogeneous composition of e-waste, fluctuations in metal concentrations, and the variability of recovery conditions, all of which introduce substantial uncertainty into extraction processes.

The development of sustainable and socially just recycling systems requires moving beyond narrowly defined technological interventions. This chapter contends that the integration of Traditional Ecological Knowledge (TEK) offers a critical complement to conventional scientific approaches (Priyadarsini and Das 2024, 2026). TEK constitutes a cumulative body of knowledge, practices, and beliefs concerning the interrelationships between living beings and their environment, maintained by Indigenous and local communities and transmitted across generations. Characterized by its holistic, place-based orientation and its emphasis on long-term ecological balance, TEK provides a framework inherently aligned with sustainability. In contrast, top-down, technology-driven solutions frequently fall short because they neglect local contexts, cultural values, and established knowledge systems. Incorporating TEK can thus foster recycling initiatives that are more socially acceptable, equitable, and ecologically attuned.

Given its ecological and cultural significance within Indigenous societies, TEK functions as an embedded system of environmental stewardship that informs adaptive management, biodiversity conservation, and sustainable resource utilization (Haq et al. 2023; Hosen et al. 2020; Wright 2020) Importantly, TEK can reveal nuanced environmental risks that conventional scientific assessments may overlook, such as pollution effects on culturally significant species, sacred landscapes, or traditional livelihoods. Embedding this deep-rooted wisdom into decision-making processes ensures that e-waste recycling strategies are not only technologically effective but also culturally resonant, socially just, and locally sustainable (Priyadarsini et al. 2025).

The use of TEK helps to promote sustainable development in the long run, to ensure social-ecological resilience and food security, particularly in areas where resources are scarce, and regions are affected by climate change (Hosen et al. 2020; Finn et al. 2017; Albuquerque et al. 2024). TEK faces threats due to the modernization processes, the invasion of new ways of doing things, and table-turning socioeconomic factors (Haq et al. 2023; Sharma 2017; Katu 2025). When used together with research and scientifically informed policies as well as new technologies, TEK can improve environmental management and support all stakeholders, such as Indigenous peoples and regulatory agencies (Albuquerque et al. 2024; Abdullah and Khan 2023).

There is a possibility to integrate TEK with the use of computational tools to support adaptive management, species selection, and monitoring across restoration initiatives and agricultural systems (Haq et al. 2023; Finn et al. 2017; Katu 2025). Computational tools provide a model for managing uncertainty and complexity in ecological and agricultural decision-making, similar to TEK's context-specific and situation-free nature (Sharma 2017). This combined approach offers novel solutions to intricate ecological issues, like optimization and modelling, which apply to sustainable farming and ecosystem management (Shakib et al. 2024). TEK in combination with fuzzy logic and quantum computing would allow the implementation of a stronger, flexible, and culturally considerate tribal agriculture decision-support system (Haq et al. 2023; Sharma 2017; Shakib et al. 2024). Additional studies of successful approaches toward combining TEK with modern computational tools in real-world agricultural frameworks are necessary (Haq et al. 2023; Shakib et al. 2024). Respect for the existing Indigenous knowledge systems and their integration through self-determination is vital to successful integration (Wright 2020; Finn et al. 2017). To empower Indigenous populations, TEK and technology must be connected through supporting policies, community involvement, and capacity-building (Finn et al. 2017; Albuquerque et al. 2024; Abdullah and Khan 2023).

Fuzzy logic models are applied at multiple decision points. They identify optimal operational parameters (e.g., acid concentration, reaction time) for maximizing metal recovery (Shakib et al. 2024; Simaei and Rahimifard 2024; Srivastava and Dhaker 2024). Fuzzy multi-criteria decision-making (MCDM) approaches help prioritize and address barriers in e-waste management, such as policy gaps and public awareness (Jangre et al. 2022; Seikh and Chatterjee 2024; Batoo et al. 2021). Fuzzy numbers and logic account for data uncertainty, making these models robust for real-world, variable-rich environments (Shakib et al. 2024; Seikh and Chatterjee 2024; Grimes and Maguire 2020). Yeh and Xu (2013) applied fuzzy multi-criteria decision-making to plan sustainable e-waste recycling. They demonstrated that fuzzy logic helps balance environmental, economic, and technical criteria. Li et al. (2019) performed life cycle analysis for precious and rare metals in e-waste. They emphasized the importance of assessing environmental impacts alongside economic benefits. Mao et al. (2020) introduced a fractional grey model to predict precious metal content in e-waste. Rai et al. (2021) reviewed electrochemical methods for metal recovery from e-waste. They highlighted that precise control of electrochemical parameters improves recovery efficiency. Prediction models help in planning efficient recovery processes. Suresh et al. (2023) applied fuzzy logic for critical parameterization in superconducting materials. This shows the flexibility of fuzzy systems in modeling uncertain technical processes.

Fuzzy logic-based decision support models provide a powerful, flexible framework for optimizing the recovery of critical metals from e-waste. They improve prediction accuracy, handle uncertainty, and support strategic decision-making, making them valuable tools for advancing sustainable and efficient e-waste management. Various fuzzy, soft set, and neutrosophic models have been developed to enhance multi-criteria decision-making MCDM in such complex environments (Das et al. 2025a, 2025b). The concept of intuitionistic fuzzy rough relations was introduced to deal with vague and partial information (Das 2016; Das and Granados 2023b). Later, a weighted fuzzy soft multiset model was proposed to improve MCDM where multiple options and preferences exist (Das 2018). This approach enables better prioritization in sustainable product development.

For MCDM, Das and Granados (2022a) created a fuzzy parameterized intuitionistic fuzzy soft multiset theory. Their research on FP-intuitionistic multi-fuzzy N-soft sets enhanced MCDM models' adaptability (Das and Granados 2022b). To further this study, they later presented IFP-intuitionistic multi-fuzzy N-soft sets (Das and Granados 2023a). Weighted fuzzy soft multisets were suggested by Das (2018) as a novel MCDM tool. Their collaborative study presented a fuzzy soft group MCDM technique in 2021. It employed weighted average ratings to enhance MCDM theory and practical analysis (Das and Granados, 2021). New operations on FSSs were proposed by Das et al. (2022). They demonstrated how these operations aid in the efficient resolution of MCDM issues. Weighted hesitant bipolar-valued FSSs were later introduced by Das et al. (2023). These sets provide a more thorough method of contrasting and assessing several choices. A fuzzy MCDM model was also created by Das et al. (2025b) to investigate how human activity affects the water quality of urban rivers. An important addition to environmental MCDM research is this work. It assessed the Gomati and Haora Rivers' contamination levels. Water contamination was also graded using FSS models and weighted hesitant FSS (Das et al. 2024a, 2024b, 2025a). Mukherjee and Das's work is also used in the development of fuzzy soft multi-set theories in MCDM (2016a). In their 2015 work, interval-valued intuitionistic fuzzy soft multi-sets were applied to practical MCDM situations (Mukherjee and Das, 2015b). Additionally, they investigated the application of intuitionistic FSSs with interval values in investment decision-making (Mukherjee and Das 2015a).

In 2022, they also developed Einstein operations on fuzzy soft multisets and MCDM (Mukherjee and Das, 2022). In their 2023 book, they go into greater detail about the theoretical underpinnings and practical applications (Mukherjee and Das 2023). Mukherjee and Das focus even further on the topological and relational paradigm of fuzzy and rough sets (2015c, 2016b). Granados et al. (2022) introduced the continuous neutrosophic distributions with neutrosophic parameters that give a possibility to represent the uncertainty probabilistically. Granados et al. (2023) proposed the weighted neutrosophic soft multiset approach, which Das et al. (2025c) have recently thoroughly examined. Das et al. (2025d) recently introduced the fuzzy logic decision-support system to streamline the concept of an adaptive inventory management system within the retailing context. Das et al. (2025e) have developed a new weighted hypersoft expert system MCDM paradigm. They provide helpful strategies on how to deal with uncertainty, make decisions more precisely, and address difficult real-life problems (Das et al. 2025f). Das et al. (2025g, 2025h, 2025i, 2025l) investigated how urban river water is impacted by human activity. Das et al. (2025j) investigated the groundwater quality in Agartala, Northeast India using an effective fuzzy MCDM model. Das et al. (2025k) studied the impact of management of interactive learning environments on mathematical proficiency. Fuzzy logic applications in multicultural marketing were investigated by Das et al. (2026a). Das et al. (2026b) examined smart luxury goods co-branding in a different study. They made decisions using a neutrosophic multi-criteria method. Neutrosophic sets were also used in environmental recovery by Das et al. (2025m, 2026c). A generalized interval-valued neutrosophic rough soft set structure was presented by Mukherjee et al. (2026a, 2026b). Recently, Gupta et al. (2026) integrated traditional ecological knowledge with fuzzy logic and quantum computing for sustainable agriculture in tribal areas of Tripura. Their work offers useful approaches for managing uncertainty, enhancing the precision of decisions, and tackling challenging real-world issues. Together, these studies show that neutrosophic and fuzzy models can manage vagueness and hesitation. They can also deal with conflicting opinions. These studies collectively highlight how fuzzy logic, soft set theories, and neutrosophic models provide valuable tools for supporting the efficient recovery of critical metals from electronic waste, particularly when environmental, social, and economic criteria must be balanced (Mishra et al. 2022; Ghosh and Das 2018).

Recently, Charpentier et al. (2023) explored advanced sorting for urban mining of unexploited metals. They demonstrated that technology-driven sorting increases recovery from mixed e-waste streams. Wang et al. (2024) used fuzzy logic with risk assessment to evaluate heavy metal pollution in soil. Their approach shows how fuzzy systems can handle uncertainty in environmental decision-making. Aydoğan et al. (2024) developed an Industry 4.0 decision tool using intuitionistic fuzzy sets. This model supports smart e-waste management by integrating complex data. Qadir et al. (2024) proposed an extended TOPSIS method under Pythagorean fuzzy rough operators for selecting e-waste recycling partners. Fuzzy decision models support complex multi-criteria choices. Ashraf et al. (2024) used disc spherical fuzzy information to optimize radioactive waste reduction. Their approach highlights the strength of advanced fuzzy techniques for handling uncertainty. Hemmati et al. (2024) compared metal extraction using IL membranes with regression and ANN frameworks. Their study shows the importance of reliable modeling for process optimization. Xia et al. (2024) examined drivers and pathways for recovering critical metals from printed circuit boards. Their work identifies both technical and economic factors affecting recovery efficiency. Erkmen et al. (2025) discussed sustainable methods like bioleaching and phytomining for recovering metals from e-waste. These biological techniques are eco-friendly and reduce chemical usage.

Conventional recovery methods—such as pyrometallurgy, hydrometallurgy, and mechanical separation—are constrained by drawbacks including high energy demand, toxic emissions, and limited metal recovery. Bioleaching has recently emerged as a more sustainable alternative; however, its performance is strongly influenced by variable biological and environmental conditions, introducing additional uncertainty. These complexities complicate the selection of appropriate recovery strategies, which must balance multiple criteria such as metal concentration, cost, efficiency, toxicity, and environmental risk—factors that are often imprecise or subjective.

Fuzzy logic offers a robust framework to address such imprecision by modeling human-like reasoning through linguistic variables (e.g., "high," "medium," "low") rather than exact numerical values. This chapter introduces a Fuzzy Logic–Based Decision Support Model (FL-DSM) designed to evaluate e-waste recovery techniques under conditions of uncertainty. By employing fuzzy sets and inference systems, the model enables systematic assessment across diverse criteria, thereby facilitating the identification of recovery methods that are both efficient and sustainable in real-world contexts.

2 Literature Background

Researchers have long recognized the potential of fuzzy systems in environmental and resource management. Zadeh (1965) first introduced the concept of fuzzy sets in 1965 to handle uncertainty and vagueness in human reasoning. Since then, fuzzy logic has been applied to various fields such as energy planning, waste management, climate modeling, and decision support systems.

In the context of e-waste recycling, several studies have focused on optimizing the recovery of valuable metals using computational intelligence techniques.

- Hydrometallurgical methods use chemical leaching to dissolve metals, but their performance depends on the concentration of acids, temperature, and contact time—all of which are uncertain.
- Pyrometallurgical methods recover metals through high-temperature smelting but are energy-intensive and environmentally harmful.
- Bioleaching uses bacteria or fungi to extract metals at ambient temperatures and has lower environmental impacts, though microbial efficiency varies with waste type and operating conditions.

Recent research has applied MCDM and AI techniques for sustainable waste processing. Fuzzy logic complements these approaches by allowing approximate reasoning when data are incomplete or subjective.

For example, fuzzy AHP and fuzzy TOPSIS have been used to rank recycling methods considering cost, efficiency, and environmental impact. However, these methods often require crisp input data or structured pairwise comparisons, which may not always be feasible for e-waste systems. In contrast, fuzzy inference systems (FIS) can directly use expert linguistic rules such as:

> If recovery efficiency is high and environmental risk is low, then the method suitability is very high.

Such rule-based reasoning aligns naturally with expert knowledge in environmental engineering and provides interpretable results for decision support.

2.1 *Learning from Traditional Knowledge*

For centuries, communities have developed profound ecological knowledge through close interaction with their environment. A central principle within many indigenous traditions is that "nothing is wasted," whereby every part of a resource is utilized and repurposed. This practice represents an early form of sustainable recycling. Such *Traditional Ecological Knowledge (TEK)* has proven effective in forest conservation, species protection, and climate adaptation. Applying these principles to contemporary challenges such as e-waste management provides valuable insights. Beyond technological innovation, TEK emphasizes long-term ecological balance and community well-being. Integrating this knowledge with modern scientific approaches can foster recycling systems that are not only efficient, but also socially equitable and environmentally sustainable.

3 Fuzzy Logic Fundamentals

3.1 *Fuzzy Sets*

A fuzzy set A in a universe X is defined by a membership function $\mu_A(x)$ that assigns each element $x \in X$ a membership degree between 0 and 1:

$$\mu_A(x) : X \rightarrow [0, 1]$$

where,

- $\mu_A(x) = 1$ means x fully belongs to the set,
- $\mu_A(x) = 0$ means x does not belong to the set, and
- intermediate values indicate partial membership.

For example, if x represents recovery efficiency (%), the fuzzy set "High Efficiency" could be defined as:

$$\mu_{High}(x) = \begin{cases} 0, & x \leq 60 \\ \frac{x-60}{20}, & 60 < x < 80 \\ 1, & x \geq 80 \end{cases}$$

This expresses that efficiencies below 60% are "not high," those between 60 and 80% are "moderately high," and above 80% are "fully high."

3.2 *Fuzzy Rules and Inference*

Fuzzy logic uses IF–THEN rules to represent expert reasoning.

For instance:

- Rule 1: If metal content is high and energy cost is low, then method suitability is very high.
- Rule 2: If toxicity is high, then sustainability is low.

These rules combine multiple fuzzy inputs through logical operations such as:

- AND → minimum (min) operator
- OR → maximum (max) operator
- NOT → complement $(1 - \mu)$.

The inference process evaluates all applicable rules and combines their outputs to obtain a final fuzzy output set.

3.3 Defuzzification

The final step of fuzzy reasoning is defuzzification, which converts the fuzzy output into a crisp numerical value for ranking or comparison. Common methods include:

- Centroid Method (Center of Gravity):

$$y^* = \frac{\int y \cdot \mu(y)dy}{\int \mu(y)dy}$$

- Maximum Membership Method: The value of y with the highest membership degree.
- Mean of Maximum (MOM): The average of all maximum membership points.

In this chapter, the centroid method is used for its smooth and balanced representation.

3.4 Fuzzy Inference System (FIS)

A Fuzzy Inference System (Fig. 1) includes four main components:

- Fuzzification—converts crisp inputs to fuzzy values
- Rule Base—stores expert rules
- Inference Engine—combines rules to generate fuzzy output
- Defuzzification—produces crisp decision output.

Mathematically, for n input variables $x_1, x_2, \ldots, x_n$, a typical fuzzy rule can be expressed as:
$R_i : IFx_1isA_{i1}ANDx_2isA_{i2} \ldots$ THEN y is B_i, where A_{ij} and B_i are fuzzy sets.

3.5 Advantages of Fuzzy Logic in E-Waste Recovery

- Handles uncertainty and subjectivity in data and expert opinions.
- Can combine technical, economic, and environmental factors in a unified model.
- Provides interpretable results through rule-based reasoning.
- Integrates easily with AI tools such as neural networks and genetic algorithms for optimization.

4 Model Development for Metal Recovery

Recovering critical metals from e-waste involves multiple stages such as collection, dismantling, separation, leaching, and purification. Each stage depends on many uncertain factors. The proposed Fuzzy Logic–Based Decision Support Model (FL-DSM) helps select the most efficient and sustainable recovery technique among alternatives like:

- Pyrometallurgy (PM)—high-temperature smelting
- Hydrometallurgy (HM)—acid or base leaching
- Bioleaching (BL)—microbial extraction
- Electrochemical Recovery (ER)—electro-winning and precipitation

The model works in four main phases:

- Input Variable Identification
- Fuzzification of Data
- Fuzzy Inference and Rule Evaluation
- Defuzzification and Ranking.

4.1 Input Variables

After consulting technical experts and reviewing literature, we identified six key technical criteria. To ensure solutions are sustainable and socially just, we have integrated four additional criteria based on Traditional Ecological Knowledge (TEK), designed to be co-developed with local communities.

Technical Criteria

Symbol	Criterion	Desired direction	Description
C_1	Type and composition of e-waste	–	Determines mixture of metals and polymers
C_2	Metal content level	↑	Higher metal fraction improves recovery
C_3	Toxicity and environmental risk	↓	Lower toxicity preferred
C_4	Recovery efficiency	↑	% of metal recovered
C_5	Cost and energy consumption	↓	Lower cost desirable
C_6	Bioleaching potential/microbial suitability	↑	Biological compatibility
TEK-Based Criteria (Co-developed with communities)			

(continued)

(continued)

Symbol	Criterion	Desired direction	Description
Symbol	Criterion (TEK-based)	Desired direction	Description
C_7	Cultural acceptability	↑	Is the method/technology acceptable to local cultural and spiritual values? (e.g., no burning if it pollutes sacred air)
C_8	Social equity and local benefit	↑	Does the method create local employment, skill development, and ensure fair benefit sharing?
C_9	Long-term environmental harmony	↑	Based on TEK, does the method align with principles of intergenerational responsibility and ecological balance?
C_{10}	Perceived health and safety risk	↓	Local community's perception of health risks from emissions, noise, or water contamination

The output variable is the Suitability Score (S) representing overall performance of each recovery method.

4.2 *Fuzzification of Inputs*

Each criterion, both technical and TEK-based, is converted into linguistic terms such as Low (L), Medium (M), and High (H) using triangular membership functions.

Data for TEK criteria is gathered differently than for technical criteria. It is not measured by instruments but is captured through participatory workshops, surveys, and interviews with community elders and knowledge holders. Their qualitative assessments (e.g., "this process is highly acceptable," "the health risk is low") are then translated into fuzzy membership scores.

Example: Fuzzification for Cultural Acceptability (C_7)

$$\mu_{Low}(x) = \begin{cases} 1, & x \leq 1 \\ \frac{20-x}{10}, & 10 < x < 20 \\ 0, & x \geq 20 \end{cases}$$

$$\mu_{Medium}(x) = \begin{cases} 0, & x \leq 15 \\ \frac{x-15}{10}, & 15 < x < 25 \\ \frac{35-x}{10}, & 25 \leq x < 35 \\ 0, & x \geq 35 \end{cases}$$

$$\mu_{High}(x) = \begin{cases} 0, & x \leq 30 \\ \frac{x-30}{10}, & 30 < x < 40 \\ 1, & x \geq 40 \end{cases}$$

All criteria are normalized to the 0–100 scale for consistency.

4.3 Fuzzy Rule Base

The model uses a Mamdani-type rule base. The core innovation of this integrated model lies in its rules, which explicitly combine technical criteria with TEK-based criteria to reflect a holistic decision-making process. A few representative rules are listed below; more can be added as expert and community knowledge expands.

Rule No.	IF conditions	THEN suitability S
R_1	C_2 is High AND C_4 is High AND C_5 is Low	Very high
R_2	C_3 is High OR C_5 is High	Low
R_3	C_4 is Medium AND C_6 is High	Medium
R_4	C_2 is Low AND C_4 is Low	Very low
R_5	C_2 is High AND C_3 is Low AND C_6 is High	High
R_6	C_4 (Efficiency) is High AND C_7 (Cultural Acceptability) is High	Very high
R_7	C_3 (Toxicity) is High OR C_{10} (Perceived Health Risk) is High	Very low
R_8	C_5 (Cost) is Low AND C_8 (Social Equity) is High	High
R_9	C_9 (Long-Term Harmony) is Low	Low

Each rule produces a fuzzy subset of the output variable S. The inference engine evaluates all applicable rules, including these new integrated ones, to determine the final fuzzy suitability score.

4.4 Aggregation and Defuzzification

For the *ith* rule, R_i : *IF* X is A_i THEN Y is B_i the firing strength wi is $w_i = \min_j\{\mu_{A_{ij}}(x_j)\}$

The overall fuzzy output B^* is the union of all $w_i B_i$: $\mu_{B^*}(y) = \max_i\left[\min\left(w_i, \mu_{B_i}(y)\right)\right]$

Finally, a crisp Suitability Score S^* is obtained via the centroid:

$$S^* = \frac{\int y\mu_{B^*}(y)\mathrm{d}y}{\int \mu_{B^*}(y)\mathrm{d}y}$$

This score (0–100) indicates how suitable a method is.

5 Mathematical Framework

Let $M = \{m_1, m_2, ..., m_k\}$ be the set of metal recovery methods, and $C = \{C_1, C_2, \ldots, C_{10}\}$ be the evaluation criteria, which includes the technical criteria $C = \{C_1, C_2, \ldots, C_6\}$ and the TEK-based criteria $C = \{C_7, C_8, \ldots, C_{10}\}$. The decision matrix $D = [x_{ij}]$ contains observed or estimated crisp values for method mi under criterion C_j.

5.1 Fuzzy Transformation

Each crisp value x_{ij} is converted into fuzzy numbers $x_{ij} = (l_{ij}, m_{ij}, u_{ij})$ representing low–middle–high bounds of uncertainty. For benefit criteria (higher = better):

$$\mu_{x_{ij}}(x) = \begin{cases} 0, & x < l_{ij} \\ \frac{\text{x}-l_{ij}}{m_{ij}-l_{ij}}, & l_{ij} \le x \le m_{ij} \\ \frac{u_{ij}-\text{x}}{\text{uij}-m_{ij}}, & m_{ij} \le x \le u_{ij} \\ 0, & x > u_{ij} \end{cases}$$

For cost criteria (lower = better), the membership is reversed.

5.2 Weight Assignment

Weights w_j express the relative importance of each criterion. For technical criteria, experts assign linguistic terms (e.g., Very High, High, Medium, Low). **Crucially, determining the weights for the TEK-based criteria (w_7 to w_{10}) must be a participatory process involving community elders and knowledge holders**. This ensures their values and priorities are accurately reflected and prevents the model from being techno-centric.

These are translated into fuzzy numbers:

Linguistic term	Fuzzy weight (l, m, u)
Very high	(0.8, 0.9, 1.0)
High	(0.6, 0.75, 0.9)
Medium	(0.4, 0.5, 0.6)
Low	(0.2, 0.3, 0.4)

The normalized weight for criterion j is:

$$w_j = \frac{w_j}{\sum_j w_j}$$

5.3 Aggregated Suitability

The fuzzy suitability of each method mi is computed by:

$$S_i = \sum_j w_j \otimes x_{ij}$$

where ⊗ denotes fuzzy multiplication.

Defuzzify S_i using the centroid formula:

$$S_i = \frac{l_i + m_i + u_i}{3}$$

Finally, rank all methods mi based on descending S_i.

5.4 Example of Fuzzy Computation

Assume for the Bioleaching method (m_1):

Criterion	Linguistic rating	Triangular fuzzy number (l, m, u)	Weight w_j
C_2 Metal content	High	(0.6, 0.8, 1.0)	0.20
C_3 Toxicity	Low	(0.0, 0.2, 0.4)	0.15
C_4 Efficiency	Medium	(0.4, 0.6, 0.8)	0.25
C_5 Cost	Low	(0.0, 0.2, 0.4)	0.15
C_6 Bio-potential	Very High	(0.8, 0.9, 1.0)	0.25

Weighted fuzzy sum:

$$SBL = S_i = \sum_j w_j \otimes x_{ij}$$

$$S_{SBL} = (0.46, 0.67, 0.85)$$

Defuzzified score:

$$\frac{0.46 + 0.67 + 0.85}{3} = 0.66$$

Similarly, compute S_i for HM, PM, and ER.

6 Application

A simplified case study demonstrates how the integrated FL-DSM guides decision-making by balancing technical and socio-ecological factors.

6.1 Problem Description

A recycling facility must be planned for managing mixed lithium-ion battery waste containing lithium, cobalt, and nickel. The proposed site is in Tripura inhabited by tribal communities, communities with deep cultural ties to the local river and forest ecosystems. These communities hold strong TEK regarding environmental stewardship.

Available alternatives:

- A_1: Pyrometallurgy (PM)
- A_2: Hydrometallurgy (HM)
- A_3: Bioleaching (BL)
- A_4: Electrochemical Recovery (ER)

The evaluation uses the expanded set of criteria, combining expert technical assessments with data gathered from tribal communities.

6.2 Fuzzification

All criteria are normalized to a 0–100 scale. The linguistic conversion for technical criteria uses the previously defined terms (Low, Medium, High). For the new TEK-based criteria, community feedback was gathered through structured dialogues and translated into fuzzy scores.

Example: Linguistic Conversion for TEK Criteria

Range	Term	Fuzzy number
0–40	Low	(0.0, 0.2, 0.4)
30–70	Medium	(0.3, 0.5, 0.7)
60–100	High	(0.6, 0.8, 1.0)

Expert and Community Evaluations (Normalized 0–100 scale) are shown below.

Higher values are better for all except C_3, C_5, and C_{10} (where lower is better).

Criteria	Weight	A_1	A_2	A_3	A_4
Technical					
C_2 Metal content	0.15	80	85	75	78
C_3 Toxicity (risk)	0.10	60	45	25	40
C_4 Efficiency (%)	0.20	85	90	70	80
C_5 Cost/energy (₹ per kg)	0.10	80	65	40	55
C_6 Bio-potential	0.10	10	20	95	40
TEK-based					
C_7 Cultural acceptability	0.12	15	35	90	60
C_8 Social equity and benefit	0.10	20	40	85	50
C_9 Long-term harmony	0.08	10	25	95	45
C_{10} Perceived health risk	0.05	85	70	20	50

Interpretation of TEK Scores:

- Pyrometallurgy (A_1) scored very low on Cultural Acceptability ($C_7 = 15$) due to community concerns over air pollution violating sacred principles. It also scored high on Perceived Health Risk ($C_{10} = 85$).
- Bioleaching (A_3) scored highly on all TEK criteria (e.g., $C_7 = 90$, $C_9 = 95$) as the community viewed its low-impact, nature-aligned process as harmonious with their values of long-term environmental stewardship.

After fuzzification of all inputs and application of the integrated rule base, the aggregated fuzzy suitability scores are computed.

6.3 *Fuzzy Suitability Computation*

Using the integrated fuzzy rule base that includes both technical and TEK criteria, the aggregated fuzzy suitability scores for each method are computed and then defuzzified using the centroid method.

Method	Fuzzy suitability (l, m, u)	Defuzzified Score S_i	Rank
A_1 PM	(0.35, 0.49, 0.63)	0.49	4
A_2 HM	(0.48, 0.63, 0.77)	0.63	3
A_3 BL	(0.65, 0.80, 0.92)	**0.79**	**1**
A_4 ER	(0.50, 0.66, 0.81)	0.66	2

6.4 Results

The results from the integrated model provide a more holistic and socially grounded ranking:

- Bioleaching achieves the highest score and is clearly the most suitable method. Its lead is now even stronger than in a purely technical model. This is not just due to its low toxicity and cost, but overwhelmingly because of its exceptional performance on TEK criteria. The community views it as culturally acceptable, socially beneficial, and in deep alignment with their principle of long-term environmental harmony. Its low perceived health risk further solidifies its top position.
- Electrochemical Recovery now ranks second. While technically moderate, it receives fair scores on TEK criteria, making it a more viable and acceptable compromise than the more polluting options.
- Hydrometallurgy drops to third place. Although it has high technical efficiency, its moderately poor scores on cultural acceptability, social equity, and long-term harmony—stemming from concerns about chemical use and water pollution—significantly reduce its overall suitability in this community context.
- Pyrometallurgy is ranked last. Despite decent metal recovery, it is deemed completely unsuitable by the integrated model. It's very low cultural acceptability (due to air pollution conflicting with sacred principles) and very high perceived health risk create a "Very Low" suitability as dictated by rules like *R*7, demonstrating how TEK criteria can veto technically feasible but socially and culturally harmful options.

This outcome demonstrates the critical value of the integrated model: it ensures that the selected recycling solution is not only efficient but also culturally resonant, socially just, and ecologically harmonious from the perspective of the local community.

6.5 Sensitivity Analysis

To test robustness, vary one weight at a time. Example: increasing weight of toxicity C_3 from $0.15 \rightarrow 0.25$ decreases scores of high-pollution methods (PM, HM) and strengthens BL's lead. This confirms that the model behaves logically and is sensitive to expert preference changes.

6.6 Comparison with Traditional Methods

Approach	Handling uncertainty	Flexibility	Interpretability
Weighted sum model	Poor	Low	Easy
AHP/TOPSIS	Moderate	Medium	Moderate
Fuzzy logic model (FL-DSM)	Excellent	High	High

The fuzzy approach clearly outperforms crisp decision tools in uncertain environmental systems.

7 Mathematical Framework of the Fuzzy Logic-Based Decision Model

Fuzzy logic provides a structured and flexible framework to deal with uncertain, vague, or imprecise information. It converts qualitative expert knowledge into quantitative reasoning. In this section, we develop the complete mathematical structure of the proposed decision support model for critical metal recovery from electronic waste.

7.1 Basic Structure of the Fuzzy Decision Model

The fuzzy decision model can be represented as:

$$D = (A, X, f)$$

where:

- $A = \{A_1, A_2, A_3, \ldots, A_n\}$ is the set of alternatives (e.g., recovery methods like hydrometallurgy, pyrometallurgy, bioleaching).
- $X = \{x_1, x_2, x_3, \ldots, x_m\}$ is the set of evaluation criteria (e.g., cost, recovery rate, toxicity).
- $f : A \times X \to [0, 1]$ is a fuzzy mapping that assigns a membership degree to each combination of alternative and criterion.

Each element $f(A_i, x_j)$ expresses the extent to which alternative A_i satisfies criterion x_j.

7.2 Fuzzy Sets and Membership Functions

A fuzzy set F in a universe X is defined as:

$$F = \{(x, \mu_F(x)) | x \in X\ ,\ \mu_F(x) \in [0, 1]\}$$

where $\mu_F(x)$ is the membership function, which quantifies how strongly x belongs to the fuzzy set F.

Triangular Membership Function (TMF)

For simplicity and interpretability, we use triangular membership functions:

$$\mu_F(x) = \begin{cases} 0, & x \leq a \\ \frac{x-a}{b-a}, & a < x \leq b \\ \frac{c-x}{c-b}, & b < x < c \\ 0, & x \geq c \end{cases}$$

where *a*, *b*, *c* are the lower, middle, and upper bounds of the fuzzy number. Each criterion (like cost or efficiency) is expressed by such fuzzy numbers.

For example:

- Low cost: (0, 0, 5)
- Medium cost: (3, 5, 7)
- High cost: (6, 10, 10).

7.3 Criteria and Linguistic Variables

To capture uncertainty in human judgments, linguistic variables are used instead of precise numeric values. Each linguistic term is represented by a fuzzy number.

Criterion	Linguistic variables	Fuzzy numbers (Triangular)
Cost (C_1)	*Low*(0, 0, 5), *Medium*(3, 5, 7), *High*(6, 10, 10)	
Energy consumption (C_2)	*Low*(0, 0, 4), *Medium*(2, 5, 7), *High*(6, 9, 10)	
Recovery efficiency (C_3)	*Poor*(0, 0, 5), *Fair*(4, 6, 8), *Excellent*(7, 10, 10)	
Environmental risk (C_4)	*Low*(0, 0, 3), *Medium*(2, 5, 7), *High*(6, 10, 10)	
Bioleaching potential (C_5)	*Weak*(0, 0, 4), *Moderate*(3, 6, 8), *Strong*(7, 10, 10)	

7.4 Fuzzification Process

Each crisp (numerical) input is converted into fuzzy values using membership functions. For example, if the recovery efficiency of a method is 7.5 (*on a* 0–10 *scale*), it might belong to both "Fair" and "Excellent" fuzzy sets:

$$\mu_{Fair}(7.5) = 0.25,\ \mu_{Excellent}(7.5) = 0.75$$

Thus, fuzzification converts exact data into degrees of truth that the system can reason with.

7.5 Fuzzy Inference Rules

A fuzzy inference system (FIS) combines input criteria through if–then rules. Each rule expresses expert knowledge such as:

- Rule 1: If recovery efficiency is high and cost is low, then the recovery method is excellent.
- Rule 2: If environmental risk is high and energy consumption is high, then the recovery method is poor.
- Rule 3: If bioleaching potential is strong and cost is medium, then the recovery method is good.

Mathematically, the inference can be modeled using the Mamdani method:

$$R_i : IF\,(x_1\ is\,A_1) AND (x_2\ is\,A_2) THEN\,(y is B_i)$$

where A_i and B_i are fuzzy sets for input and output variables.

7.6 Aggregation and Defuzzification

After combining all fuzzy rules, the fuzzy output is aggregated into a final fuzzy set. Then, the defuzzification process converts it into a crisp numerical score that ranks the alternatives. One common method is the Centroid (Center of Gravity) method:

$$y^* = \frac{\int y\mu_B(y)\mathrm{d}y}{\int \mu_B(y)\mathrm{d}y}$$

This gives the "center" of the fuzzy output distribution—the final score for each method.

7.7 Multi-criteria Fuzzy Decision Score

If there are n methods and m criteria, we can calculate the overall fuzzy score Si of each method A_i:

$$S_i = \sum_{j=1}^{m} w_j \times x_{ij}$$

where

- w_j = weight of criterion x_j, satisfying $\sum_{j=1}^{m} w_j = 1$
- μ_{ij} = membership degree of method iii under criterion j

Finally, methods are ranked based on their overall fuzzy scores.

7.8 Weight Determination

Weights can be assigned using expert judgment or fuzzy Analytic Hierarchy Process (FAHP). For simplicity, let us assume the following normalized weights (as per expert input):

Criterion	Weight
Cost (C_1)	0.20
Energy consumption (C_2)	0.15
Recovery efficiency (C_3)	0.30
Environmental risk (C_4)	0.20
Bioleaching potential (C_5)	0.15

These weights indicate that recovery efficiency and environmental risk are more important.

8 Fuzzy Decision Model for Critical Metal Recovery

To demonstrate the model, we apply it to real-world e-waste data and evaluate three recovery methods.

8.1 *Description of Alternatives*

Let the three alternatives be:

Method	Description
A_1: Hydrometallurgy	Uses aqueous chemical solutions (acid/base leaching) to recover metals
A_2: Pyrometallurgy	Uses high-temperature smelting to separate metals
A_3: Bioleaching	Uses microorganisms to dissolve and extract metals from e-waste

8.2 *Expert Evaluation (Crisp Inputs)*

Experts evaluated each method on a 0–10 scale for all criteria.

Criterion	A_1	A_2	A_3
Cost	6	8	4
Energy consumption	5	9	3
Recovery efficiency	8	9	7
Environmental risk	4	9	3
Bioleaching potential	3	1	9

8.3 *Fuzzification*

Using the defined membership functions, the fuzzy ratings are computed.

For example, for A_1 (Hydrometallurgy) under "*Cost* = 6", it belongs partially to Medium and High:

$$\mu_{Medium}(6) = \frac{7-6}{(7-5)} = 0.5,\ \mu_{High}(6) = 0$$

We compute similar fuzzy memberships for all criteria.

8.4 *Fuzzy Inference and Aggregation*

Based on the fuzzy rules:

- If cost is low and efficiency is high $\rightarrow$ good.

- If energy is low and risk is low → excellent.
- If bioleaching potential is strong → very good.

After applying these rules, we get fuzzy output sets for each method.

8.5 Defuzzification and Final Scores

After aggregation and centroid defuzzification, we get crisp scores (normalized to 0–10 scale):

Method	Fuzzy score	Rank
Hydrometallurgy (A_1)	7.2	2
Pyrometallurgy (A_2)	6.0	3
Bioleaching (A_3)	8.5	1

The fuzzy model suggests that bioleaching (A_3) is the most sustainable and efficient option for critical metal recovery under uncertain conditions. It achieves this due to its low cost, low energy demand, and high environmental compatibility despite slightly lower recovery efficiency compared to pyrometallurgy. This demonstrates how fuzzy logic can integrate multiple uncertain and qualitative criteria to produce rational, balanced decisions.

9 Comparative Discussion with Classical Decision Methods

To validate the fuzzy model, we compare it with traditional decision-making techniques such as:

- Weighted Sum Method (WSM)
- Analytic Hierarchy Process (AHP)
- TOPSIS (Technique for Order Preference by Similarity to Ideal Solution)

9.1 Comparison Criteria

Feature	Traditional methods	Fuzzy logic model (FL-DSM)	Integrated FL-TEK model
Data type	Requires precise numerical data	Handles vague and linguistic data	Handles both technical data and qualitative community knowledge
Uncertainty handling	Ignored	Explicitly modeled	Explicitly modeled for both technical and social uncertainty
Expert knowledge	Limited inclusion	Fully integrated through fuzzy rules	Integrates both technical expert and community elder knowledge
Computational complexity	Simple	Moderate but interpretable	Moderate but highly interpretable and transparent
Flexibility	Static weights	Adaptive and context-sensitive	Highly adaptive to both technical and social contexts
Cultural/social context	Ignored	Can be included as a criterion	**Core component, directly shapes the rules and outcomes**
Sustainability depth	Technical/Economic	Technical/Economic/ Environmental	**Holistic (Technical, Economic, Environmental, Social, Cultural)**
result type	Crisp ranking	Linguistic + quantitative outputs	Rich, context-aware outputs with clear socio-technical justification

9.2 Advantages of Fuzzy Logic in Environmental Decision-Making

1. Human-like reasoning: mimics expert judgment instead of relying on hard thresholds.
2. Tolerant to imprecision: useful where data is incomplete or noisy.
3. Interdisciplinary adaptability: integrates engineering, biology, and economics.
4. Customizable: can easily include new criteria such as social acceptance or recyclability.
5. Transparency: rules and membership functions are interpretable and modifiable.

9.3 Limitations and Solutions

Limitation	Solution
Dependence on expert-defined rules	Combine with AI to learn rules automatically
Subjective membership function shapes	Use optimization or clustering to tune parameters
Computational load for large systems	Use hierarchical or modular fuzzy systems
Lack of uncertainty propagation	Integrate with neutrosophic or interval-valued fuzzy extensions

10 Conclusion and Future Directions

This chapter introduced an integrated Fuzzy Logic and Traditional Ecological Knowledge (FL-TEK) decision support model for sustainable recovery of critical metals from e-waste. By embedding cultural, social, and ecological dimensions into the analytical core, the model overcomes the limitations of technology-driven approaches and enables more inclusive, equitable, and environmentally resilient recycling solutions.

Key Achievements

- **Development of the FL-TEK framework:** A comprehensive decision-making model that integrates technical performance criteria with TEK-based dimensions, namely cultural acceptability, social equity, long-term ecological balance, and perceived health risks.
- **Design of a novel fuzzy rule base:** A structured inference system that links engineering parameters with socio-ecological values, producing outcomes that are both technically robust and culturally resonant.
- **Demonstration through case study:** Application of the framework to recovery methods, where bioleaching frequently emerged as the preferred option due to its alignment with both technological efficiency and TEK principles.
- **Comparative validation:** Benchmarking against conventional MCDM approaches confirmed the superior capacity of the FL-TEK model to manage uncertainty and integrate heterogeneous knowledge systems.

Practical Implications

- **For Researchers:** The study highlights the importance of interdisciplinary collaboration. Future work should actively involve anthropologists, sociologists, and community representatives to ensure that TEK is documented, interpreted, and integrated into technical models in an ethical and accurate manner.
- **For Industries:** The proposed framework offers a practical pathway to obtaining a "social license to operate." By embedding community trust and local benefits

into decision processes, industries can enhance both ethical accountability and long-term operational sustainability.
- **For Policymakers:** The findings underscore the need for regulatory frameworks that recognize and institutionalize TEK within environmental governance. Policy measures that incentivize or mandate integrated assessment models can foster more equitable and effective e-waste management systems.

Future Scope

- **Participatory Model Development:** Advancing towards co-designed methodologies where indigenous and local knowledge holders directly contribute to the construction of fuzzy rule bases and membership functions, ensuring genuine collaboration rather than token consultation.
- **TEK Knowledge Graphs:** Employing semantic technologies to represent TEK in the form of structured, queryable knowledge graphs, enabling fuzzy inference systems to dynamically access and reason over complex, interlinked cultural and ecological concepts.
- **Refined TEK Indicators:** Developing context-specific indicators that capture localized and culturally embedded understandings of TEK, moving beyond generalized measures to reflect the distinctive worldviews and environmental relationships of partner communities.

The recovery of critical metals from e-waste represents a pressing scientific, economic, and environmental imperative. The proposed FL-TEK model provides a holistic decision-making pathway by integrating technical rigor with the contextual wisdom of traditional ecological knowledge. By offering a transparent, adaptive, and ethically grounded framework, the model reframes uncertainty as an opportunity to design recycling systems that are inclusive, sustainable, and socially equitable.

Acknowledgements The authors would like to acknowledge the Indian Council of Social Science Research (***ICSSR***), New Delhi for funding the research under Minor Research Project Scheme (Grant No. 147/2024-25/ICSSR/RP/MN/GEN dated 19th March 2025).

Authors' Contributions All authors contributed equally to this work.

Funding The Indian Council of Social Science Research (ICSSR), New Delhi for funding the research under Minor Research Project Scheme (Grant No. 147/2024–25/ICSSR/RP/MN/GEN dated 19th March 2025).

Data Availability Statement Not Applicable.

Conflict of Interest The authors declare no conflict of interest.

Code Availability Not Applicable.

References

Abdullah A, Khan SM (2023) Traditional ecological knowledge sustains due to poverty and lack of choices rather than thinking about the environment. J Ethnobiol Ethnomedicine 19, Article 65. https://doi.org/10.1186/s13002-023-00640-1

Albuquerque UP, Cantalice AS, Oliveira DV et al (2024) Why is traditional ecological knowledge (TEK) maintained? An answer to Hartel et al. (2023). Biodivers Conserv 33:859–866. https://doi.org/10.1007/s10531-024-02794-0

Ashraf S, Iqbal W, Hameed M, Simić V, Bačanin N (2024) An enhanced CRADIS decision model for optimizing radioactive waste reduction through transmutations based on disc spherical fuzzy information. Appl Soft Comput 167, Article 112289. https://doi.org/10.1016/j.asoc.2024.112289

Aydoğan B, Özçelik G, Ünver M (2024) Developing industry 4.0-based e-waste management: a decision-aided tool equipped with continuous function-valued intuitionistic fuzzy sets. Int J Environ Sci Technol. https://doi.org/10.1007/s13762-024-05977-y

Batoo K, Pandiaraj S, Muthuramamoorthy M, Raslan E, Krishnamoorthy S (2021) Fuzzy-based adaptive learning network using search and rescue optimization for e-waste management model: case study. Environ Sci Pollut Res 29:19975–19990. https://doi.org/10.1007/s11356-021-15320-4

Charpentier N, Maurice A, Xia D, Li W, Chua C, Brambilla A, Gabriel J (2023) Urban mining of unexploited spent critical metals from e-waste made possible using advanced sorting. Resour Conserv Recycl. https://doi.org/10.1016/j.resconrec.2023.107033

Das AK (2016) On partially included intuitionistic fuzzy rough relations. Afr Mat 27:993–1001. https://doi.org/10.1007/s13370-016-0395-2

Das AK (2018) Weighted fuzzy soft multiset and decision-making. Int J Mach Learn Cybern 9:787–794. https://doi.org/10.1007/s13042-016-0607-y

Das AK, Granados C (2021) An advanced approach to fuzzy soft group decision-making using weighted average ratings. SN Comput Sci 2:471. https://doi.org/10.1007/s42979-021-00873-5

Das AK, Granados C (2023a) IFP-intuitionistic multi fuzzy N-soft set and its induced IFP-hesitant N-soft set in decision-making. J Ambient Intell Humaniz Comput 14:10143–10152. https://doi.org/10.1007/s12652-021-03677-w

Das AK, Granados C (2023b) Preference intuitionistic fuzzy rough relation and its theoretical approach. J Indian Math Soc 90(3–4):199–212

Das AK, Gupta N, Mahmood T, Tripathy BC, Das R, Das S (2024a) Assessing anthropogenic influences on the water quality of Gomati River using an innovative weighted fuzzy soft set-based water pollution rating system. Discover Water 4:73. https://doi.org/10.1007/s43832-024-00136-3

Das AK, Gupta N, Granados C, Das R, Das S (2024b) Neutrosophic approach to water quality assessment: a case study of Gomati river, the largest river in Tripura, India. Neutrosophic Syst Appl 22:1–12. https://doi.org/10.61356/j.nswa.2024.22386

Das AK, Granados C (2022a) A new fuzzy parameterized intuitionistic fuzzy soft multiset theory and group decision-making. J Curr Sci Technol 12(3):547–567. https://doi.org/10.14456/jcst.2022.42

Das AK, Granados C (2022b) FP-intuitionistic multi fuzzy N-soft set and its induced FP-Hesitant N soft set in decision-making. Decis Mak: Appl Manag Eng 5(1):67–89. https://doi.org/10.31181/dmame181221045d

Das AK, Granados C, Bhattacharya J (2022) Some new operations on fuzzy soft sets and their applications in decision-making. Songklanakarin J Sci Technol 44(2):440–449. https://doi.org/10.14456/sjst-psu.2022.61

Das AK, Gupta N, Granados C (2023) Weighted hesitant bipolar-valued fuzzy soft set in decision-making. Songklanakarin J Sci Technol 45(6):681–690. https://sjst.psu.ac.th/journal/45-6/10.pdf

Das AK, Gupta N, Mahmood T, Tripathy BC, Das R, Das S (2025a) An efficient water quality evaluation model using weighted hesitant fuzzy soft sets for water pollution rating. In: Kumar

A, Kumar P, Rathee S, Kumar B (eds) Mechatronics: concepts, tools, applications, and new trends, 1st edn. CRC Press, p 17. https://doi.org/10.1201/9781003494478-10

Das AK, Gupta N, Mahmood T, Tripathy BC, Das R, Das S (2025b) An innovative fuzzy multi-criteria decision making model for analyzing anthropogenic influences on urban river water quality. Iran J Comput Sci 8:103–124. https://doi.org/10.1007/s42044-024-00211-x

Das AK, Das R, Das S, Debnath BK, Granados C, Shil B, Das R (2025c) A comprehensive study of neutrosophic SuperHyper BCI-semigroups and their algebraic significance. Trans Fuzzy Sets Syst 4(2):80–101. https://doi.org/10.71602/tfss.2025.1198050

Das AK, Mahmood T, Das R, Das S (2025d) Optimizing inventory levels in retail: fuzzy logic-based decision support systems for adaptive inventory management. In: Jana S, Basu Mallik B, Sarkar A, Jana C (eds) Applications of fuzzy logic in decision making and management science. Information systems engineering and management, vol 29. Springer, Cham. https://doi.org/10.1007/978-3-031-77719-6_8

Das AK, Das R, Datta R, Granados C (2025e) An innovative approach to group decision-making based on weighted hypersoft expert system. In: Smarandache F, Jdid M, Leyva-Vazquez M (eds) Neutrosophic and Plithogenic inventory models for applied mathematics. IGI Global Scientific Publishing, pp 483–518. https://doi.org/10.4018/979-8-3693-3204-7.ch019

Das AK, Patra S, Granados C (2025f) An innovative approach to hesitant bipolar fuzzy soft sets in multi-criteria group decision-making. Adv Comp Int 5:4. https://doi.org/10.1007/s43674-025-00082-0

Das AK, Gupta N, Mahmood T, Tripathy BC, Das R, Das S (2025g) Anthropogenic impacts on urban river water quality for rating water pollution using an effective fuzzy soft multi criteria group decision making model. Iraqi J Sci 66(9):3921–3937. https://doi.org/10.24996/ijs.2025.66.9.32

Das AK, Gupta N, Mahmood T, Tripathy BC, Das R, Das S (2025h) Analyzing seasonal changes and anthropogenic impacts on urban river water quality using a weighted fuzzy soft set-based multi-criteria decision-making model: a case study of the Manu River, India. Arab J Geosci 18:228. https://doi.org/10.1007/s12517-025-12323-0

Das AK, Gupta N, Das R, Mallik R, Das S (2025i) Interval-valued neutrosophic rough soft set based intelligent multi-criteria decision-making framework. Boletim da Sociedade Paranaense de Matemática (3rd Series) 43:1–26. https://doi.org/10.5269/bspm.76016

Das AK, Gupta N, Mahmood T, Das R, Das S, Granados C (2025j) Assessment of groundwater quality in Agartala, Tripura, Northeast India using an effective fuzzy multicriteria decision-making model. In: Das J, Pradhan B, Mitra R, Chouhan P, Lee C, Chatterjee U (eds) Developments in environmental science, vol 19. Elsevier, pp 311–333. https://doi.org/10.1016/B978-0-443-33787-1.00023-X

Das S, Mallik R, Shil B, Bardhan R, Das AK (2025k) The impact of management of interactive learning environments on mathematical proficiency: a neutrosophic overview. Boletim da Sociedade Paranaense de Matemática (3rd Series) 43:1–6. https://doi.org/10.5269/bspm.76458

Das AK, Gupta N, Das S, Mallik R, Granados C, Das R, Sinha K, Fujita T (2025l) AI-driven decision-making for sustainable water quality management: a fuzzy multi-criteria approach to handling uncertainty. In: Kadry S, Mahajan S, Jain V (eds) Exploring generative AI with computational intelligence: navigating the uncertainty of creative machines, 1st edn. CRC Press, pp 336. https://doi.org/10.1201/9781003680192-3

Das AK, Gupta N, Das R, Granados C, Das S (2025m) An innovative approach to neutrosophic rough soft set in intelligent multi-criteria decision-making. OPSEARCH. https://doi.org/10.1007/s12597-025-01067-w

Das AK, Gupta N, Das R, Das S, Das R, Granados C, Debnath BK (2026a) Fuzzy logic-driven strategies for diversity and inclusion in multicultural marketing. In: Tarnanidis T, Sousa B, Vlachopoulou M, Papachristou E, Kittu Manda V (eds) Diversity and inclusion-driven marketing for multicultural marketplaces. IGI Global Scientific Publishing, pp 189–236. https://doi.org/10.4018/979-8-3373-2111-0.ch009

Das AK, Gupta N, Fujita T, Das S, Granados C, Debnath BK, Das R (2026b) Advanced co-branding strategies for smart luxury products: a multi-criteria decision-making approach using neutrosophic sets. In: Tarnanidis T, Sousa B, Vlachopoulou M, Papachristou E, Kittu Manda V (eds) Co-branding strategies for smart luxury products. IGI Global Scientific Publishing, pp 203–258. https://doi.org/10.4018/979-8-3373-3010-5.ch006

Das AK, Gupta N, Das R, Granados C, Das S (2026c) Neutrosophic sets for decision-making in natural disasters and environmental recovery. In: Ezaier Y, Gaamouche R, Lahby M (eds) Computational intelligence and optimization methods for sustainable water management. IGI Global Scientific Publishing, pp 119–162. https://doi.org/10.4018/979-8-3373-2700-6.ch004

Erkmen A, Ulber R, Jüstel T, Altendorfner M (2025) Towards sustainable recycling of critical metals from e-waste: Bioleaching and phytomining. Resour, Conserv Recycl. https://doi.org/10.1016/j.resconrec.2024.108057

Finn S, Herne M, Castille D (2017) The value of traditional ecological knowledge for the environmental health sciences and biomedical research. Environ Health Perspect 125(8):085006. https://doi.org/10.1289/EHP858

Ghosh S, Das AP (2018) Metagenomic insights into the microbial diversity in manganese-contaminated mine tailings and their role in biogeochemical cycling of manganese. Sci Rep 2018 8:1, 8(1), 1–12. https://doi.org/10.1038/s41598-018-26311-w

Granados C, Das AK, Osu B (2023) Weighted neutrosophic soft multiset and its application to decision making. Yugosl J Oper Res 33(2):293–308. https://doi.org/10.2298/YJOR220915034G

Granados C, Das AK, Das B (2022) Some continuous neutrosophic distributions with neutrosophic parameters based on neutrosophic random variables. Adv Theory Nonlinear Anal Appl 33(2):293–308. https://doi.org/10.31197/atnaa.1056480

Grimes S, Maguire D (2020). Assessment of priorities in critical material recovery from waste electrical and electronic equipment. Resour Policy 68, Article 101658. https://doi.org/10.1016/j.resourpol.2020.101658

Gupta N, Das AK, Granados C (2026) Integrating traditional ecological knowledge with fuzzy logic and quantum computing for sustainable agriculture in tribal areas of Tripura. In: Horng M, Khalaf O, Shieh C, Jain V (eds) Revolutionizing sustainable food production with quantum computing. IGI Global Scientific Publishing, pp 177–214. https://doi.org/10.4018/979-8-3373-3957-3.ch006

Haq SM, Pieroni A, Bussmann RW, Abd-ElGawad A, El-Ansary H (2023) Integrating traditional ecological knowledge into habitat restoration: Implications for meeting forest restoration challenges. J Ethnobiol Ethnomedicine 19, Article 33. https://doi.org/10.1186/s13002-023-00606-3

Hemmati A, Asadollahzadeh M, Torkaman R (2024) Assessment of metal extraction from e-waste using supported IL membrane with reliable comparison between RSM regression and ANN framework. Sci Rep 14, Article 54591. https://doi.org/10.1038/s41598-024-54591-y

Hosen N, Nakamura H, Hamzah A (2020) Adaptation to climate change: does traditional ecological knowledge hold the key? Sustainability 12(2), Article 676. https://doi.org/10.3390/su12020676

Jangre J, Prasad K, Patel D (2022) Analysis of barriers in e-waste management in developing economy: an integrated multiple-criteria decision-making approach. Environ Sci Pollut Res 29:72294–72308. https://doi.org/10.1007/s11356-022-21363-y

Katu Amina H (2025) Traditional ecological knowledge: Sustainable use of medicinal plants for health. Res Inven J Res Med Sci 4(1):29–35. https://doi.org/10.59298/RIJRMS/2025/412935

Li Z, Diaz L, Yang Z, Jin H, Lister T, Vahidi E, Zhao F (2019) Comparative life cycle analysis for value recovery of precious metals and rare earth elements from electronic waste. Resour, Conserv Recycl. https://doi.org/10.1016/j.resconrec.2019.05.025

Mao S, Kang Y, Zhang Y, Xiao X, Zhu H (2020) Fractional grey model based on non-singular exponential kernel and its application in the prediction of electronic waste precious metal content. ISA Transactions. https://doi.org/10.1016/j.isatra.2020.07.023

Mishra S, Dash D, Al-Tawaha ARMS, Das AP (2022) A review on heavy metal ion adsorption on synthetic microfiber surface in aquatic environments. Appl Biochem Biotechnol, 194(10), 4639–4654. https://doi.org/10.1007/S12010-022-04029-W/METRICS

Mishra S, Ghosh S, van Hullebusch ED, Singh S, Das AP (2023) A critical review on the recovery of base and critical elements from electronic waste-contaminated streams using microbial biotechnology. Appl Biochem Biotechnol 2023:1–30. https://doi.org/10.1007/S12010-023-04440-X

Mukherjee A, Das AK (2015a) Application of interval valued intuitionistic fuzzy soft set in investment decision making. In: 2015 Fifth international conference on advances in computing and communications (ICACC), Kochi, India, 2015, pp 61–64. https://doi.org/10.1109/ICACC.2015.37

Mukherjee A, Das AK (2015b) Interval valued intuitionistic fuzzy soft multi set theoretic approach to decision making problems. In: 2015 International conference on computer, communication and control (IC4), Indore, India, pp 1–5. https://doi.org/10.1109/IC4.2015.7375640

Mukherjee A, Das AK (2015c) Relations on intuitionistic fuzzy soft multi sets. In: Kim K (eds) Information science and applications. Lecture notes in electrical engineering, vol 339. Springer, Berlin, Heidelberg. https://doi.org/10.1007/978-3-662-46578-3_71

Mukherjee A, Das AK (2016a) Application of fuzzy soft multi sets in decision-making problems. In: Nagar A, Mohapatra D, Chaki N (eds) Proceedings of 3rd international conference on advanced computing, networking and informatics. Smart innovation, systems and technologies, vol 43. Springer, New Delhi. https://doi.org/10.1007/978-81-322-2538-6_3

Mukherjee A, Das AK (2016b) Topological structure formed by intuitionistic fuzzy rough relations. J Indian Math Soc 83(1–2):135–144

Mukherjee A, Das AK (2022) Einstein operations on fuzzy soft multisets and decision making. Boletim Da Sociedade Paranaense De Matematica 40:1–10. https://doi.org/10.5269/bspm.32546

Mukherjee A, Das AK (2023) Essentials of fuzzy soft multisets: theory and applications, Springer Nature Singapore Pte Ltd., pp 1–151. ISBN: 978-981-19-2759-1, https://doi.org/10.1007/978-981-19-2760-7

Mukherjee A, Das AK, Gupta N, Patra S, Mahmood T, Das S, Das R (2026a) A novel generalized interval-valued neutrosophic rough soft set framework for enhanced decision-making: application in water quality assessment. Trans Fuzzy Sets Syst 5(1):1–25. https://doi.org/10.71602/tfss.2026.1194037

Mukherjee A, Mukherjee A, Das R, Das AK (2026b) A new approach of generalized interval-valued neutrosophic rough set and its application in medical diagnosis. Trans Fuzzy Sets Syst 5(1):115–134. https://doi.org/10.71602/tfss.2026.1204833

Priyadarsini S, Das AP (2024) Analytical and structural characterization of waste lithium-ion batteries for their effective recycling strategy. Environ Chem Ecotoxicol. https://doi.org/10.1016/j.enceco.2024.12.004

Priyadarsini S, Alok, Das P, Journal G, Das AP (2025) Isolation and identification of native lithium-resistant bacteria from electronic waste contaminated sites: a focused investigation on tolerance assessment with potential lithium bio-recovery application. Geomicrobiol J. https://doi.org/10.1080/01490451.2025.2539509

Priyadarsini S, Das AP (2026) Lithium in waste water: impacts on pollution and human health. Decarbonization of Wastewater Pollutants as a Sustainable Solution, 353–364. https://doi.org/10.1016/B978-0-443-33062-9.00010-2

Qadir A, Abdullah S, Khan S, Khan F (2024) A new extended TOPSIS method for e-waste recycling partner selection under complex Pythagorean fuzzy rough Dombi aggregation operator. IEEE Access 12:193520–193551. https://doi.org/10.1109/ACCESS.2023.3338446

Rai V, Liu D, Xia D, Jayaraman Y, Gabriel J (2021) Electrochemical approaches for the recovery of metals from electronic waste: a critical review. Recycling 6(3), Article 53. https://doi.org/10.3390/recycling6030053

Seikh M, Chatterjee P (2024) Identifying sustainable strategies for electronic waste management utilizing confidence-based group decision-making method in an interval-valued Fermatean fuzzy

environment. Eng Appl Artif Intell 135, Article 108701. https://doi.org/10.1016/j.engappai.2024.108701
Shakib B, Lee J, Kang H, Petranikova M (2024) Uncertainty evaluation associated with the removal of molybdenum from industrial effluents as environmental impacts using solvent extraction with LIX 63. Process Saf Environ Prot. https://doi.org/10.1016/j.psep.2024.12.005
Sharma SB (2017) The relevance of traditional ecological knowledge (TEK) in agricultural sustainability of the semi-arid tropics. In: Rakshit A, Abhilash P, Singh H, Ghosh S (eds) Adaptive soil management: From theory to practices. Springer, pp 335–349. https://doi.org/10.1007/978-981-10-3638-5_20
Simaei E, Rahimifard S (2024) AI-based decision support system for enhancing end-of-life value recovery from e-wastes. Int J Sustain Eng 17:80–96. https://doi.org/10.1080/19397038.2024.2306293
Srivastava S, Dhaker K (2024) Data-driven approach for Cu recovery from hazardous e-waste. Process Saf Environ Prot.https://doi.org/10.1016/j.psep.2024.01.013
Suresh N, Sadeghi A, Yazdani-Asrami M (2023) Critical current parameterization of high temperature superconducting tapes: a novel approach based on fuzzy logic. Superconductivity, Article 100036. https://doi.org/10.1016/j.supcon.2023.100036
Wang X, Zhao C, Li Z, Huang J (2024) Modeling risk assessment of soil heavy metal pollution using partial least squares and fuzzy logic: a case study of a gully-type coal-based solid waste dumpsite. Environm Pollut. Article 124147. https://doi.org/10.1016/j.envpol.2024.124147
Wright R (2020) Review of traditional ecological knowledge: learning from Indigenous practices for environmental sustainability. J Study Relig, Nat Cult 14(2):304–306. https://doi.org/10.1558/jsrnc.39475
Xia D, Lee C, Charpentier N, Deng Y, Yan Q, Gabriel J (2024) Drivers and pathways for the recovery of critical metals from waste printed circuit boards. Adv Sci 11, Article e202309635. https://doi.org/10.1002/advs.202309635
Yeh C, Xu Y (2013) Sustainable planning of e-waste recycling activities using fuzzy multicriteria decision making. J Clean Prod 52:194–204. https://doi.org/10.1016/j.jclepro.2013.03.003
Zadeh LA (1965) Fuzzy sets. Inf Control 8(3):338–353

Evaluating the Environmental Footprint of E-Waste Recycling Methods

Suchinnata S. Sardar, Smruti Smaranika Nayak, and Ajaya K. Biswal

Abstract The rapid proliferation of electronics development has been fuelling an exponential increase in e-waste, with the projected 59.4 million metric tonnes of e-waste produced globally in 2022 expected to exceed 74 million metric tonnes by 2030. This poses extreme environmental and health conditions due to the improper disposal and recycling of toxic chemicals. This chapter explores the environmental impact assessment (EIA) and life cycle analysis (LCA) of some of the major e-waste recycling processes like manual dismantling, mechanical separation, pyrometallurgy, hydrometallurgy, and new bioleaching methods. In Low- and middle-income country, EIA reports suggest that informal recycling greatly contaminates the air, land, and water and causes heavy metal pollution (lead, cadmium, mercury) 2–5 times more than acceptable levels in the environment. Alternatively, controlled recycling facilities with emissions control reduce greenhouse gas emissions by 60% compared to uncontrolled recycling. The LCA outcome indicates that mechanical recycling has the lowest environmental footprint among current industrial processes and has a global warming potential (GWP) of approximately 0.8 kg CO_2-eq per kg treated e-waste, while pyrometallurgy has 3.2 kg CO_2-eq. Recycling of e-waste can also have the ability to limit net emissions by another 25–30% by virtue of e-waste recycling industries using renewable energy sources. The chapter highlights the importance of regulation, technology innovation, and impact assessment tools in ensuring sustainable and circular management of e-waste.

S. S. Sardar (✉)
Department of Botany, Rama Devi Women's University, Bhubaneswar, Odisha, India
e-mail: ssardar@rdwu.ac.in

S. S. Nayak
Department of Industrial Microbiology, Rama Devi Women's University, Bhubaneswar, Odisha, India
e-mail: smrutismaranika2003@gmail.com

A. K. Biswal
Department of Biochemistry and Molecular Biology, University of Georgia, Athens, GA, USA
e-mail: biswalajaya@ccrc.uga.edu

A. P. Das and S. Priyadarsini (eds.), *Recovery and Recycling of Critical Metals from Electronic Waste for Environmental Sustainability*, Earth and Environmental Sciences Library, https://doi.org/10.1007/978-3-032-29064-9_13

2.1 Formal and Informal Recycling Markets

Due to mounting concern for sustainability, more than 60% of the world's e-waste is being treated informally (Baldé et al. 2022; Liu and Zhang 2023b). Electronics dismantling during the informal process by informal recyclers in China, India, Ghana, and Nigeria will be conducted with crude processes, open burning, acid leaching, and manual sorting—without emission control. Wong et al. (2009) approximated that more than 80% of the urban housing in Guiyu conducted informal recycling, resulting in severe environmental pollution and occupational lead, dioxin, and PCB exposure.

Conversely, however, Japan, South Korea, and the EU, respectively, have official recycling facilities with mechanized shredding, solvent extraction, and green metal refining technology. According to Yla Mella and Pongracz (2016), EU facilities recycle precious metals by 95%, as compared to informal facilities with recovery below 30%. Extended Producer Responsibility (EPR) programs have also enhanced traceability and reduced landfilling in such countries. International and regional law have led the way in sustainable e-waste management development. The Basel Convention of 1989 is the global leading convention governing transboundary movement of hazardous waste, including e-waste (Priyadarsini and Das 2024, 2026). This has been followed by a succession of regional and national legislations:

European Union—Waste Electrical and Electronic Equipment (WEEE) Directive (2012/19/EU) imposes the cost on producers for recovery and recycling.

India.—E-Waste (Management) Rules (2022) set EPR goals and sanctioned collection channels. China—Circular Economy Promotion Law (2008) and Solid Waste Pollution Prevention Law (2020) control extended producer responsibility (Table 1).

Empirical data of Dwivedy and Mittal (2012) indicated that e-waste can be redirected to the formal sector by ~ 30% from the informal sector with the proper implementation of policy action. Baldé et al. (2022) referenced that worldwide, it is just

Table 1 Global e-waste generation and recycling trends (2010–2024)

Year	Global e-waste generated (million tons)	E-waste recycled (million tons)	Recycling rate (%)	Major contributing regions
2010	33.8	7.2	21.3	Europe, North America
2014	44.7	9.5	21.2	East Asia, Europe
2018	53.6	11.8	22.0	China, India, USA
2020	57.4	12.5	21.8	Asia, Europe
2022	62.0	13.3	21.4	China, EU, USA, India
2024	67.1 (Projected)	14.9 (Projected)	22.1 (Projected)	Asia-Pacific, Africa rising

17.4% of the e-waste that was officially collected and recycled in 2020 in the face of a huge data and regulation gap.

There is also dependence on process technology and policy-making for efficient recycling (Osborne 2024). Hydrometallurgical recycling will require 35% less energy and produce 25% less toxic effluent than pyrometallurgical treatment, and carbon emission will be cut by up to 40% with mechanical pre-treatment, Hossain and Sahajwalla (2024) assert.

At the global level, metal recycling of traditional streams (i.e., palladium, gold, silver) is 90–95%, whereas for plastic recycling it is 15–20% (Liu et al. 2023). In Japanese "urban mining" plants, 98% of e-waste recycling is diverted from landfills and reaches near-zero-waste quality (Oguchi et al. 2012). But in developing countries still the lack of advanced plants and integration of EIA-LCA keeps leaving tremendous environmental footprints and resource loss.

3 Environmental Impact Assessment (EIA) of E-Waste Recycling

Environmental Impact Assessment (EIA) is a systematic analytical instrument that is aimed at forecasting, assessing, and mitigating potential environmental and social effects before project construction. For e-waste recycling, EIA guarantees that facilities comply with environmental requirements, reduce emissions, and protect worker health (Glasson et al. 2012). The process offers a scientific and legal context to incorporate environmental aspects into decision-making, particularly in countries where informal recycling is prevalent.

According to United Nations Environment Programme (2023), the absence of pre-project environmental impact assessments in more than 65% of developing countries has resulted in uncontrolled releases of heavy metals and persistent organic pollutants. Effective EIA frameworks, therefore, are crucial to sustaining economic benefits of recycling alongside sustainable environmental performance (Liu and Zhang 2023a).

3.1 EIA Framework and Methodology

The typical EIA procedure includes a sequence of key steps: screening, scoping, baseline study, impact prediction, mitigation planning, public consultation, and monitoring. Jay et al. (2007) emphasized that the success of an EIA relies on early stakeholder involvement and openness of data.

Baseline studies of e-waste recycling facilities usually assess:

Air emissions ($PM_{2.5}$, heavy metal aerosols, dioxins/furans),
Water pollution (acidic leachates, dissolved metals like Pb, Cd, Ni),

Soil quality (residual hydrocarbons, brominated flame retardants), and.
Socio-economic parameters (employment, public health).

Post-operational impact predictive models of air dispersion and surface-water pollution such as AERMOD and QUAL2K are being used more commonly to model post-operational impacts (Chakraborty et al. 2022; Michael and Kumar 2024). Integration of GIS and remote sensing in the mapping of EIA has also increased spatial representation of pollution hotspots.

An EIA of the Bangalore E-waste Recycling Park, commissioned in 2016, estimated expected emissions of 4.5 kg CO_2-eq per tonne of e-waste processed. Emissions were minimized by 35% following post-implementation monitoring following the commissioning of mechanical pre-treatment and wet scrubbing equipment (Ramachandra and Varghese 2018).

On earth's largest e-waste informal dump site, $PM_{2.5}$ averaged 2400 μg/m^3 by air sampling, with dioxin concentrations in excess of WHO guidelines by nearly 60 times (Song et al. 2021). The absence of EIA and open burning created years of soil contamination and workers having elevated blood lead levels. Remediation based on EIA, which commenced in 2015, also reflected quantifiable outcomes Lead concentration in surface soil decreased from 3000 mg/kg to below 500 mg/kg within five years (Leung et al. 2020). This indicates that systematic assessment together with enforcement by regulatory bodies can provide noteworthy environmental rehabilitation.

Implementation of EIA varies significantly among various regions. Developed nations such as Japan and European Union nations apply strong EIA procedures infused with life-cycle indicators and continuous auditing (Yla Mella and Pongracz 2016). Developing countries, on the other hand, face data inadequacies, lack of technical expertise, and informal sector dominance, leading to surface-level or retrospective analysis (Dwivedy and Mittal 2012).

In the EU, EIA compliance exceeds 90% in recycling facilities, while in South and Southeast Asia, fewer than 25% of operational units have official EIAs (UNEP 2023). The disparity emphasizes the necessity for capacity building, uniform guidelines for evaluation, and integration of socio-environmental equity in emerging economies.

3.2 *Environmental Hotspots and Impact Indicators Identified*

From numerous EIAs globally, the major environmental hotspots of e-waste recycling are:

Air pollution: Volatile metals and dioxin and furan emission by open burning and smelting.

Water pollution: Leachates acidic with Pb, Cd, and Cr in hydrometallurgical process effluents.

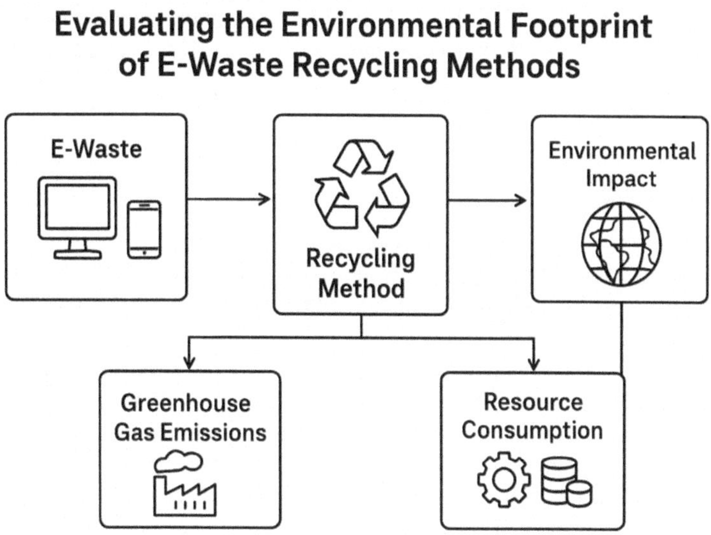

Fig. 1 Evaluating the environmental footprint of E-waste recycling methods

Soil pollution: Deposition of heavy metals and persistent organic pollutants by fallout of dust.

Noise and vibration: Shredding and mechanical activities in excess of 85 dB in non-regulated plants.

Quantitative factors such as Air Quality Index (AQI), Soil Pollution Index (SPI), and Water Quality Index (WQI) are included in EIA reports to measure cumulative impact levels today (Sivaranjani et al. 2021). Incorporation of occupational exposure monitoring, community health studies, and cumulative environmental indices makes modern EIA a necessary tool for sustainable recycling of e-waste (Fig. 1).

4 Environmental Impact Assessment (EIA) Structure for E-Waste Recycling

Environmental Impact Assessment (EIA) is a formal process used to study probable environmental impacts of development or industrial activities prior to their commencement. For the recycling of e-waste, EIA seeks, anticipates, and quantifies direct and indirect environmental consequences of dismantling plants, metal extraction plants, and waste treatment plants (Rathi 2017). EIA ensures recycling plant establishment and decision-making regarding operation incorporating sustainability factors and measures to avert pollution. The framework has incorporated the monitoring of the environment, stakeholders, and mitigation measures in order to ensure that recycling processes minimize damage to air, water, and land quality (Kiddee et al. 2013).

4.1 Methodological Approach in EIA for E-Waste Facilities

The EIA of e-waste recycling usually adopts the following method: screening, scoping, impact prediction, assessment, mitigation, and monitoring (Sinha-Khetriwal et al. 2005).

Screening determines whether a project shall have a full EIA, and scoping establishes limits and members of analysis such as emissions, leachates, and public exposure. Impact prediction applies qualitative and quantitative methods such as dispersion modeling and toxicity indices. Evaluation is where the anticipated effects are synthesized for decision-making, while monitoring is carried out to guarantee compliance in utilization. These emerging trends include the use of GIS-based mapping for site location, and risk matrices to provide an estimate of potential ecological hazards (Dutta and Goel 2021). Combination of EIA with life cycle-based tools facilitates a complete analysis of upstream and downstream environmental influences (Rautela et al. 2022).

4.2 Significant Environmental Indicators

A majority of parameters assessed in EIA for e-waste recycling are air emissions ($PM_{2.5}$, VOCs, SO_2), soil contamination (Pb, Cd, Hg), water quality parameters (COD, BOD, heavy metals), and level of noise pollution. Experiments show that open burning and acid leaching operations enhance concentration of airborne particulate matter up to 5 times the WHO norm, and untreated leachates in informal recycling elevate groundwater lead concentration to 0.8 mg/L, higher than the WHO permissible level of 0.01 mg/L (Awasthi et al. 2018).

Successful implementation of EIA depends on policy integration and public involvement. Interest groups like citizens, environmental protection organizations, and recycling entrepreneurs play an important role in identifying potential social and environmental risks (Wang et al. 2019). In countries like India and China, EIA regimes are being harmonized with Extended Producer Responsibility (EPR) laws to ensure that producers remain responsible for the end-of-life products.

Despite its demonstrated usefulness, EIA in the e-waste sector is bedeviled by weak data, poor regulation, and dominance of the informal economy (Awasthi et al. 2016a, b). Future emphasis is placed on integrating real-time monitoring of the environment, machine learning algorithms to forecast consequences, and digital transparency technologies to facilitate public reporting of environmental performance

(Zeng et al. 2022). EIA, incorporating circular economy discourse and sustainability assessment, can be a strategic tool to facilitate low-impact, resource-saving recycling regimes (Dutta and Goel 2021).

5 Life Cycle Assessment Structure for Recycling Processes of E-Waste

Life Cycle Assessment (LCA) is a normative evaluation process that is used to estimate the environmental impact of each stage of the life cycle of a product—raw material extraction to production, end-use, and end-of-life treatment through disposal or recycling (ISO 14040/44 system). LCA quantifies in e-waste recycling the energy and emissions inputs, and resource consumption in recycling processes like mechanical, hydrometallurgical, and pyrometallurgical processes (Wäger et al. 2011). LCA enables decision-makers to identify the most environmentally friendly recycling alternatives by contrasting their relative environmental trade-offs.

LCA has also been reported to provide vital insights on greenhouse gas reduction and toxic metal releases inherent in informal recycling activities (Wang et al. 2011). Further, LCA is a strategic instrument for achieving circular economy strategies and developing Extended Producer Responsibility (EPR) policy (Ardente et al. 2018) (Table 2).

5.1 Phases of LCA Applied for E-Waste

LCA studies typically entail four major phases:

Table 2 Life cycle assessment (LCA) results of e-waste management scenarios

Scenario	Functional unit	Impact on climate change (kg CO_2 eq)	Human toxicity potential (kg 1.4-DB eq)	Freshwater ecotoxicity (kg 1.4-DB eq)	Energy demand (MJ)
Landfilling	1 kg of e-waste	3.6	0.042	0.015	12.3
Informal recycling	1 kg of e-waste	2.9	0.080	0.038	8.1
Mechanical + hydrometallurgical	1 kg of e-waste	1.5	0.025	0.010	9.7
Full formal recycling chain	1 kg of e-waste	0.9	0.018	0.007	10.5
Circular economy integration	1 kg of e-waste	0.7	0.015	0.005	11.0

1. Definition of Goal and Scope—identification of the goal, system boundaries, and functional unit (usually per ton of treated e-waste).
2. Life Cycle Inventory (LCI)—creation of quantitative data on material flows, energy consumption, and emissions.
3. Life Cycle Impact Assessment (LCIA)—determination of probable impacts such as global warming potential, acidification, eutrophication, and human toxicity (Cucchiella et al. 2015).
4. Interpretation—conclusion drawing and improvement description.

Inventory data for e-waste recycling can comprise electricity use (~450–750 MJ/ton), CO_2 emissions (0.8–1.2 tons CO_2 eq/ton), and efficiencies of metal extraction (Li et al. 2019). LCIA results reveal that copper and gold hydrometallurgical metal extraction preserves energy up to 35% over pyrometallurgy (Xue et al. 2020).

5.2 Comparative Study of Recycling Processes using LCA

Reduced size and separation of low emission materials but metal recovery efficiency is mechanical recycling (Cui and Forssberg 2017). Pyrometallurgical recycling is with high metal recovery efficiency (95% for Cu, Au, and Ag) but air pollution and high energy usage. Leaching and solvent extraction hydrometallurgical processes are utilized more with lower environmental impact (Tuncuk et al. 2012) but with difficulty in dealing with chemicals (Song et al. 2018).

New LCA indicates that the most sustainable option is a combined method of a mechanical-hydrometallurgical process that strikes a balance between recovery performance and environmental performance (Rautela et al. 2022).

5.3 LCA-Based Decision Support in Policy and Industrial Design

LCA outcomes are being used more and more in policy appraisals, and industries and governments are being steered towards sustainable recycling facilities. The European Union WEEE Directive, for instance, is guided by LCA outcomes and takes recycling efficiency into consideration as a factor and minimizes industry energy intensity (Ardente et al. 2018). LCA-based analysis has directed a stakeholder in resource efficiency plans and green design guides in China and India (Gu et al. 2020). Businesses utilize LCA in order to design recyclable products and select low-impact processing pathways in a bid to reduce the environmental burden by 20–30% (Chatterjee et al. 2021).

Even with all these benefits, certain limitations of LCA are heterogeneity of data, lack of standard inventories for e-waste, and minimal geographical databases (Kumar and Samadder 2022). The future direction is to wedge dynamic models of LCA, environmental monitoring in real-time, and AI-driven simulations so that the ability to foresee improves (Zeng et al. 2022). Furthermore, wedging LCA with S-LCA and LCC frameworks gives the holistic sustainability analysis (Cucchiella et al. 2015).

6 Comparative Case Studies of EIA and LCA Use for World Recycling of E-Waste

The European Union (EU) boasts one of the globe's most controlled and prepared e-waste management networks, with Environmental Impact Assessment (EIA) and Life Cycle Assessment (LCA) to ready recycling plants. EU member nations must assess environmental elements of e-waste collection, transport, disassembly, and waste treatment and material recovery facilities under the Waste Electrical and Electronic Equipment (WEEE) Directive (2012/19/EU).

Ardente et al. (2018) performed an LCA of hydrometallurgical and mechanical recycling of Italian printed circuit board (PCB) waste. In their study, mechanical pre-treatment with chemical recovery was above 90% efficient for the recovery of precious metals and the global warming potential (GWP) decreased by approximately 25% compared to pyrometallurgical recycling conditions. Complementary EIA reports completed LCA results with locale-specific analysis of air emissions, effluent discharge, and likely community health impact (Yla Mella and Pongracz 2016). Such a combined methodology has upgraded the environmental sustainability of EU recycling processes (Gui et al. 2021).

China, the former global largest informal e-waste recycling center, has consistently drifted towards institutionalized EIA regimes and LCA methods to restore environments with pollution as well as improve recycling facilities (Wang and Zhang 2021). Guiyu, being the largest e-waste center, had the highest contamination with up to 3000 mg/kg lead in soils (Leung et al. 2020). Under national post-2015 interventions, under regulation and site remediation based on EIA, the site experienced a decrease in the soil lead level by close to 85% and dioxins in the atmosphere by 60% within five years.

In China, Xue et al. (2020) carried out an LCA of conventional pyrometallurgical processes and mechanochemical-hydrometallurgical recycling. The study recorded 30–35% energy savings and sufficient reduction in water consumption for the mechanochemical route. Therefore, use of EIA and LCA sequentially accomplished environmental rejuvenation along with process improvement, reflecting the importance of conjoining regulatory assessment with sustainability methods.

India's E-Waste (Management) Rules 2022 have placed greater emphasis on formal recycling facilities and imposing EIA and LCA-based environmental studies. At Bangalore E-Waste Park, pre-construction EIA estimated emissions at 4.5 kg CO_2-equivalent per ton of processed e-waste. Post-performance tests confirmed the inclusion of mechanical pre-treatment units and wet scrubbing systems to reduce the CO_2

emissions up to 35% and heavy metal discharge to acceptable limits (Ramachandra and Varghese 2018).

Conversely, LCA research involving informal recycling shops in Delhi presented significant environmental loads, having a maximum 1.2 tons CO_2-equivalent greenhouse gas emissions per ton of e-waste treated and heavy build-up of persistent organic pollutants in surface soils near the area (Awasthi et al. 2018). These relative findings are the impetus for moving past informal processes to controlled recycling practices tested scientifically.

The following lessons of great value are extracted from these local case studies.

1. Monitoring and formalization: The European Union demonstrates that integrated science of LCA and EIA leads to superior material recycling efficiency and lower environmental impact.
2. Remediation capacity: China demonstrates how intervention after impact with EIA coupled with technology upgrade with LCA can significantly minimize pollution in what was once polluted areas.
3. Issues in developing economy: India's experience reflects the ubiquity of informal recycling, non-harmonized adoption of EIA standards, and absence of LCA data as hurdles to sustainable development.

Comparative evaluation is also in consensus that the complementarity of EIA and LCA provides an integrated approach to environmental hotspots, ensuring evidence-based decision-making, application of cleaner technology, and effective use of resources needed for circular economy growth.

6.1 Global E-Waste Management Recommendations

1. Compulsory integration of EIA and LCA processes in all large e-waste recycling facilities to improve overall environmental responsibility.
2. Incentives through the medium of mechanochemical-hydrometallurgical recycling technology with improved energy efficiency, lowered emissions, and metal recovery.
3. Capacity development and training exercises for developing countries in relation to the concurrent application of environmental appraisal.
4. GIS-based maps, remote sensing, and real-time monitoring systems to monitor air, land, and water pollution.
5. Harmonization of national policies to balance Extended Producer Responsibility (EPR), EIA, and LCA policies to make global e-waste regulation consistent.

7 Sustainable E-Waste Recycling Strategies

International policies and conventions are of utmost importance in global e-waste management with regard to governing toxic wastes and promotion of recycling in a sustainable manner. Export of hazardous wastes, including e-waste, is banned by the Basel Convention (1989, 2019 amended), and environmentally sound management (ESM) is encouraged among export nations. Baldé et al. (2022) state that prior to the implementation of the Basel Amendment, 50% of global e-waste was supposedly smuggled into developing countries. Other global standards such as Stockholm Convention (2001) of persistent organic pollutants and Minamata Convention (2013) of mercury control e-waste recycling practice indirectly through the control of toxic emission of treatment plants (Li et al. 2019). The conventions are the bases of mandatory environmental monitoring, removal standards of pollutants, and recovery practice of materials safely across the globe (Li et al. 2020) (Table 3).

7.1 Regional Regulatory Frameworks

European Union (EU):

The WEEE Directive (2012/19/EU) and the RoHS Directive (2011/65/EU) have implemented producer responsibility, regulated collection, and e-waste treatment.

Table 3 Environmental impacts of major e-waste recycling methods

Recycling method	Major process	Energy consumption (MJ/kg)	GHG emission (kg CO_2 eq/kg)	Toxic waste generation (kg/kg)	Remarks
Mechanical recycling	Shredding, magnetic and eddy separation	8–15	0.45	0.02	Low emission; high metal recovery efficiency
Pyrometallurgical	Smelting, furnace refining	25–40	1.8	0.09	High recovery but emits heavy metals
Hydrometallurgical	Leaching, solvent extraction	15–25	1.0	0.05	Eco-friendly; moderate cost
Bioleaching	Microbial recovery	10–20	0.8	0.03	Sustainable; low scalability
Informal recycling	Open burning, acid stripping	5–8	2.5	0.12	Extremely hazardous and uncontrolled

The EU has an official rate of collection of 54% and precious-metal recovery efficiency higher than 90% (Ardente et al. 2018).

China:

China's 2008 Law on Encouraging the Circular Economy and 2020 Prevention of Pollution by Solid Waste Law controls the formal and informal waste recycling markets. Post-policy activities in Guiyu and Dali factories, after LCA and EIA surveys, treated 60% of air pollutants and lowered soil heavy metal contamination by 50–80% (Leung et al. 2020; Xue et al. 2020).

India:

2016 and 2022 E-Waste (Management) Rules implemented Extended Producer Responsibility (EPR) with collection, recycling, and recovery programs being compulsory for manufacturers. Implementation at Bangalore E-Waste Park was effective, resulting in a 35% reduction in CO_2 emissions and toxic leachates reduction (Ramachandra and Varghese 2018).

7.2 Technological Strategies

Pre-treatment mechanical achieves value from separation of material using low energy (Cui and Forssberg 2017).

Hydrometallurgical recovery is metal recovery efficiently with low emissions relative to pyrometallurgy (Song et al. 2018).

Integrated recycling is the union of mechanical and chemical processes in an attempt to employ all efficiency with low environmental impact (Rautela et al. 2022).

7.3 Policy-Based Strategies

Extended Producer Responsibility (EPR) inspires producers to create recyclable products and end-of-life take-back electronics.

Tax subsidies and incentives for formal sector development stimulate environmentally sustainable recycling facilities (Awasthi et al. 2018).

Public participation and campaign approaches bring down unorganized dumping and cause accurate collection.

Integration of LCA and EIA in design through to recycling sustainably, blockchain and digital tracing for transparent collection and recycling pathways, and circular economy deployment for achieving optimal resource recovery and alleviation of the environment are possibilities (Zeng et al. 2022).

8 Integration of EIA and LCA towards Sustainable E-Waste Management

Environmental Impact Assessment (EIA) and Life Cycle Assessment (LCA) are two converging green management practices which, when integrated, provide an overall perception of the environmental impacts of e-waste management. EIA considers site-level and project-level impacts, i.e., ecological and social impacts, while LCA considers environmental performance of the system from collection to transport, treatment to disposal or recycling at end-of-life stage (Gu et al. 2020).

The hybridization of EIA and LCA enables policymakers, managers, and planners to identify hotspots of the environment, attribute economic values to contending recycling technologies against each other, and adopt strategies that will be sustainable at both local and global levels. Zeng et al. (2022) further suggested that hybridization of EIA and LCA will be energy efficient with higher resource recovery and lower emission production, especially in nations where there are large-scale informal recycling industries.

8.1 *Integration Methodological Framework*

In general, integration of LCA and EIA comprises the following steps:

1. Definition of purpose and scope, which determines site-specific and life cycle system boundaries.
2. Inventory analysis, which gathers data on emissions, energy use, material flow, and waste generation.
3. Environmental impact assessment, whereby LCA is used to obtain global burdens such as global warming potential, toxicity, acidification, and resource depletion, and EIA is used to obtain local air, land, and water impacts.
4. Interpretation and optimization, combining results to inform technology choice, regulation compliance, and environmentally favorable management practices.

Rautela et al. (2022) applied the hybrid framework to a hybrid mechanical-hydrometallurgical e-waste recycling process in India. The study achieved 92% metal recovery efficiency and reduced CO_2 emission by 28% when compared to conventional processes, which proved the practical applicability of the integration of the two methods.

In China, integrated EIA-LCA research was carried out at Guiyu and Dali e-waste recycling plants for guiding remediation and technology upgradation. Interventions resulted in 50–85% reduction in lead in soil and around 60% reduction in air contaminants (Leung et al. 2020; Xue et al. 2020).

In Germany and Italy, as members of the European Union, integration of EIA in LCA allowed policy-makers to evaluate recycling plants, increase efficiency in terms of energy and material, and reduce environmental impact per ton of e-waste by about 30% without losing regulatory compliance (Ardente et al. 2018).

Integrated EIA-LCA method was used in India by Bangalore E-Waste Park to decide the most suitable pre-treatment and wet scrubbing technologies. The process resulted in 35% reduced CO_2 emissions as well as significant light metal leachates reduction (Ramachandra and Varghese 2018).

Despite the advantages, challenges still dominate. Variability and scarcity of information are key issues in informal networks of recycling. There is no common framework to integrate site-specific EIA data with system-level LCA indicators. Technical data and high monitoring, modelling, and computational analysis costs can limit its large-scale application. Stakeholder involvement in informal and decentralization networks is also limited in implementing it effectively (Awasthi et al. 2018).

The future direction is the development of dynamic EIA-LCA models capable of tracking and modelling recycling in real time. Artificial intelligence and big data provide the potential for improving predictive emission, energy use, and material output models. Harmonization of international policy can simplify Extended Producer Responsibility, EIA, and LCA systems in an effort to stem illegal flows of e-waste. Combining EIA-LCA with social and economic studies, such as social life cycle assessment and life cycle costing, can ensure overall sustainability (Cucchiella et al. 2015, 2021). Emerging economies should strengthen themselves for better technical capacity and more environmentally sound management practices.

9 New Technologies and New Innovations for Sustainable E-Waste Recycling

Mechanic pre-treatment is central to e-waste recycling but innovation develops it into effective and eco-friendly form. New technologies such as eddy current separation, air classification, and sensor-based separation enhance the recovery rate of metal and plastic by 95% for copper and aluminum (Cui and Forssberg 2017).

Zeng et al. (2019) illustrated how robotic destruction and AI-driven sorting can recover 20–25% more precious metals and minimize human exposure to toxic fumes and dust by significant margins.

9.1 Hydrometallurgical Innovations

Hydrometallurgical processes selectively utilize leaching, solvent extraction, and electrowinning for metal separation. Recent research involves bioleaching with genetically engineered microorganisms for copper, gold, and palladium separation at ambient pressures of atmosphere, reducing acid usage by 40% (Li et al. 2021). Ionic liquids and deep eutectic solvents are utilized for metal separation with high selectivity and low energy requirement and low wastewater production (Song et al. 2018).

These hydrometallurgical processes are now coming with a recovery ratio of up to 95% along with 50% lower greenhouse gas emissions than those that are produced in the conventional pyrometallurgical processes (Xue et al. 2020).

9.2 Pyrometallurgical and Thermal Treatment Innovations

Pyrometallurgical treatment is still the champion of metal content e-waste streams. Conservation of energy and reducing emissions are the flavour of the day in pyrometallurgical treatment. Microwave smelting conserves energy by 25–30% and lowers dioxin emissions (Liang et al. 2020). Closed-loop off-gas dust cleaning and heavy metals capture with zero space for on-site air pollution.

9.3 Digital and AI-Based Innovations

Machine learning and artificial intelligence are implemented for predictive modeling of metal recovery, energy usage, and emission savings (Zeng et al. 2022). Blockchain is used to ensure traceability and transparency along the e-waste supply chain and guaranteed Extended Producer Responsibility policy enforcement. IoT sensors provide real-time tracking of air, wastewater, and energy use so that recycling facilities can reduce environmental impacts by 10–15% per ton of e-waste (Gu et al. 2020).

9.4 Circular Economy Directions

Emerging technologies are increasingly circular economy-oriented. Recycling design expands material recycling rates. Closed-loop recycling allows recovered metals to directly attain reuse in new electronic devices. Resource efficiency optimization minimizes environmental effects while yielding maximum economic advantages (Chatterjee et al. 2021).

There are limitations even with the technology advances. There is great cost of capital and operation with relatively lower application in developing nations. It is difficult to measure environmental benefit of new processes due to a lack of data on single regions' life cycle assessment. Methods must be established to ensure environmental sustainability of chemical-based recycling and bioleaching processes.

Future directions are high-efficiency metal recovery hybrid biological-mechanical-chemical systems and low emissions. Life cycle assessment and environmental impact assessment in real time will allow for ongoing process improvement.

Harmonization of international policy is also needed to facilitate the platform for technology transfer as well as sustainable e-waste management (Baldé et al. 2022; Zeng et al. 2022).

10 Conclusion

The wide international distribution of electronic devices has triggered unprecedented volumes of e-waste generation, generating serious environmental, health, and management issues. Environmental Impact Assessment and Life Cycle Assessment have been emphasized in the chapter to assess and compensate for the environmental impact of e-waste recycling. EIA provides a site-specific method for evaluating the possible effects of recycling plants on ambient air, water, and soil quality and on human health, while LCA provides a system-based, holistic comparison of energy use, greenhouse gas emissions, resource depletion, and toxic releases throughout the entire life cycle of an electronic product. The integration of the devices allows decision-makers to track hotspots in the environment, maximize recycling technologies, and apply sustainable management practices that integrate resource recovery and environmental protection.

The chapter has demonstrated that recycling activities, i.e., mechanical, hydrometallurgical, and pyrometallurgical, vary considerably in environmental as well as operating performance. Mechanical separation processes, though saving energy and low direct emissions, offer poor recovery of precious metals. Hydrometallurgical processes, with selective leaching and chemical recovery, are energy-efficient and low in energy consumption but require strict control of chemical effluents. Pyrometallurgical processes are energy-consuming and high metal recovery but are inclined to be with air pollutant release. Technological choice, infrastructure formalization, and regulation control make huge contributions in European Union, Chinese, and Indian case studies to the environmental impacts. For example, metal recycling facilities in Europe have low ecological impact and high recovery rates and less developed countries are confronted with informal sector supremacy, inadequate monitoring, and absence of standard operation procedures.

Extended Producer Responsibility, international agreements, and e-waste management regulations are among the policy regimes that have been hailed as green recycling leaders. The policies mentioned above encourage formality in recycling facilities, green environmental administration, and design for recyclability. But there are some problems such as inefficiency in the enforcement process, there is no data available, and low participation of stakeholders in illegal e-waste recycling chains. The integration of EIA and LCA is an efficient policy choice decision system that can assist policymakers to research alternative technologies, make an estimation of environmental implications, and set priorities on interventions to have maximum recovery of resources at the lowest emissions and release of toxics. New and new emerging technologies are revolutionizing the e-waste recycling industry.

Mechanical separation, robot stripping, hydrometallurgical advancements, bioleaching, and computerized monitoring make the recovery more efficient, minimize energy usage, and reduce environmental risks. IoT-based real-time monitoring and blockchain-based supply chain monitoring add further transparency, traceability, and compliance with environmental legislation. Aggregation of these technology solutions with circular economy principles guarantees that the precious resources are harvested, reprocessed, and recycled to production lines again with less use of virgin materials and in total less environmental footprint. Effective e-waste management in the future is a call for an integrated process.

There must be institutionalization of recycling industries, especially in developing economies, in order to avoid pollution as well as occupational health hazards. Training and capacity building and technical assistance will enable the local stakeholders with green practices. Policy alignment at national and international levels will ensure prevention of illegal transboundary e-waste movement and fair recovery of the materials. Incorporation of dynamic and real-time methods of assessment, blending EIA, LCA, social, and economic research will offer end-to-end solutions for sustainability measures to maintain continuous development. Finally, e-waste sustainable management relies on concerted action of technology, policy, and environmental assessment tool.

EIA and LCA, each separately or as a double application, are proper instruments for the environmental risk identification, resource exploitation optimization, and assistance for informed decision-making. Through the application of advanced recycling technologies, embracing circular economy concepts, and imposing firm policy interventions, it is achievable to minimize the environmental impact of e-waste and recover valuable resources for future world electronics production. The future management of e-waste will be technology-led, policy-facilitated, and one solution that protects the environment, guarantees safe human health, and guarantees sustainable resource use for current and future generations.

References

Ádám B, Kovács L (2021) From inequitable to sustainable e-waste processing for resource recovery. Sci Total Environ 755:142568. https://doi.org/10.1016/j.scitotenv.2020.142568

Ardente F, Mathieux F, Recchioni M, Blengini GA (2018) Integration of life cycle assessment in the European e-waste policy framework. Environ Sci Policy 86:1–10. https://doi.org/10.1016/j.envsci.2018.04.007

Awasthi AK, Zeng X, Li J (2018) Environmental pollution from electronic waste recycling in India: a critical review. Environ Sci Pollut Res 25:16215–16235. https://doi.org/10.1007/s11356-018-2147-5

Awasthi AK, Zeng X, Li J (2016a) Relationship between e-waste recycling and human health risk in India: a critical review. Environ Sci Pollut Res 23:11509–11532. https://doi.org/10.1007/s11356-016-6085-7

Awasthi AK, Zeng X, Li J (2016b) Environmental pollution of electronic waste recycling in India: a critical review. Environ Pollut 211:259–270. https://doi.org/10.1016/j.envpol.2015.11.027

Baldé CP, Angelo ED, Luda V, Deubzer O, Kuehr R (2022) Global transboundary e-waste flows monitor-2022. United Nations Institute for Training and Research (UNITAR), Bonn, Germany
Chakraborty SC, Zaman MWU, Hoque M (2022) Metals extraction processes from electronic waste: constraints and opportunities. Environ Sci Pollut Res 29:32651–32669. https://doi.org/10.1007/s11356-022-19322-8
Chatterjee S, Kumar A, Basu S (2021) Role of LCA in sustainable design and e-waste recycling optimization. Cleaner Engineering and Technology 4:100210. https://doi.org/10.1016/j.clet.2021.100210
Cucchiella F, D'Adamo I, Rosa P (2015) Life cycle analysis of e-waste management systems: environmental and economic assessment. Renew Sust Energ Rev 51:263–272. https://doi.org/10.1016/j.rser.2015.06.049
Cucchiella F, Gastaldi M, Koh SL (2021) Environmental and economic impacts of e-waste recycling: A systematic review. Renew Sust Energ Rev 135:110235. https://doi.org/10.1016/j.rser.2020.110235
Cui J, Forssberg E (2017) Mechanical separation of metals from e-waste: Environmental and technical assessment. Waste Manag 61:135–145. https://doi.org/10.1016/j.wasman.2017.01.006
Dutta D, Goel S (2021) Understanding the gap between formal and informal e-waste recycling facilities in India. Waste Manag 125:163–171. https://doi.org/10.1016/j.wasman.2021.02.045
Dwivedy M, Mittal RK (2012) An investigation into e-waste flows in India. J Clean Prod 37:229–242. https://doi.org/10.1016/j.jclepro.2012.07.017
Evans ML, Kumar S (2023) Critical material recovery from e-waste. United States Energy Association. https://usea.org/sites/default/files/USEA633-2023-004-01_CMfromEwaste_FINAL_REPORT.pdf
Forti V, Baldé CP, Kuehr R, Bel G (2023) The global e-waste monitor 2020: quantities, flows, and the circular economy potential. United Nations University, ITU & ISWA. https://ewastemonitor.info/wp-content/uploads/2020/11/GEM_2020_def_july1_low.pdf
Glasson J, Therivel R, Chadwick A (2012) Introduction to environmental impact assessment. ISBN: 978–0–415–66468–4 (hbk) ISBN: 978–0–415–66470–7 (pbk)
Grant K, Goldizen FC, Sly PD, Brune MN, Neira M, van den Berg M, Norman RE (2013) Health consequences of exposure to e-waste: a systematic review. Lancet Glob Health 1(6):e350–e361. https://doi.org/10.1016/S2214-109X(13)70101-3. PMID: 25104600
Gu F, Zhang Y, Li J (2020) Life cycle assessment of e-waste recycling: a case study of mobile phone recycling in China. J Clean Prod 258:120688. https://doi.org/10.1016/j.jclepro.2020.120688
Gui L, Chen Y, Chen X (2021) Advanced recycling technologies for printed circuit boards: a review. J Environ Manag 280:111825. https://doi.org/10.1016/j.jenvman.2020.111825
Hossain R, Sahajwalla V (2024) Current recycling innovations to utilize e-waste in sustainable green metal manufacturing. Philos Trans A Math Phys Eng Sci 382(2284):20230239. https://doi.org/10.1098/rsta.2023.0239. PMID: 39489178; PMCID: PMC11531906
Jay S, Jones C, Slinn P, Wood C (2007) Environmental impact assessment: retrospect and prospect. Environ Impact Assess Rev 27(4):287–300. https://doi.org/10.1016/j.eiar.2006.12.001
Kiddee P, Naidu R, Wong MH (2013) Electronic waste management approaches: an overview. Waste Manag 33(5):1237–1250. https://doi.org/10.1016/j.wasman.2013.01.006. PMID: 23402807
Kumar A, Samadder SR (2022) Assessment of energy recovery potential and analysis of environmental impacts of waste to energy options using life cycle assessment. J Clean Prod 365:132854. https://doi.org/10.1016/j.jclepro.2022.132854
Leung AOW, Wong MH, Luo Q, Deng W (2020) Remediation effectiveness at Guiyu, China: post-intervention assessment of soil and air quality. Environ Res 188:109783. https://doi.org/10.1016/j.envres.2020.109783
Li J, Xu L, Liu L (2021) Bioleaching of metals from e-waste: advances and challenges. J Hazard Mater 414:125527. https://doi.org/10.1016/j.jhazmat.2021.125527
Li J, Zhang Y, Gu F (2020) Recovery of valuable metals from e-waste: a review. Waste Manag 102:1–18. https://doi.org/10.1016/j.wasman.2019.10.014

Li Z, Diaz LA, Yang Z, Jin H, Lister TE, Vahidi E, Zhao F (2019) Comparative life cycle analysis for value recovery of precious metals and rare earth elements from electronic waste. Resour Conserv Recycl 149:20–30. https://doi.org/10.1016/j.resconrec.2019.05.025

Liang Y, Wang H, Zhang X (2020) Microwave-assisted pyrometallurgical processing of e-waste: Energy efficiency and emission control. J Clean Prod 260:121062. https://doi.org/10.1016/j.jclepro.2020.121062

Liu K, Zhang Y (2023a) A global perspective on e-waste recycling: challenges and opportunities. Environ Sci Technol 57(2):123–135. https://doi.org/10.1021/acs.est.2c06234

Liu Z, Zhang Y (2023b) Informal electrical and electronics waste recycling and its health impacts: a systematic review. Environ Health Perspect 131(2):027001. https://doi.org/10.1289/EHP7266

Liu K, Tan Q, Yu J, Wang M (2023) A global perspective on e-waste recycling. Circ Econ 2:100028. https://doi.org/10.1016/j.cec.2023.100028

Michael LK, Kumar S (2024) Factors influencing the behavior in recycling of e-waste: a study in the Indian context. J Environ Manag 300:113567. https://doi.org/10.1016/j.jenvman.2021.113567

Oguchi M, Sakanakura H, Terazono A, Takigami H (2012) Fate of metals contained in waste electrical and electronic equipment in a municipal waste treatment process. Waste Manag 32(1):96–103. https://doi.org/10.1016/j.wasman.2011.09.012

Osborne VO (2024) Strategies used in United States e-waste recycling businesses. Walden dissertations and doctoral studies. https://scholarworks.waldenu.edu/cgi/viewcontent.cgi?article=17285&context=dissertations

Panwar R, Ahmed S (2018) Assessment of contamination of soil and groundwater due to e-waste handling. Curr Sci 114(1):166–173. https://doi.org/10.18520/cs/v114/i01/166-173

Parvez SM, Rahman M (2021) Health consequences of exposure to e-waste: an updated review. Environ Health Perspect 129(4):045001. https://doi.org/10.1289/EHP7266

Priyadarsini S, Das AP (2024) Analytical and structural characterization of waste lithium-ion batteries for their effective recycling strategy. Environ Chem Ecotoxicol. https://doi.org/10.1016/j.enceco.2024.12.004

Priyadarsini S, Das AP (2026) Lithium in waste water: impacts on pollution and human health. Decarbonization of Wastewater Pollutants as a Sustainable Solution, 353–364. https://doi.org/10.1016/B978-0-443-33062-9.00010-2

Ramachandra TV, Varghese S (2018) E-waste management: mitigation strategies and assessment of Bengaluru E-waste Park. Curr Sci 115(4):657–664. https://www.currentscience.ac.in/Volumes/115/04/0657.pdf

Rathi AKA (2017) Evaluation of project-level environmental impact assessment and SWOT analysis of EIA process in India. Environ Impact Assess Rev 67:31–39. https://doi.org/10.1016/j.eiar.2017.08.004

Rautela R, Singh P, Arya S (2022) Comparative assessment of recycling processes for waste electrical and electronic equipment: a life cycle perspective. Sustain Mater Technol 32:e00373. https://doi.org/10.1016/j.susmat.2022.e00373

Shahabuddin M, Hossain M (2023) A review of the recent development, challenges, and prospects of electronic waste (e-waste) recycling. Environ Sci Pollut Res 30(5):1234–1248. https://doi.org/10.1007/s11356-022-04274-w

Sinha-Khetriwal D, Kraeuchi P, Schwaninger M (2005) A comparison of electronic waste recycling in Switzerland and in India. Environ Impact Assess Rev 25(5):492–504. https://doi.org/10.1016/j.eiar.2005.04.006

Sivaranjani S, Marimuthu R, Shanthi M, Aramvith S (2021) Smart waste management model for effective disposal of waste management through technology. In: Maheswar R, Balasaraswathi M, Rastogi R, Sampathkumar A, Kanagachidambaresan GR (eds) Challenges and solutions for sustainable smart city development. EAI/Springer Innovations in Communication and Computing. Springer, Cham. https://doi.org/10.1007/978-3-030-70183-3_9

Song Q, Li J, Zeng X (2018) Comparative life cycle assessment of e-waste recycling technologies. Sci Total Environ 626:1359–1368. https://doi.org/10.1016/j.scitotenv.2018.01.163

Song Q, Liu Y, Zhang L, Xu Z (2021) Selective electrochemical extraction of copper from multi-metal e-waste leaching solution and its enhanced recovery mechanism. J Hazard Mater 407:124799. https://doi.org/10.1016/j.jhazmat.2020.124799

Tuncuk A, Stazi V, Akcil A, Yazici EY, Deveci H (2012) Aqueous metal recovery techniques from e-scrap. Hydrometallurgy 118(1–2):1–10. https://doi.org/10.1016/j.hydromet.2011.11.004

UNEP (2023) Annual report. https://www.unep.org/resources/annual-report-2023

Wäger PA, Hischier R, Eugster M (2011) Environmental impacts of the Swiss collection and recovery systems for waste electrical and electronic equipment (WEEE): a follow-up. Sci Total Environ 409(10):1746–1756. https://doi.org/10.1016/j.scitotenv.2011.01.050

Wang F, Zhang Y (2021) Life cycle assessment of e-waste recycling: a case study of mobile phone recycling in China. J Clean Prod 278:123876. https://doi.org/10.1016/j.jclepro.2020.123876

Wang MYL, Chi X, Streicher-Porte M, Reuter MA (2011) Informal electronic waste recycling: a sector review with special focus on China. Waste Manag 31(4):731–742. https://doi.org/10.1016/j.wasman.2010.11.006

Wang B, Ren C, Dong X, Zhang B, Wang Z (2019) Determinants shaping willingness towards online recycling behaviour: an empirical study of household e-waste recycling in China. Resour Conserv Recycl 143:218–225. https://doi.org/10.1016/j.resconrec.2019.01.005

Xue M, Chen W, Li B (2020) Life cycle assessment of hydrometallurgical and pyrometallurgical recycling of waste printed circuit boards. Environ Sci Pollut Res 27:44259–44272. https://doi.org/10.1007/s11356-020-10425-2

Yla Mella J, Pongracz E (2016). Drivers and constraints of critical materials recycling: the case of Indium. Resources 5(4):34. https://doi.org/10.3390/resources5040034

Zeng X, Chen S, Li J (2019) AI and robotics in e-waste recycling: opportunities for automation and recovery optimization. Resour Conserv Recycl 141:102–110. https://doi.org/10.1016/j.resconrec.2018.10.034

Zeng X, Yang C, Li J (2022) AI-driven life cycle analysis for e-waste recycling optimization: emerging trends. J Environ Manag 316:115306. https://doi.org/10.1016/j.jenvman.2022.115306

Critical Metal Bioleaching from E-Waste: Global Case Studies and Circular Economy Integration

Santoshi Mohanta, Chinmayee Mohapatra, Don Adrian Meshan Dassanayake, and Lipika Parida

Abstract The increasing trend in the production of electronic waste (e-waste) has been of critical concern in terms of depletion of resources and environmental pollution. The e-waste carries substantial amounts of critical metals like nickel, gold, copper, cobalt and rare earth elements, which are highly important to the current technologies. Nonetheless, conventional techniques of metal recovery are, on most occasions, energy-intensive and costly as well as hazardous to the environment. Microbial-based process bioleaching has become a sustainable and low-impact process that can be applied to metal extraction found in e-waste and industry residues. This chapter discusses a research series of various case studies of various countries like China, India, Germany, etc., to show the practical example of bioleaching technology implementation into various surroundings of the earth. Each case commentates on the kind of waste used, the microbial consortium applied, the process factors, percentage recovery of metals, and the environmental advantages. Particular emphasis was put on the role of acidophilic and chemolithotrophic bacteria, including *Acidithiobacillus ferrooxidans* and *Leptospirillum* spp., in facilitating metal solubilization. Another aspect covered in the chapter is the policy structures, technological preparedness, and socio-economic forces that define the incorporation of bioleaching practices in various regions. Incorporating bioleaching into a circular economy management strategy, nations will be able to mitigate environmental impact, save natural resources, and reclaim useful resources in waste streams.

Keywords Electronic waste · Bio-leaching · Metal recovery · India · China · Germany

S. Mohanta · C. Mohapatra · D. A. M. Dassanayake
School of Chemical Engineering, KIIT Deemed University, Bhubaneswar, Odisha, India
e-mail: santoshi.mohanta@kiitbiotech.ac.in

L. Parida (✉)
Chemical Engineering Department, Veer Surendra Sai University of Technology, Burla, Odisha, India
e-mail: lparida_chemical@vssut.ac.in

A. P. Das and S. Priyadarsini (eds.), *Recovery and Recycling of Critical Metals from Electronic Waste for Environmental Sustainability*, Earth and Environmental Sciences Library, https://doi.org/10.1007/978-3-032-29064-9_14

1 Introduction

Rapid technological advancement and digitization have led to an extraordinary surge in the production of electronic waste, or "e-waste." Discarded electrical and electronic equipment (EEE), including cell phones, computers, communication devices, and household appliances, is commonly known as e-waste. Any socioeconomic area has seen a sharp increase in the handling of electronic devices because to the worldwide trends of automation, smart technology, and the Internet of Things (IoT). Thus, the global supply of obsolete electronics has become a significant environmental issue (ITU 2022). According to Fig. 1, just 23.3% (13.8 billion kg) of the 62 billion kg (or 7.8 kg/person) of e-waste produced worldwide in 2022 was collected and recycled formally. Since 2010, the quantity of e-waste generated year has climbed by over 2.3 billion kg, which is nearly five times faster than the growth in formal recycling (Baldé et al. 2024a, b). This growing disparity highlights how ineffective the current global e-waste management methods are and how urgently the world needs sustainable and circular solutions.

The majority of e-waste consists of metals (31 billion kg) and plastics (17 billion kg), with the remainder made up of glass, ceramics, and composite materials. Unpleasant recycling procedures, especially in the chaotic regions of low- and middle-income countries, put the environment and human health at risk by releasing toxic materials into the atmosphere. Ineffective e-waste management leads to significant greenhouse gas emissions and mercury contamination. The Global E-waste

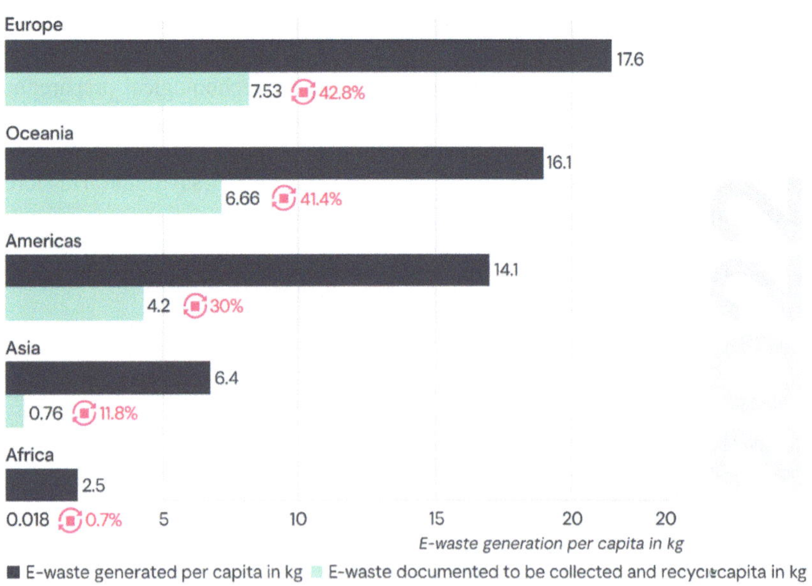

Fig. 1 Amount of e–waste produced and collected (Baldé et al. 2024a, b

Monitor 2024 reports that about 95% of mercury emissions from e-waste come from discarded lamps, releasing around 58 tonnes of mercury in 2022 due to poor recycling. The Minamata Convention on Mercury (UNEP 2019) aims to phase out such mercury-added products. Additionally, improper disposal of temperature-exchange devices, such air conditioners and refrigerators, emits refrigerants that contribute to global warming and the thinning of the ozone layer.

The efficient e-waste management systems are the key to recovering valuable secondary raw materials, preventing environmental pollution, and reducing greenhouse gas (GHG) emissions. By recycling e-waste in an appropriate manner 52 billion kg of CO_2-equivalent emissions and the digging of approximately 900 billion kg of ore in primary mining activities were avoided. Urban mining is increasingly being embraced as a viable and sustainable way of recovering resources because it focuses on acquiring useful materials of anthropogenic origin waste rather than on virgin ores (UNDP and UNEP 2018). Also, compared to traditional mining, urban mining mitigates the loss of biodiversity and land degradation, air and water pollution caused by the other activities, and social challenges such as child labor and illegal mining (Urban Mines, n.d.). The metal waste contained in e-waste around the world in 2022, when including copper, gold, and iron, has a total value of about USD 91 billion. This is however the case since the collection and recycling efficiencies were low, which meant that USD 28 billion of the secondary raw materials could only be collected and recycled and this implies that a lot of money and resources were wasted (IPCC 2023).

It was estimated that e-waste generation will keep on growing exponentially with an extrapolation of up to 82 billion kg of e-waste worldwide by the year 2030. The developed world such as the European Union and North America generates disproportionately large amounts of e-waste in comparison to the developing world. An example can be given of Pakistan, where the scrap is imported of over 50 kg and generates about 38 kg of e-waste annually, making it more difficult to handle at home (Sajid et al. 2018). It is estimated that annually the entire world produces 2.6 million tonnes of e-waste and this is why there is a need to ensure that effective policies are implemented and that the international community works together. E-waste disposal remains a significant threat to the human health and environment despite improving the product design and using materials that are less toxic (Lee et al. 2017). Tough waste management strategies that are based on the 3Rs principle that include reduce, reuse, and recycle are thus instrumental in moving to the global circular economy.

The e-waste management technology that is the most recent one is bioleaching. It is also an economical and ecofriendly method of extracting metal. Microorganisms have been applied in bioleaching by biochemical oxidation-reduction reactions to solubilize metals and form soluble metal complexes. This biotechnological method has several merits compared to conventional hydrometallurgical and pyrometallurgical techniques that are usually costly, energy-consuming, and dangerous in terms of their chemical characteristics (Yaashikaa et al. 2022; Mohanta et al. 2026). Furthermore, bioleaching has been recognized as an essential component of environmentally responsible e-waste recycling as well as the application of the circular economy idea of resource recovery (Parida et al. 2023).

The procedure has proven especially promising in the treatment of low-grade ores and secondary resources that are unprofitable under standard methods. Other study results show that bioleaching has the potential to be utilized to recover REEs from red mud, fly ash, and electronic debris, among other things, while also reducing health hazards associated with waste storage (Joshi et al. 2025). On the same note, research on goldmine tailings shows that bioleaching of some of these metals, including praseodymium, europium, and cerium, can be effectively mobilized using selected microbial systems, which can be combined with pretreatments to reduce environmental risks associated with waste storage. Its industrial-level uses underscore its relevance to key supply networks. Nickel bioleaching is one example of a low-carbon method of producing battery-grade material for electric cars and renewable energy storage. According to the comparative analysis, bio-heap leaching has the potential to reduce cradle-to-gate carbon emissions by several orders of magnitude when compared to pyrometallurgical refining (Panyushkina and Muravyov 2025). Thus, bioleaching is more than a technological advancement; it symbolizes a paradigm change toward low-impact, cyclical resource recovery. In a period of resource scarcity and environmental urgency, it allows for metal extraction from both primary ores and various waste streams, therefore decarbonizing mining operations, reducing environmental risks, and ensuring access to essential raw materials.

3.1 Principles of Bioleaching

Bioleaching is a hydrometallurgical process in which microorganisms dissolve and mobilize metals in solid matrices. It works by combining microbial metabolism with mineral substrates to recover both base and vital metals in an ecologically benign and energy efficient manner (Harshan and Rajan 2025). Sulfur compounds are oxidised by chemolithoautotrophic microbes to sulfuric acid, which reduces the pH and increases the proton attack on mineral lattices. This holds in particular in the case of oxide and carbonate matrix (Shokoohi et al. 2025). The oxidation of ferrous iron to ferric iron is catalyzed by many acidophilic bacteria (e.g., *Acidithiobacillus* spp.) which in turn acts as a strong oxidant of sulfide mineral. This indirect control is important in sulfide ores, though it may be applied to waste streams with Fe-sulfide phases (Vardanyan et al. 2025). Organic acids (citric, oxalic, gluconic), siderophores and in certain instances biogenic cyanide are produced by heterotrophic organisms, where the phytonid group binds and solubilizes the metals. The process is particularly suitable to REEs and valuable metals in secondary wastes such as phosphorus, slag, fly ash, and printed circuit boards (Belov et al. 2025).

The efficiency is regulated by optional settings. Pulp density and particle size control lixiviant accessibility, aeration controls a sufficient number of electron acceptors to support oxidative bioleaching, and chloride, heavy metal load, and other inhibitors directly affect microbial viability and thus dissolution kinetics. Temperature also has an effect on the efficiency of leaching-the mesophilic and thermophilic acidophiles exhibit variances. Proteomic studies have revealed that microbial adaptation to stress (e.g. incineration residues) involves up-regulation of proteins involved with metal tolerance and redox balance that's also associated with a higher yield of leaching. Bioleaching is related to the circular economy paradigm if combined with subsequent separation process and in lifecycle assessment methods. Both Munonye et al. and Mugion et al. highlight that the integration of microbial processes within closed loop value chains optimizes sustainability because of the lower consumption of energy as compared to pyrometallurgy and lower secondary waste streams (Munonye et al. 2025; Mugion et al. 2025).

3.2 Microorganisms Used in Bioleaching

The activity and adaptability of microorganisms that are able to oxidize metal sulfides is important to effectiveness of bioleaching. *Leptospirillum* species, *Sulfobacillus* species, *Acidithiobacillus ferrooxidans*, and *Acidithiobacillus thiooxidans* are required in sulfide bioleaching. These bacteria accelerate the dissolution of the minerals by oxidizing ferrous iron and reduced sulfur compounds resulting in ferric iron and sulphuric acid (Vardanyan et al. 2025). Their success in industrial biomining may be owed to their resilience to high acidity and heavy metal conditions. There is proteomic evidence of adaptive changes in *Acidithiobacillus ferridurans* under high metal stress with more expression of tolerance and redox enzymes to support an efficient bioleaching in incineration residues. Fungi (*Aspergillus*, *Penicillium*, microscopic fungi) and yeasts are widely applied on the use of oxide and REE bearing matrices. Belov et al. showed that the dissolution of poorly soluble neodymium compounds has been promoted by microscopic fungi (Belov et al. 2025) which secrete oxalic and citric acids.

3.3 Applications of Bioleaching

Bioleaching is a technique that offers innovative solutions to mining, waste valorisation and circular resource recovery applications, alongside the original one of sulfide ore processing. Recent research has proven that microbial systems can potentially be adapted to meet economic and environmental concerns. For example the Bacillus cereus has been found to extract residual gold from tailings and metabolize cyanide as a nutrition, thus it combines metal recovery and detoxification. Closed-loop bioleaching is a sustainable approach to recover copper, gold and lithium from discarded mobile phones in the framework of electronic waste, and the regenerative leaching process is especially essential for the process viability (Morell et al. 2025).

Apart from mining, bioleaching technique also has found an application in the agricultural sector, in which microbial communities in natural pig manure produce organic acids that dissolve copper and zinc, redefining waste treatment as resource recovery. The application of autochthonous microbial consortia from Peruvian tailings provides evidence of the strength and place-specific adaptability of indigenous communities whereas the quorum sensing manipulation of *Acidithiobacillus thiooxidans* with synthetic analogs has been shown to boost the efficiencies of sulfur oxidation and leaching substantially. These applications are subsequently ventured into by advancements in microbial engineering. In more complex matrices, oxidative bioleaching has allowed for the recovery of nickel from ores containing high arsenic levels whilst stabilizing hazardous byproducts, and optimization of *Acidithiobacillus ferroxidans* has allowed for selective ability in the recovery of precious metals from discharged lithium-ion batteries 50. Complementary strategies have also been implemented to overcome chalcopyrite passivation including mechanical activation which facilitates the extraction of copper. Collectively these applications demonstrate the versatility of bioleaching both as a green extraction technology and a remediation tool, although reviews of the field continue to stress the need for overcoming the limitations on kinetics and scaling problems before widespread industrial deployment can be achieved (Nnaemeka et al. 2025).

4 Management of E-Waste in India

Regarding the production of e-waste, India is among the top nations worldwide. Around 3.23 million tons of e-waste was generated nationwide in 2019, translating to an annual e-waste output per person of about 2.4 kg. To control e-waste, India has put in place a regulatory framework. Below are some of them.

- E-waste Management and Handling Rules 2016 (amended in 2022)
- 2010 E-waste Management and Handling Rules (effective 2012)
- 2008 Hazardous and Waste Management Rules
- 2004 Municipal Solid Waste Management Rules.

Regulations in India place a strong focus on Extended Producer Responsibility (EPR), which mandates that electronics manufacturers set up collecting and recycling facilities. However, due to idiosyncrasies in the e-waste management system, they have been difficult to execute (Phogat et al. 2025). 90% of India's e-waste is recycled in the unorganized sector by a variety of individuals, including door-to-door scrap collectors known as Kawariwalas. Workers without proper training handle and handle over 95% of e-waste. It can be poisonous and dangerous for these personnel to handle such garbage in these situations (Andeobu et al. 2021).

Some unique market dynamics such as, customers in India are eager to sell their obsolete electronic garbage to door-to-door collectors, unlike developed countries where producers and consumers pay recycling/disposal fees and the recycling process being operated as a market driven industry can be challengers when it comes to implementing sustainable recycling habits (Andeobu et al. 2021). However, despite being faced by many challenges India has emerged as a significant contributor to e-waste management related research. India was able to release 116 publications, second only to China with 124 publications. These articles mostly focus on economical and ecologically friendly ways to counteract unofficial recycling methods and the rising quantity of e-waste (Phogat et al. 2025).

Even though India faces these challenges considering the research done on e-waste it's safe to say that there could be a promising future regarding e-waste in India. The research done could lead to tailor made solutions which can be more effective than solutions which were implemented in other countries. Additionally, funding organizations are essential to the continued study of metal recovery from e-waste, which promotes creative sustainable practices and other technological developments. The main organization in India that provides funding for research initiatives with an economic and environmental impact is the Department of Science and Technology (DST). With the support received from DST, 12 research publications happened in 2025 on advancement of metal extraction from e-wastes. Also, when research is done at this scale it attracts international collaboration. Which then would bring in more advanced technologies into the country.

In conclusion India's e-waste management presents a complex challenge due to rapid generation growth, lack of implementation of regulations, high dependence on informal sector and the significant health and environmental risks. However, considering the research done, India can use it to develop tailer made solutions to the country's unique socio-economic conditions.

5 Case Studies of E-Waste Management in Various Countries

Globally, the handling of e-waste has developed into an vital component of sustainable resource recovery techniques. Because various nations have varied tactics and techniques depending on their technical prowess, policy formations, and socioeconomic situations, the level of formal e-waste management development differs amongst them. Table 1 displays the most recent e-waste generation figures by country and area, which were gathered by The Global E-waste Monitor 2024 (Baldé et al. 2024a, b).

In 2022, Asia contributed 30 billion kilograms of e-waste, close to half the global production of it, and Europe and the Americas are coming behind (13 and 14 billion kg), with Africa and Oceania trailing the formal recycling rates (Baldé et al. 2024a, b).

5.1 E-Waste Bioleaching in China

China is the largest e-waste generating country in the world as well as one of the top countries in the development of bioleaching type of recycling. Traditionally, China

Table 1 Global e-waste generation by country and region, 2022 (Baldé et al. 2024a, b)

Rank	Country	Region	E-waste produced (million kg)	Per capita (kg)
1	China	Asia	12,000	8
2	United States of America	Americas	7200	21
3	India	Asia	4100	2.9
4	Japan	Asia	2600	21
5	Brazil	Americas	2400	11
6	Germany	Europe	1767	21.2
7	Indonesia	Asia	1900	6.9
8	Russia	Europe	1900	13
9	United Kingdom	Europe	1700	24
10	France	Europe	1445	22.4
11	Italy	Europe	1100	18
12	Turkey	Asia (Western)	1100	13
13	Mexico	Americas	1500	12
14	Iran (Islamic Republic of)	Asia	820	9.3
15	Canada	Americas	770	20
16	South Africa	Africa	530	7.8
17	Spain	Europe	930	19
18	Egypt	Africa	690	6.3
19	Saudi Arabia	Asia	620	14
20	Australia	Oceania	580	22

was a big importer of foreign e-waste, especially in such areas as Guiyu, Guangdong Province, which was associated with informal and ecologically dangerous recycling operations. Since the policy on importing foreign e-waste called the National Sword was introduced, the attention was shifted to the management of waste stream produced domestically (Fu et al. 2018). In 2022, China's domestic e-waste production grew by over 12 million tons, and by 2030, it is expected to reach 28.4 million tonnes (Fu et al. 2018; Baldé et al. 2024a, b). Although 109 licensed recycling businesses exist, merely one-fourth of all e-waste is recycled by formal channels, and a large part of it goes to informal recyclers (Fu et al. 2018). According to Grant et al. (2013), formal recycling has harmed the ecosystem and human health by contaminating food products, soil, and the air with heavy metals. China in turn has increased the study of bioleaching an environmentally friendly biotechnological substitute of conventional hydrometallurgical and pyrometallurgical operations. The Chinese Academy of Sciences and Beijing University of Chemical Technology have conducted bioleaching studies on acidophilic bacteria, such as *Leptospirillum ferrooxidans, Acidithiobacillus ferrooxidans*, and *Thiobacillus caldus* (Wang et al.

2022). Using organic acid, other fungal species as *Penicillium simplicissimum* and *Aspergillus niger* are also utilized to dissolve metals (Li et al. 2018; Parida 2025).

In China, there are optimized bioleaching processes that attain over 95 and 90 percent recovery efficiencies of copper (Cu) and zinc (Zn) in controlled conditions (Zhu et al. 2017). New piloting small-scale applications made by the Chinese Academy of Sciences have shown semi-continuous systems capable of recovering Cu, Zn, Ni, and Au at efficiencies over 70 percent, with the applications consuming a small amount of chemicals and generating minimal secondary pollution (Wang et al. 2022). They are currently incorporated into eco-industrial zones, including the Guiyu Circular Economy Industrial Park, as part of the Circular Economy Promotion Law (2008) of China and the 14th Five-Year Plan (2021–2025). The future direction of microbial biomineralization is the genetic engineering of microbial strains, enhanced metal tolerance, the combination of bioleaching with bioelectrochemical system (BES), and the valorization of the bioleached residues to create zero-waste outcomes (Li et al. 2018; Liu et al. 2019). The bioleaching innovations of China are a clear example of the way in which biotechnology can support the circular resource recovery process in the national systems of sustainability.

5.2 E-Waste Bioleaching in the United States

The bioleaching programs in the United States are still in the laboratory and pilot phase, unlike the case of China where bioleaching programs are in the advanced stage. Bioleaching has been investigated by national laboratories and universities as well as federal agencies, including the U.S. Department of Energy (DOE) and the United States Energy Association (USEA) in the context of recovering from certain e-waste sources, including phosphors, magnets and printed circuit boards (Reed 2016; Battelle 2023). A team of researchers at the DOE in the Oak Ridge National Laboratory demonstrated microbial leakage of REE phosphors used in fluorescent lamps, and revealed that microbes and microbial-produced organic acids might be useful in mobilizing critical elements. Other studies have successfully recovered metals of magnets in laboratory scale with the help of acidophilic and heterotrophic microbial systems (Tapia et al. 2022).

The Battelle/USEA (2023) report highlighted that despite the environmentally friendly benefits of bioleaching as compared to chemical extraction, the major challenges to commercialisation are long reaction times, feedstock heterogeneity, and microbial inhibition caused by complex additives in e-waste. The research of the U.S. is aimed at hybrid processing pathways, i.e., the mechanical pretreatment, mild chemical leaching, and microbial metal recovery (USEA/Battelle 2023). Nevertheless, the technical issues notwithstanding, specific niche applications, like NdFeB magnets, and high-value printed circuit boards are discussed as possible candidates to pilot-scale demonstration. The DOE and Department of Defense (DoD) are also funding scaling up of low-carbon recovery technologies through federal funds (Kwok 2019). The situation in the U.S. bioleaching, therefore, has a strategic interest in biologically aided critical material recovery, in keeping with national supply-chain security objectives.

5.3 Bioleaching and E-Waste Management Germany

Germany is one of the strongest countries, with regard to e-waste management and the implementation of a circular economy, producing nearly 1.77 million tonnes of e-waste (21.2 kg per person) in 2022 (Baldé et al. 2024a, b). Its system is being run in accordance with the Waste electrical and electronic equipment (WEEE) Directive (2012/19/EU) and national Elektro- und Elektronikgerraetegesetz (ElektroG), with the focus on EPR and high recycling levels (BMUV, 2021). Formal collection rates are above 45 and there is effective collection of ferrous and non-ferrous metals by using the sophisticated recycling facilities.

The Circular Economy Act (KrWG) and the Raw Materials Strategy (2020) of Germany facilitate the idea of sustainable resource recovery and the minimization of primary mining (BMUV 2021). Bioleaching studies have increased as part of such a strategy, and other studies by the Helmholtz Centre for Environmental Research and TU Freiberg show that the following acids can be used to extract Cu, Zn, and Au out of e-waste: *Acidithiobacillus ferrooxidans* and *Leptospirillum ferrooxidans* (Brandl et al. 2001; Rohwerder et al. 2008). More recent initiatives like Biohydrometall 2020 and ReBioMetal funded by the German Federal Ministry of Education and Research (BMBF) recover over 90% of the Cu and almost 80% of the Au under favourable bioreactor conditions (Hein et al. 2022). These projects show how Germany has embraced the concept of biotechnology and the circular economy to minimize the environmental effects and enhance the recovery of metal sustainably.

6 Circular Economy Integration

Bioleaching mostly comes into play in areas such as repairing, refurbishing and most importantly Recycling. Our focus here mostly has been on extracting metals and metallic materials from E-waste using bioleaching rather than conventional methods like pyrometallurgy and hydrometallurgy. The final goal in the circular economy model is to close the loop (Murmu, et al. 2025). This means utilizing waste to extract materials which will be used back in production. Methods like pyrometallurgy and hydrometallurgy were fundamentally used in extracting virgin materials from ores. Moreover, bioleaching being a more modern technology is being focused on material/metal recovery. This supports the circular economy goals of conserving virgin materials, sustainability and resource efficiency (Phogat et al. 2025). Table 2 shows the fraction of metals recovered through bioleaching from spent lithium-ion phone batteries using *Aspergillus niger* (Bahaloo-Horeh et al. 2018). Some of these yields are even higher than the yield possible from ores.

Table 2 Metal fraction recovered from spent Li-ion batteries by bio-leaching

Metal	Percentage (%)
Lithium	100
Manganese	72
Nickel	45
Copper	94
Aluminium	62
Cobalt	38

Apart from the said contribution to the circular economy bioleaching can be greener than conventional metal recovery methods. While conservation of virgin materials is already a step towards being greener, bioleaching also eliminates the harsh chemicals and high energy demand for the process (Adetunji et al. 2023). This technique uses microorganisms which are essentially living organisms rather than chemicals like in hydrometallurgy, thus producing environmentally safer effluents. High energy demands also can be environmentally unfriendly since we are yet unable to produce pure green energy. These characteristics of bioleaching reduces overall carbon footprint and increases profit margins which ultimately leads to achieving goals in the circular economy where sustainability is prioritised (Erkmen et al. 2025; Phogat et al. 2025; Adetunji et al. 2023; Brown et al. 2022).

Bioleaching also carves solutions for a very important process; Waste Valorization (Adetunji et al. 2023), which is one of the primary goals in the circular economy where it closes the loop in the circular economy (Fig. 4). However, direct implementation of bioleaching can be a major challenge. Hence using hybrid methods where bioleaching is incorporated with other methods would be far better in initial implementations (Erkmen et al. 2025). Importantly it can be considered as a key technology in achieving a circular economy. It delivers cost-effective metal recovery, decreases waste, lowers environmental pollution, promotes closed-loop material flows, and reduces virgin resource extraction, all of which are sustainability objectives that are addressed concurrently.

7 Conclusions

E-waste management has grown to be a significant environmental and social issue worldwide as a result of the increasing popularity of consumer electronics. The ways in which different countries handle e-waste differ, with an emphasis on environmental protection, resource recovery, and rubbish reduction. While wealthier nations use sophisticated recycling techniques and well-organized collection networks, underdeveloped countries often resort to uncontrolled, health-hazardous activities. Bioleaching, which uses microorganisms, is a sustainable method of extracting metal from e-waste. To ensure success with implementation, operational characteristics

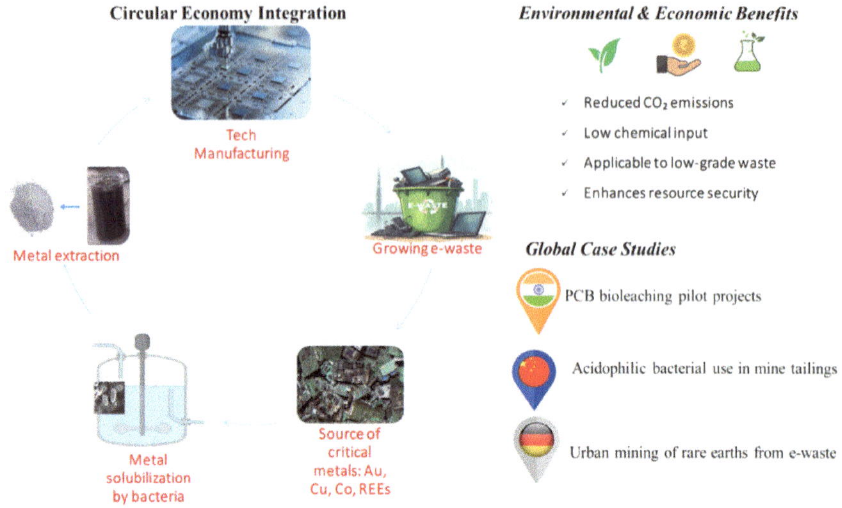

Fig. 4 Circular bioleaching approach for through microbial metal solubilization and recycling

must be modified for better efficiency. These processes can be incorporated into a circular economy to promote resource sustainability while reducing environmental effects. To transition to a sustainable electronic economy, bio-based technological progress and effective e-waste disposal need strong legislation, public participation, and multidisciplinary collaboration.

References

Adetunji AI, Oberholster PJ, Erasmus M (2023) Bioleaching of metals from e-waste using microorganisms: a review. Minerals 13(6):828. https://doi.org/10.3390/min13060828

Andeobu L, Wibowo S, Grandhi S (2021) A systematic review of e-waste generation and environmental management of Asia Pacific countries. Int J Environ Res Public Health 18(17):9051. https://doi.org/10.3390/ijerph18179051

Bahaloo-Horeh N, Mousavi SM, Shojaosadati SA (2018) Use of adapted metal tolerant *Aspergillus niger* to enhance bioleaching efficiency of valuable metals from spent lithium-ion mobile phone batteries. J Clean Prod 197(1):1546–1557. https://doi.org/10.1016/j.jclepro.2018.06.299

Baldé CP et al (2024a) The Global E-waste Monitor 2024. ITU & UNITAR

Baldé CP, Kuehr R, Yamamoto T, McDonald R, D'Angelo E, Althaf S, Bel G, Deubzer O, Fernandez-Cubillo E, Forti V, Gray V, Herat S, Honda S, Iattoni G, Khetriwal DS, Luda di Cortemiglia V, Lobuntsova Y, Nnorom I, Pralat N, Wagner M (2024b) The Global E-Waste Monitor 2024. International Telecommunication Union (ITU) and United Nations Institute for Training and Research (UNITAR), Geneva, Bonn. https://www.itu.int/itu-d/sites/environment

BMUV (2021) Federal Ministry for the Environment, Nature Conservation, Nuclear Safety and Consumer Protection

Battelle (2023) Critical material recovery from e-waste—Final report (prepared for United States Energy Association). USEA, Battelle, USEA, United States Energy Association. https://usea.org/sites/default/files/USEA633-2023-004-01_CMfromEwaste_FINAL_REPORT.pdf

Belov DV, Belyaev SN, Razov EN, Sorokoletova NA, Serebrov EI, Mosyagin PV (2025) Green strategy for bioleaching of poorly soluble neodymium compounds by microscopic fungi. Appl Biochem Microbiol 61(4):761–776. https://doi.org/10.1134/S0003683825040021

Brandl H, Bosshard R, Wegmann M (2001) Computer-munching microbes: metal leaching from electronic scrap by bacteria and fungi. Hydrometallurgy 59:319–326

Rohwerder T, Gehrke T, Kinzler K, Sand W (2008) Bioleaching review: Part A. Appl Microbiol Biotechnol 63:239–248

Brown RM, Mirkouei A, Reed D, Thompson V (2022) Current nature-based biological practices for rare earth elements extraction and recovery: bioleaching and biosorption. Renew Sust Energ Rev 173:113111. https://doi.org/10.1016/j.rser.2022.113111

Erkmen AN, Ulber R, Jüstel T, Altendorfner M (2025) Towards sustainable recycling of critical metals from e-waste: bioleaching and phytomining. Resour Conserv Recycl 215:108057. https://doi.org/10.1016/j.resconrec.2024.108057

Fu J, Zhang H, Zhang A, Jiang G (2018) E-waste recycling in China: a challenging field. Environ Sci Technol 52(12):6727–6728. https://doi.org/10.1021/acs.est.8b02329

Ghosh S, Das AP (2018) Metagenomic insights into the microbial diversity in manganese-contaminated mine tailings and their role in biogeochemical cycling of manganese. Sci Rep 2018 8:1, 8(1), 1–12. https://doi.org/10.1038/s41598-018-26311-w

Grant K, Goldizen FC, Sly PD, Brune MN, Neira M, van den Berg M, Norman RE (2013) Health consequences of exposure to e-waste: a systematic review. Lancet Glob Health 1(6):e350–e361

Harshan K, Rajan AP (2025) Unveiling the potential of microbial biominers in bioleaching for heavy metal recovery from e-waste—A comprehensive review. J Hazardous Mater Adv 18:100691. https://doi.org/10.1016/j.hazadv.2025.100691

Hein M, Rohwerder T, Sand W (2022) Biotechnol Adv 58:107938

Intergovernmental Panel on Climate Change (2023) Summary for policymakers. In: Lee H, Romero J (eds) Climate change 2023: Synthesis report. Contribution of Working Groups I, II, and III to the Sixth Assessment Report of the Intergovernmental Panel on Climate Change. IPCC, pp 1–34

ITU (2022) Facts and figures 2022. ITU Subscriptions

Jia X, Zhan Y, Tian X (2025) Recent advances in treatment technologies and resource utilization of mine tailings in Hunan Province, China. Processes 13(9):2957. https://doi.org/10.3390/pr13092957

Joshi K, Magdouli S, Brar SK (2025) Bioleaching for the recovery of rare earth elements from industrial waste: A sustainable approach. Resour Conserv Recycl 215:108129. https://doi.org/10.1016/j.resconrec.2025.108129

Kwok R (2019) How bacteria could help recycle electronic waste. Proc Natl Acad Sci (commentary/reporting on research). PNAS. https://www.pnas.org/doi/10.1073/pnas.1820329116?utm_source=chatgpt.com

Lee D, Offenhuber D, Duarte F, Biderman A, Ratti C (2017) Monitour: tracking global routes of electronic waste. Waste Manag 72:362–370. https://doi.org/10.1016/j.wasman.2017.11.014

Li J, Zeng X, Chen M (2018) Sustainable e-waste management in China: the bioleaching approach. J Hazard Mater 360:196–208

Liu Y, Zhang S, Huang D (2019) Synergistic bacterial consortia for enhanced bioleaching of copper and zinc from printed circuit boards. Bioresour Technol 278:34–42

Mishra S, Dash D, Al-Tawaha ARMS, Das AP (2022) A Review on Heavy Metal Ion Adsorption on Synthetic Microfiber Surface in Aquatic Environments. Appl Biochem Biotechnol, 194(10), 4639–4654. https://doi.org/10.1007/S12010-022-04029-W/METRICS

Mohanta S, Parida L, Sarkar A, Dassanayake DAM (2026) Bioleaching of heavy metals from electronic waste. In: Das AP, Priyadarsini S (eds) Electronic waste and environmental pollution. Sustainable environmental waste management strategies. Springer, Cham. https://doi.org/10.1007/978-3-031-90287-1_11

Morell J, Lao-Luque C, Solé-Sardans M, Ribas D, Guimerà X, Dorado AD (2025) Closed-loop bioleaching for mobile phones from the bioregeneration of the leaching agent to the recovery of metal: a review. J Ind Eng Chem 151:90–111. https://doi.org/10.1016/j.jiec.2025.03.012

Mugion RG, Elmo GC, Ungaro V, Di Pietro L, Martucci O (2025) A systematic literature review of selected aspects of life cycle assessment of rare earth elements: Integration of digital technologies for sustainable production and recycling. Sustainability 17(9):5825. https://doi.org/10.3390/su17095825

Munonye WC, Ajonye GO, Ahonsi SO, Munonye DI, Chigozie OI, Akinloye OA (2025) Advancing circularity in battery systems for renewable energy: technologies, barriers and future directions. Adv Energy Sustain Res 6(2):2500255. https://doi.org/10.1002/aesr.202500255

Murmu AK, Parida L, Prakash V (2025) Sustainable management and advanced techniques of micropollutants through circular economy. In: Behera ID, Das AP, Bal M (eds) Emerging micropollutants. Sustainable environmental waste management strategies. Springer, Cham. https://doi.org/10.1007/978-3-032-03530-1_19

Nguyen VK, Lee J-U (2015) A comparison of microbial leaching and chemical leaching of arsenic and heavy metals from mine tailings. Biotechnol Bioprocess Eng 20:91–99. https://doi.org/10.1007/s12257-014-0223-1

Nnaemeka IC, Chime TO, Nonso UC, Ikechukwu OM, Anezichukwu AF, Ikechukwu NA, Omotioma M, Chukwudi EB, Chisom MK, Ifeanyichukwu OT, Chinelo OS (2025) Examining the efficiency of microbe-assisted metal extraction: A review of bio-hydrometallurgical leaching techniques. Hybrid Adv 9:100407. https://doi.org/10.1016/j.hyaadv.2025.100407

Panyushkina A, Muravyov M (2025) Biomining nickel for a greener low-carbon future: Progress in bioleaching technologies and acidophilic microbiology. Rev Environ Sci Biotechnol 24:753–803. https://doi.org/10.1007/s11157-025-09798-5

Parida L (2025) Fungal bioremediation: a sustainable solution to petroleum hydrocarbon contamination. In: Behera ID, Das AP (eds) Environmental hydrocarbon pollution and zero waste approach towards a sustainable waste management. Environmental science and engineering. Springer, Cham. https://doi.org/10.1007/978-3-031-81615-4_9

Parida L, Murmu AK, Prakash V (2023) Circular economy model for petroleum waste and its implementation in India. In: Behera ID, Das AP (eds) Impact of petroleum waste on environmental pollution and its sustainable management through circular economy. Environmental science and engineering. Springer, Cham. https://doi.org/10.1007/978-3-031-48220-5_12

Phogat P, Kumar S, Wan M (2025) A scientometrics study of advancing sustainable metal recovery from e-waste: processes, challenges, and future directions. RSC Sustainability 3:2434–2454. https://doi.org/10.1039/d5su00049a

Priyadarsini S, Das AP (2024) Analytical and structural characterization of waste lithium-ion batteries for their effective recycling strategy. Environ Chem Ecotoxicol. https://doi.org/10.1016/j.enceco.2024.12.004

Priyadarsini S, Das AP (2026) Lithium in waste water: impacts on pollution and human health. Decarbonization of Wastewater Pollutants as a Sustainable Solution, 353–364. https://doi.org/10.1016/B978-0-443-33062-9.00010-2

Reed DW (2016) Bioleaching of rare earth elements from waste phosphors. U.S. DOE/OSTI report. OSTI. https://www.osti.gov/servlets/purl/1357607?utm_source=chatgpt.com

Rinne M, Lappalainen H, Lundström M (2025) Evaluating the possibilities and limitations of the pyrometallurgical recycling of waste Li-ion batteries using simulation and life cycle assessment. Green Chem 27(9):2522–2537. https://doi.org/10.1039/D4GC05409A

Roy S, Vishwakarma A, Hait S (2025) Enhanced bioleaching of spent lithium-ion battery cathode powder by *Acidithiobacillus ferrooxidans* for selective metal recovery: influence of ferrous iron and elemental sulfur as energy sources. Sep Purif Technol 376:134165. https://doi.org/10.1016/j.seppur.2024.134165

Sajid M, Syed JH, Iqbal M, Abbas Z, Hussain I, Baig MA (2018) Assessing the generation, recycling and disposal practices of electronic/electrical-waste (e-waste) from major cities in Pakistan. Waste Manag 84:394–401. https://doi.org/10.1016/j.wasman.2018.11.026

Shokoohi R, Vosough RN, Shabanloo A, Torkshavand Z, Mahmoudi MM, Abasi MA (2025) Removing heavy metals and improving the dewaterability of sewage sludge with the bioleaching process by *Thiobacillus ferrooxidans* bacteria. Appl Water Sci 15:34. https://doi.org/10.1007/s13201-025-02334-8

Stegemann L, Gutsch M (2025) Environmental impacts of pyro- and hydrometallurgical recycling for lithium-ion batteries: a review. J Bus Chem 22(1):62–76. https://doi.org/10.17879/43998523309

Tapia J, et al (2022). Bioleaching of heavy metals from printed circuit boards—Review and experimental studies (peer-reviewed articles summarizing lab methods relevant to U.S. Research

UNDP, UNEP (2018) Managing mining for sustainable development: a sourcebook. United Nations Development Programme, Bangkok, Thailand

UNEP 2019 Minamata convention on mercury—Text and annexes. Nairobi, Kenya

Vardanyan N, Khachatryan A, Melkonyan Z, Abrahamyan N, Barseghyan S, Zhang R, Vardanyan A (2025) Bacterial tolerance and bioleaching in the presence of chloride. Materials 18:4407. https://doi.org/10.3390/ma18134407

Velis CA, Cook E (2021) Mismanagement of plastic waste through open burning with emphasis on the global south: a systematic review of risks to occupational and public health. Environ Sci Technol 55(11):7186–7207. https://doi.org/10.1021/acs.est.0c08536. Epub 2021 May 18. PMID: 34003007

Wang J, Xu M, Li H, Zhao Y, Zhang X (2022) Bioleaching of metals from E-waste: recent advances and mechanisms. Int J Environ Res Public Health 19(7508):1–25. https://doi.org/10.3390/ijerph1907508

Wu W, Liu X, Zhang X, Zhu M (2018) Bioleaching of copper from waste printed circuit boards by bacteria-free cultural supernatant of iron–sulfur-oxidizing bacteria. Bioresour Bioprocess 5(10). https://doi.org/10.1186/s40643-018-0196-6

Yaashikaa PR, Priyanka B, Senthil Kumar P, Karishma S, Jeevanantham S, Indraganti S (2022) A review on recent advancements in recovery of valuable and toxic metals from e-waste using bioleaching approach. Chemosphere 287:132230. https://doi.org/10.1016/j.chemosphere.2021.132230

Zhu N, Xiang Y, Zhang T, Wu P, Zhang Y (2017) Bioleaching of metals from waste printed circuit boards by *Acidithiobacillus ferrooxidans* and its kinetics. Waste Management 60:357–365. US

Sustainability Assessment of E-Waste Recycling Technologies: A Life Cycle Perspective

Sweta Sinha, Ravi Ranjan, Abhishek Kumar, and Sumira Malik

Abstract The exponential growth of electronic waste presents significant environmental and economic challenges globally. This manuscript examines the sustainability of various e-waste recycling technologies through a comprehensive life cycle assessment framework. By analysing environmental impacts, economic viability, and social implications across different recycling methods, this study provides insights into optimizing e-waste management strategies. The findings reveal that hydrometallurgical processes combined with mechanical pre-treatment offer the most balanced approach to sustainability, though context-specific factors significantly influence optimal technology selection.

Keywords Environmental challenges · E-waste · Environmental sustainability · LCA

1 Introduction

As technology advances at an unprecedented pace, it brings with it a growing global challenge: electronic waste (e-waste). E-waste has rapidly become one of the largest and fastest-growing sources of pollution worldwide, driven by the expanding electronics manufacturing industry and the constant demand for newer, more sophisticated devices. Approximately 62 million tonnes of e-waste were produced in 2022,

S. Sinha (✉) · A. Kumar · S. Malik
Amity University Jharkhand, Ranchi, Jharkhand, India
e-mail: sweta.sinha2203@gmail.com

R. Ranjan · A. Kumar
Centre of Engineering and Technology (CET), Steel Authority of India (SAIL), Ranchi Unit, Jharkhand, India
e-mail: ravi.rjn@gmail.com

A. Kumar
Research Scientist, Department of Land, Air and Water Resources, University of California, Davis, CA 95616, US

A. P. Das and S. Priyadarsini (eds.), *Recovery and Recycling of Critical Metals from Electronic Waste for Environmental Sustainability*, Earth and Environmental Sciences Library, https://doi.org/10.1007/978-3-032-29064-9_15

with projections indicating an alarming increase to 82 million tonnes by 2030 (Serpe et al. 2025). The management of waste electrical and electronic equipment (WEEE) presents significant challenges due to its complex and heterogeneous composition, combining metals, plastics, ceramics, and hazardous materials (de Oliveira Neto et al. 2023). The frequent release of upgraded electronic models and the shortening of product life cycles have led to a surge in device disposal. Consumer demand for the latest smartphones, laptops, and wearable devices has accelerated the turnover rate of electronics, thereby intensifying the global e-waste crisis (Priyadarsini and Das 2025). Despite the technological feasibility of recycling electronic products, the lack of efficient waste management practices remains a major obstacle. In the United States, only about 25% of WEEE is effectively recycled, while the rest is often illegally exported, discarded to developing countries that lack the infrastructure for safe handling (Quinto et al. 2025). In these regions, valuable materials such as copper, gold, and palladium are informally recovered, but improper recycling practices lead to severe environmental contamination and health hazards (Izatt and Hagelüken 2016; Frazzoli et al. 2024). Although e-waste recycling is technically achievable, it is often economically burdensome, requiring high initial investments and advanced facilities for large-scale operations. Exporting e-waste to developing countries is viewed as a cheaper alternative, offering short-term economic gains for both parties, developed nations offload their waste, while developing nations extract valuable metals for income. However, this system is deeply flawed, as it prioritizes cost efficiency over environmental safety, ethical responsibility, and sustainable development (Izatt and Hagelüken 2016).

E-waste typically contains up to 60% valuable metals (such as copper, aluminum, gold, and iron) alongside approximately 3% toxic substances, including heavy metals and persistent organic pollutants (Frazzoli et al. 2024). While metal extraction provides economic incentives, the residual fractions particularly plastics and brominated flame-retardant materials pose severe disposal challenges. Inadequate waste treatment often results in open burning, landfilling, or dumping into water bodies, leading to long-term contamination of air, soil, and aquatic ecosystems (Al-Wabel et al. 2022). To address the e-waste challenge, a comprehensive and multifaceted strategy is essential. This includes enhancing recycling infrastructure, developing eco-design products for easier disassembly, and promoting reuse and material recovery (Sarkhoshkalat et al. 2024; Sinha 2026a). No single intervention can solve the issue entirely; rather, a coordinated approach that combines technological innovation, policy reform, and international collaboration is required for sustainable global management (Sinha 2026b; Sinha et al. 2026; Mishra et al. 2024).

Global consumption of electronics has surged dramatically over the past few decades, resulting in parallel increases in e-waste generation from sectors such as household appliances, medical devices, office equipment, and industrial electronics. Major contributors include the USA, China, Russia, and Japan, with U.S. states like Minnesota, Oregon, and Washington reporting the highest per capita e-waste collection rates (5.9–6.4 pounds/person) (Quinto et al. 2025). However, the high cost of domestic recycling continues to drive illegal exports to developing nations, where poor management practices cause significant environmental degradation and public health risks. E-waste contains numerous hazardous chemicals, including lead, mercury, cadmium, and arsenic, which contribute to soil, air, and water pollution (Houessionon et al. 2021; Sinha 2018, 2022, 2026b). Effective and sustainable disposal methods are thus critical, especially in high-generation regions. Despite the economic potential of metal recovery, large-scale recycling remains limited by inadequate policy enforcement and the absence of comprehensive federal regulations on e-waste handling and recovery technologies. Many electronic devices also contain components that retain economic value even after disposal, offering opportunities for reuse and recycling. However, in countries like India, improper disposal practices have led to severe health and ecological concerns. The Associated Chambers of Commerce of India reported that India contributes nearly 25% of global e-waste generation annually (Houessionon et al. 2021). Although recycling strategies such as pyrometallurgical and bio-metallurgical processes enable recovery of valuable metals from electronic scrap (Sinha and Hamdan 2023; Magoda and Mekuto 2022; Nagarajan and Panchatcharam 2023), these methods vary in efficiency and environmental impact. For instance, physical separation techniques, though simple and low-cost, often result in high metal loss and the release of brominated flame-retardant dust, contributing to air pollution. Thus, advanced recycling technologies aim to convert e-waste into secondary raw materials, reducing the environmental footprint and supporting circular economy goals.

Efficient management of WEEE remains challenging due to insufficient collection systems, low public awareness, and limited economic incentives for recycling (Yong et al. 2019). Many manufacturers still view e-waste recycling as a cost burden rather than an environmental responsibility. The introduction of Extended Producer Responsibility (EPR) policies has begun to transform this mindset by mandating manufacturers to take responsibility for the entire lifecycle of their products. EPR encourages eco-friendly design, product durability, and efficient material recovery, ultimately contributing to a sustainable and circular electronics industry (Sinha and Dey 2025; Sarkhoshkalat et al. 2024).

This study investigates the sustainability of e-waste recycling technologies, focusing on the integration of Life Cycle Assessment (LCA) to evaluate environmental performance. It explores the harmful impacts of improper e-waste disposal on ecosystems and human health, compares international recycling systems and policies, and identifies technological innovations including AI-driven sorting and process optimization for improved recycling efficiency. Furthermore, the study emphasizes the importance of EPR and policy-driven frameworks for establishing robust and sustainable e-waste management systems, particularly in developing countries (Priyadarsini and Das 2024, 2026).

2 Global Electronic Waste Generation and Material Profile

The accelerated growth of technology has been accompanied by an equally dramatic rise in electronic waste (e-waste), now recognized as the fastest-growing solid waste stream worldwide (Kaya 2019). Contrary to the popular belief that e-waste mainly originates from large industrial machinery, most of it comes from short-lived consumer electronics and small industrial devices. The surge in digital innovations, including the development of cryptocurrencies such as Bitcoin, has intensified this issue. Mining cryptocurrencies requires powerful, specialized computers that are frequently upgraded or replaced due to rapid hardware obsolescence (Shuaib et al. 2022). Once outdated, these components become a major contributor to global e-waste accumulation, a trend similarly observed across numerous consumer devices that undergo constant technological upgrades. Understanding global e-waste trends requires examining the key factors behind its generation. Continuous technological progress, economic development, and consumer demand for the latest devices have created a cycle of rapid obsolescence. For instance, in 2019, the global volume of e-waste was estimated at 53.6 million tonnes, marking a 21% increase since 2015. Regional variations are significant Asia alone accounted for 46.4% of total e-waste generation, followed by the Americas with 24.4% (Quinto et al. 2025).

E-waste is composed of both valuable and hazardous elements. Valuable components include metals, glass, and plastics that can be recovered through proper recycling. Approximately 60% of e-waste is made up of recoverable metals, and about 15% consists of plastics. Despite this, only around 17.4% of global e-waste is formally collected and recycled, mainly because recycling requires complex and costly processes. Techniques such as acid leaching, thermal treatment, and incineration are often necessary to extract metals like gold, copper, and palladium, which makes recycling energy- and resource-intensive (Quinto et al. 2025). Consequently, many companies find landfilling more economically convenient than investing in sustainable recycling systems. Furthermore, the heterogeneous composition of e-waste complicates the separation of valuable resources from hazardous components, exposing recyclers and the environment to toxic chemicals.

The dual challenge of managing e-waste responsibly while recovering valuable materials is central to achieving sustainability in modern production systems. Precious and base metals such as copper, silver, and gold can be extracted and reintegrated into manufacturing supply chains, reducing dependence on virgin resources. However, disparities persist between developed and developing regions. Nations with advanced economies generally maintain stricter environmental standards and e-waste regulations, whereas developing nations often struggle with weak legal frameworks and inadequate infrastructure. Research by Chatterjee and Abraham (Lepawsky 2025) emphasizes that global e-waste volumes are increasing much faster than municipal solid waste, and many countries face difficulties managing both legal and illegal e-waste flows. The implementation of Extended Producer Responsibility (EPR) policies has been identified as a key solution to this issue.

Globally, countries have adopted different strategies to address e-waste management. Switzerland was among the pioneers in establishing formal recycling systems, supported by organizations such as the Swiss Foundation for Waste Management and the Swiss Association for Information, Communication, and Organization Technology (Lepawsky 2025). These bodies operate under a Producer Responsibility Organization (PRO) framework to ensure efficient waste handling. Despite such models, e-waste management practices still vary widely across nations. While some developing countries have begun adopting eco-friendly recycling methods, the presence of formal recycling facilities remains minimal, and the informal sector continues to dominate e-waste processing.

Government intervention or the absence of it plays a decisive role in shaping e-waste outcomes. In many developing and low-income nations, a lack of effective legislation results in improper disposal methods such as open dumping and backyard recycling. Informal recyclers, often lacking protective equipment, manually handle hazardous components, exposing themselves to toxic metals and harmful fumes (Mishra et al. 2022; Ghosh and Das 2018). In most cases, the informal sector manages the majority of e-waste collection and recovery, while formal systems remain underdeveloped. This mismanagement not only harms local ecosystems but also contributes to transboundary pollution, as e-waste can be transported through air and ocean currents. Some countries, such as Nigeria, have started to implement stricter e-waste policies and promote domestic recycling initiatives to mitigate these risks (Bimir 2020).

In Nigeria, the government has taken measures to enhance reuse and recycling practices and reduce dependence on second-hand imports from developed nations. By investing in the information and communication technology (ICT) sector, Nigeria aims to transition toward more sustainable and durable technological solutions (Bimir 2020). However, despite these improvements, challenges persist due to the rapid pace of technological innovation, which continues to shorten product lifespans and increase disposal rates. Since the widespread adoption of electricity, technological innovation has revolutionized daily life but also created persistent waste management challenges. Nearly every modern invention operates using electrical components, raising the question of what happens once these products reach obsolescence. While some electronic devices are resold or donated, the majority eventually become waste. This problem has escalated as product life cycles have shortened drastically over the decades. For example, a 1999 National Safety Council study showed that personal computers had an average lifespan of 4.5 years in 1992, which was projected to decline to two years by 2005 (Lepawsky 2025). Moreover, the increasing prevalence of inexpensive and short-lived devices, such as kitchen gadgets, calculators, and small electronics, further compounds the global e-waste burden.

The growing e-waste problem cannot be effectively mitigated through policy reform or recycling initiatives alone. A broader, more systemic approach is required one that rethinks the entire life cycle of electronic products, from design to disposal. This necessity led to the emergence of the Life Cycle Assessment (LCA) framework in the early 1990s, developed to guide manufacturers and engineers in designing products that are easier to recycle, reuse, and safely dispose of. LCA evaluates the environmental footprint of a product across its lifespan from raw material extraction to end-of-life treatment promoting sustainability-oriented innovation in product design (Yadav et al. 2025). Although LCA adoption has been most prominent in Western Europe, its application has demonstrated measurable success in minimizing waste generation and encouraging circular manufacturing systems.

3 Recycling Technologies of E-Waste

Electronic waste (e-waste) recycling technologies play a crucial role in managing the rapidly increasing volume of discarded electronic devices across the globe. These technologies aim to recover valuable metals and components while minimizing the environmental impacts associated with improper disposal (Inobeme et al. 2025). The recycling process generally begins with the collection and segregation of e-waste into different categories such as consumer electronics, IT equipment, and household appliances. Once sorted, the materials undergo dismantling, where reusable parts and hazardous components like batteries, circuit boards, and cathode ray tubes are separated for specialized treatment (Adil et al. 2025). Subsequently, mechanical processing techniques such as shredding, crushing, and magnetic or eddy current separation are employed to isolate different material fractions metals, plastics, and glass. Pyrometallurgical methods, involving high-temperature smelting, are used to recover precious and base metals like gold, copper, and silver (Li et al. 2025). However, these processes can release toxic gases if not properly controlled. Hydrometallurgical techniques, on the other hand, use chemical leaching solutions to selectively extract metals with higher precision and lower energy consumption, though they require careful management of chemical residues (Ostellari and Gross 2025).

Emerging biotechnological approaches, such as bioleaching, utilize microorganisms to recover metals from e-waste in an eco-friendly manner, offering a sustainable alternative to chemical extraction (Magoda and Mekuto 2022). Additionally, cryogenic and electrochemical methods are being explored for cleaner and more efficient recovery of valuable materials (Laftah and Rahman 2025). With the growing emphasis on the circular economy, advanced automated and AI-assisted recycling systems are also being developed to improve sorting accuracy and process efficiency (Adil et al. 2025). Therefore, the integration of modern technologies in e-waste recycling not only enhances material recovery and reduces landfill burden but also supports resource conservation and environmental protection. However, the success of these technologies depends on establishing efficient collection networks, ensuring worker safety, and implementing strong environmental regulations to promote sustainable recycling practices worldwide (Table 1).

4 Health and Environmental Impact of Electronic Waste

Electronic waste (e-waste) poses serious environmental and public health challenges worldwide. Improper disposal practices such as open dumping, burning, and landfilling release toxic chemicals into the air, soil, and water, leading to long-term contamination of natural resources. Metals constitute the major component of e-waste (Prasad 2025), including valuable elements such as gold, copper, and aluminium, as well as toxic heavy metals like lead, cadmium, nickel, and mercury (Fatima et al.

Table 1 Details of e-waste recycling technologies

Technology	Process	Key recovered materials	Advantages	Limitations	Sustainability aspect
Manual dismantling	Hand separation of components	Reusable parts, metals	Low cost, simple	Labor-intensive, unsafe	Sustainable if formalized
Pyrometallurgical	High-temp smelting	Cu, au, ag	High recovery rate	High energy use, emissions	Low; needs emission control
Bio metallurgical	Microbial metal extraction	Cu, Zn, au	Eco-friendly, low energy	Slow rate	High; greenest method
Mechanical	Shredding and physical separation	Metals, plastics	Scalable, efficient	Dust, noise pollution	Moderate; needs filters
Hydrometallurgical	Chemical leaching	Au, ag, cu	Selective, less energy	Toxic liquid waste	Moderate; greener solvents help
Cryogenic	Low-temp brittle separation	Metals, plastics	Clean process	High setup cost	Sustainable if renewable energy used
Electrochemical	Electric current for recovery	Cu, Ni, au	High purity	Energy-intensive	Sustainable with renewables
AI-assisted	Automated sorting/ dismantling	All types	Efficient, safe	Costly setup	Supports circular economy

2025). These hazardous substances, along with polybrominated flame retardants and lithium, can severely damage vital human organs, including the brain, kidneys, liver, lungs, and heart (Sinha 2026c; Marconi et al. 2025). Continuous exposure through inhalation, ingestion, or dermal contact has been associated with neurological disorders, reproductive problems, developmental delays, and even cancer (Coradduzza et al. 2025).

Communities near informal e-waste recycling sites face the greatest risks due to direct exposure to contaminated air, dust, and water. Pregnant women and children are particularly vulnerable, with documented effects including low birth weight, respiratory illness, reduced cognitive function, and impaired growth (Pal et al. 2025). Studies have also linked high levels of heavy metals such as nickel and thallium to increased cancer risk and mitochondrial DNA damage (Coradduzza et al. 2025). Burning e-waste generates fine particulate matter and toxic fumes, contributing to respiratory diseases and premature mortality, which have been estimated to increase from 300,000 to over one million early deaths annually due to e-waste pollution (Jones et al. 2025). E-waste contamination extends beyond human health to ecosystems. Toxic metals leach from discarded electronics into the soil and groundwater, disrupting agricultural productivity and contaminating food chains (Sinha 2024a; Thakur et al. 2025). In regions like Guiyu, China, soil and water near recycling hubs contain lead, cadmium, and copper levels exceeding safety standards by more than tenfold (Quinto et al. 2025). Wildlife consuming contaminated vegetation suffer from organ damage and reproductive failure, leading to biodiversity loss and ecosystem imbalance (Ghulam and Abushammala 2023). Groundwater near e-waste sites in South China has been reported to contain heavy metals at concentrations up to 140 times higher than national safety limits, posing severe risks to communities relying on these sources for drinking and irrigation (Quinto et al. 2025).

Air pollution from the burning or melting of e-waste releases carcinogenic compounds, acids, and flame retardants into the atmosphere, degrading air quality and contributing to global warming. Manufacturing and decomposition of electronic devices also emit large quantities of carbon dioxide (CO_2), further accelerating climate change. For instance, the production of a single ton of laptops can emit nearly 10 tons of CO_2, and only 17.4% of global e-waste recycling in 2020 prevented approximately 15 million tons of CO_2 emissions (Quinto et al. 2025). So, e-waste pollution represents a growing environmental hazard linked to rapid technological advancement and inadequate waste management systems. Nearly 80% of global e-waste is mismanaged, especially in developing nations such as India, Ghana, and Vietnam, where unsafe recycling and disposal practices release heavy metals and persistent organic pollutants into the environment (Bajpai 2025). Addressing these impacts requires urgent global action strengthening regulations, promoting safe recycling technologies, and raising public awareness to safeguard both human health and ecological stability (Fig. 1; Table 2).

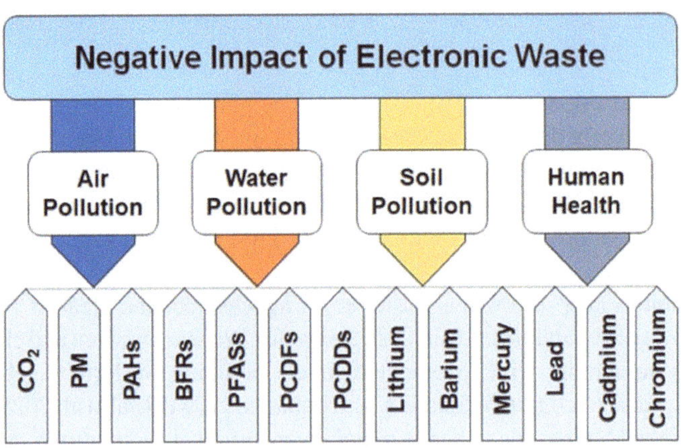

Fig. 1 Metals constitute the major component of e-waste [40]

Table 2 Major metal components of e-waste

Category	Metal	Common source in E-waste	Environmental concern/ economic value
Toxic heavy metals	Lead (Pb)	Cathode ray tubes, solder, batteries	Highly toxic; causes soil and water contamination
	Cadmium (cd)	Batteries, semiconductors	Carcinogenic; bioaccumulative in ecosystems
	Nickel (Ni)	Batteries, plating, alloys	Toxic at high levels; may cause respiratory issues
	Mercury (hg)	Switches, fluorescent lamps, displays	Persistent pollutant; affects nervous system
Valuable metals	Gold (au)	Printed circuit boards, connectors	High economic value; used in recovery and refining
	Copper (cu)	Wires, cables, transformers, motors	Valuable for recycling; essential industrial metal
	Aluminum (Al)	Casings, frames, capacitors	Lightweight and recyclable; reduces energy demand

5 Sustainable Solutions and Future Directions

Adopting circular economy principles in electronic waste (e-waste) management offers a viable pathway toward sustainability by emphasizing resource efficiency, waste minimization, and long-term value retention. The circular model promotes reuse, repair, remanufacture, and recycling of electronic components, extending product lifespans while reducing dependence on raw material extraction (Ghulam and Abushammala 2023). Circular systems create closed loops, minimizing greenhouse gas emissions and resource depletion through efficient recovery and reuse processes. Regulatory frameworks play a crucial role in operationalizing circular economy strategies across all stages of a product's life cycle from resource extraction and manufacturing to consumption and end-of-life management. Effective implementation requires harmonized policies at local, national, and global levels to close regulatory gaps and encourage shared responsibility among producers, consumers, and recyclers (Soni and Soni 2025). Eco-design principles are vital to this transition, emphasizing sustainable material selection, modular design, and integration of digital identifiers that provide product composition data to facilitate recycling and reuse (Sinha 2024b; Pietroni et al. 2025). Embedding such principles into product development can substantially reduce environmental impacts and support the efficient recovery of valuable materials.

Given the complexity of e-waste management, Multi-Criteria Decision-Making (MCDM) approaches are increasingly used to evaluate and prioritize circular economy strategies based on environmental, economic, social, and technical factors (Morganti 2025). This method helps identify the most effective interventions aligned

biomining & leachate treatment: sustainable solid waste management: waste to wealth. Springer, pp 65–93

Bajpai A (2025) Categorisation of technological waste. In: Beneficiation and management of technological waste. CRC Press, pp 14–45

Bimir MN (2020) Revisiting e-waste management practices in selected African countries. J Air Waste Manage Assoc 70(7):659–669

Coradduzza D, Sanna A, Di Lorenzo B, Congiargiu A, Marra S, Cossu M, Medici S (2025) Associations between plasma and urinary heavy metal concentrations and the risk of prostate cancer. Sci Rep 15(1):14274

de Oliveira Neto JF, Candido LA, de Freitas Dourado AB, Santos SM, Florencio L (2023) Waste of electrical and electronic equipment management from the perspective of a circular economy: a review. Waste Manage Res 41(4):760–780

Fatima G, Raza AM, Dhole P (2025) Heavy metal exposure and its health implications: a comprehensive review. Ind J Clin Biochem:1–29

Frazzoli C, Bocca B, Battistini B, Ruggieri F, Rovira J, Amadi CN, Offor SJ, Orisakwe OE (2024) Rare earth and platinum group elements in sub-Saharan Africa and global health: the dark side of the burgeoning of technology. Environ Health Insights 18:11786302241271553

Ghulam ST, Abushammala H (2023) Challenges and opportunities in the management of electronic waste and its impact on human health and environment. Sustainability 15(3):1837. https://doi.org/10.3390/su15031837

Ghosh S, Das AP (2018) Metagenomic insights into the microbial diversity in manganese-contaminated mine tailings and their role in biogeochemical cycling of manganese. Sci Rep 2018 8:1, 8(1), 1–12. https://doi.org/10.1038/s41598-018-26311-w

Houessionon MGK, Ouendo EMD, Bouland C, Takyi SA, Kedote NM, Fayomi B, Fobil JN, Basu N (2021) Environmental heavy metal contamination from electronic waste (e-waste) recycling activities worldwide: a systematic review from 2005 to 2017. Int J Environ Res Public Health 18(7):3517

Inobeme A, Mathew JT, Devolli A, Adetunji CO, Sharma N, Maliki M, Ajai A et al (2025) Metal components in industrial wastes and methods for metal ions recovery. In: Metal value recovery from industrial waste using advanced physicochemical treatment technologies. Elsevier, pp 1–15

Izatt RM, Hagelüken C (2016) Recycling and sustainable utilization of precious and specialty metals. In: Metal sustainability: global challenges, consequences, and prospects, pp 1–22

Jones BA, Goodkind AL, Berrens RP (2025) Energy use in bitcoin mining: the environmental impact of cryptocurrencies. Taylor & Francis

Kaya, M. (2019). Electronic waste and printed circuit board recycling technologies

Laftah WA, Rahman WAWA (2025) A comprehensive review of tire recycling technologies and applications. Mater Adv 6(15):4992–5010

Lepawsky J (2025) Electronic waste: a reference handbook. Bloomsbury Publishing, USA

Li Y, Baker J, Fang Y, Cao H, Pleydell-Pearce C, Watson T, Chen S, Zhao G (2025) Comparative environmental impacts analysis of technologies for recovering critical metals from copper anode slime: insights from LCA. Environ Chem Ecotoxicol 7:275–285

Magoda K, Mekuto L (2022) Biohydrometallurgical recovery of metals from waste electronic equipment: current status and proposed process. Recycling 7(5):67

Marconi G, Di Resta C, Naclerio A, Banfi G, Tomaiuolo R (2025) Heavy metals and human reproductive toxicity: mechanisms, pregnancy outcomes, and mitigation strategies. Reprod Toxicol 109104

Mishra S, Dash D, Das AP (2024) Microbial degradation of polyester microfibers using indigenously isolated bacterial strain *Exiguobacterium* sp. CLEAN—Soil Air Water 52(12):e202300343. https://doi.org/10.1002/CLEN.202300343

Mishra S, Dash D, Al-Tawaha ARMS, Das AP (2022) A Review on Heavy Metal Ion Adsorption on Synthetic Microfiber Surface in Aquatic Environments. App Biochem Biotechnol, 194(10), 4639–4654. https://doi.org/10.1007/S12010-022-04029-W/METRICS

Morganti, L. (2025). Lifecycle environmental and circular assessment in building product manufacturing: systemic approach for data optimization of custom prefabricated façade modules
Nagarajan N, Panchatcharam P (2023) A comprehensive review on sustainable metal recovery from e-waste based on physicochemical and biotechnological methods. Eng Sci 22(2):844
Ostellari P, Gross S (2025) A mature tool to address new challenges: harnessing coordination chemistry for the sustainable copper recovery from industrial and e-waste in the age of energy transition. Eur J Inorg Chem 28(4):e202400704
Pal D, Tabia T, Konar A, Banerjee P, Ghosh P (2025) Ecological effects and management of e-waste: a review
Pietroni G, Marconi M, Favi C, Tamborrini PM, Marino C, Fiore E (2025) Circular economy for electrical and electronic equipment: a life cycle design and management platform. Int J Interact Des Manuf (IJIDeM):1–19
Priyadarsini S, Das AP (2024) Analytical and structural characterization of waste lithium-ion batteries for their effective recycling strategy. Environ Chem Ecotoxicol. https://doi.org/10.1016/j.enceco.2024.12.004
Priyadarsini S, Das AP (2025) Isolation and identification of native lithium-resistant bacteria from electronic waste contaminated sites: a focused investigation on tolerance assessment with potential lithium bio-recovery application. Geomicrobiol J. https://doi.org/10.1080/01490451.2025.2539509
Priyadarsini S, Das AP (2026) Lithium in waste water: impacts on pollution and human health. Decarbonization of Wastewater Pollutants as a Sustainable Solution, 353–364. https://doi.org/10.1016/B978-0-443-33062-9.00010-2
Prasad KVS (2025) E-waste: a looming hazard to the environment and health—A critical review. In: AIP conference proceedings, vol 3298, no 1, p 040044
Quinto S, Law N, Fletcher C, Le J, Antony Jose S, Menezes PL (2025) Exploring the e-waste crisis: strategies for sustainable recycling and circular economy integration. Recycling 10(2):72
Sarkhoshkalat MM, Afkham A, Bonyadi Manesh M, Sarkhosh M (2024) Circular economy and the recycling of e-waste. In: New technologies for energy transition based on sustainable development goals. Springer, pp 319–354
Serpe A, Purchase D, Bisschop L, Chatterjee D, De Gioannis G, Garelick H, Kumar A et al (2025) 2002–2022: 20 years of e-waste regulation in the European Union and the worldwide trends in legislation and innovation technologies for a circular economy. RSC Sustain 3(3):1039–1083
Shuaib M, Badotra S, Khalid MI, Algarni AD, Ullah SS, Bourouis S, Iqbal J, Bharany S, Gundaboina L (2022) A novel optimization for GPU mining using overclocking and undervolting. Sustainability 14(14):8708
Sinha S (2018) Derivatization of milk protein using poly (acrylamide): Its characterization and application. In: Macromolecular symposia, vol 381, no 1, p 1800120
Sinha S (2022) Textile wastewater clarification using milk-based formulations as an effective dye absorbent and flocculant. Mater Today Proc 67:1304–1309
Sinha S, Hamdan A (2023) E-plastic waste—A sustainable construction material: an Indian perspective. In: Handbook of sustainable materials: modelling, characterization, and optimization. CRC Press, pp 213–232
Sinha S (2024a) Food security challenges and microplastics: a comprehensive review. Microplastics Afr Asian Environ Influencers Challenges Solutions 361–371
Sinha S (2024b) Advancements in biopolymer-derived graft copolymers: a comprehensive review on their application in sustainable wastewater treatment. Environ Qual Manage 34(2):e22320
Sinha S, Dey KP (2025) Advancements and diverse applications of waterborne epoxy coatings: a comprehensive review. Recent Adv Waterborne Epoxy Coat Synth Prop Appl 245–256
Sinha S (2026a) Seaweed polysaccharide-derived biopolymers: a sustainable alternative for food packaging. Polym Renew Resour 20412479261436665
Sinha S (2026b) Review on e-waste plastics in construction: a circular economy solution. In: Electronic waste and environmental pollution: impacts and sustainable management through circular economy. Cham, Springer, pp 139–154

Sinha S (2026c) Marine ecosystem restoration through different bioremediation tools. Bioremediat Greener Solution Mar Plast Pollutants 179–190

Sinha S, Singh R, Lima S, Srivastava M, Garg N, Kumari S (2026) Waste management and its impact on environment and health: addressing emerging waste challenges in a post-pandemic world. In: Synergizing sustainability for integrated waste management: artificial intelligence, economic stability and energy recovery. Cham, Springer, pp 53–68

Soni SK, Soni R (2025) Policy and regulatory frameworks for green biorefinery: driving sustainable waste valorization. In: Green biorefinery solutions: transforming biodegradable waste into resources. Springer, pp 315–349

Thakur S, Chandra A, Kumar V, Bharti S (2025) Environmental pollutants: endocrine disruptors/pesticides/reactive dyes and inorganic toxic compounds metals, radionuclides, and metalloids and their impact on the ecosystem. In: Biotechnology for environmental sustainability. Springer, pp 55–100

Yadav A, Lata K, Saurabh S, Garg RK (2025) Sustainable transformation of manufacturing small and medium-sized enterprises via life cycle assessment: insights from a hybrid SEM-ANN model. Rapid Prototyp J:1–17

Yong YS, Lim YA, Ilankoon IMSK (2019) An analysis of electronic waste management strategies and recycling operations in Malaysia: challenges and future prospects. J Clean Prod 224:151–166

Artificial Intelligence in Critical Metal Bio-circularity: A New Paradigm for Sustainable Resource Recovery

Hasan Volkan Oral and Hasan Saygin

Abstract This section discusses the revolutionary impact of artificial intelligence (AI) in creating a new paradigm for key metals' sustainable bio-circularity. The demand for key metals is continuously increasing as new technologies become more prevalent, while the main sources are gradually decreasing. This situation highlights the need for novel approaches to resource recovery. Conventional methods are frequently inefficient, energy-intensive, and environmentally harmful. In this respect, it is proposed that AI, with its powers in data analysis, predictive modeling, optimization, and automation, can transform all stages of crucial metal bio-circularity. These stages address a wide range of topics, from intelligent waste stream sorting and classification to improved biolysis and biorecovery procedures. AI algorithms can forecast the best operational parameters for optimum yield and effectiveness, optimize microbial consortia for targeted metal extraction, and allow for real-time bioreactor monitoring and control. Furthermore, machine learning approaches can speed the discovery of new bio-mining agents and pathways, allowing for more efficient and environmentally friendly biological processes. AI also enables the continuous management of secondary essential metal supply chains, promoting closed-loop systems and lowering reliance on raw materials. This integration not only improves economic sustainability but also reduces environmental effects, paving the way for a sustainable future in which resources move continuously inside technological systems. Furthermore, this section examines current AI applications through case studies and evaluates future potential. Additionally, challenges encountered during the transition to a circular economy are discussed.

H. V. Oral (✉)
Department of Civil Engineering (English), Faculty of Engineering, İstanbul Aydin University, Florya Campus, Çekmece, İstanbul, Türkiye
e-mail: volkan.oral@gmail.com

H. Saygin
Application and Research Center for Advanced Studies, İstanbul Aydin University, Florya Campus, K. Çekmece, İstanbul, Türkiye
e-mail: hasansaygin@aydin.edu.tr

A. P. Das and S. Priyadarsini (eds.), *Recovery and Recycling of Critical Metals from Electronic Waste for Environmental Sustainability*, Earth and Environmental Sciences Library, https://doi.org/10.1007/978-3-032-29064-9_16

Keywords Artificial intelligence · Biocircularity · Sustainability · Industrial symbiosis · Circular economy · Advanced waste management methods

1 Introduction

Sustainable resource recovery is a strategic approach to material and energy management that strives to reduce waste while maximizing resource use in a closed-loop system. It goes beyond typical waste management, which is frequently concerned with disposal (e.g., landfills), and instead views waste as a useful resource (Climate Sustainability 2024). The relationship between sustainable resource recovery and metal biocircularity serves as a means to a goal. Sustainable resource recovery is the intended outcome, and metal biocircularity is a green and efficient technology that helps to achieve it. To extract vital minerals like copper, lithium, nickel, cobalt, and rare earths, significant investments in new mines and refining capacity will be needed as the transition to a sustainable energy system quickens, particularly in geographically varied places.

For sustainable energy transitions, recycling is essential to the sustainability and security of vital mineral supplies. Recycling produces a valuable secondary supply source that lessens dependency on new mines and improves supply security for nations importing minerals, even while it does not completely replace the requirement for mining investment. Additionally, increasing recycling reduces the negative social and environmental effects of mining and refining while keeping trash from end-of-use technology out of landfills (IEA 2025). Moreover, the global transition to low-carbon energy depends heavily on the critical metal minerals. The worldwide low-carbon energy transition and the climate goals of "carbon peaks" and "carbon neutrality" are constrained by the sustainability of the critical metal mineral supply (Zhang et al. 2021a, b).

Industrial Symbiosis (IS) refers to a (regional) coordinated effort in the industrial sector. The underlying principle is that byproducts or waste from one business can be used as raw materials or inputs in another. This enables more sustainable use of materials and resources, thereby contributing to the development of a circular economy. (CE) In practical terms, it cuts energy consumption and pollution (Cluste rcollaboration.eu 2025). IS is currently being used in the field of critical metal and mineral supply (Jung et al. 2025; Scafa et al. 2025).

According to Holden et al. (2023), the bioeconomy must embrace circularity for the purpose of producing renewable feedstocks while also preserving important natural capital. The concept of biocircularity is proposed as an integrated systems approach to the sustainable production of renewable biological materials, with a focus on extended use, maximum reuse, recycling, and design for degradation from polymers to monomers, while avoiding "failure" at the end of life and minimizing energy demand and waste. Existing and emerging circular bioeconomy methods can serve as excellent examples of system integration. These approaches enable the connectivity of complicated biomass supply chain networks with other networks that include feedstock-independent renewable power, hydrogen, carbon dioxide (CO_2), water, and other biotic, abiotic, and intangible resources. Raising the profile of these connectors will enable policymakers to direct the amplification of synergies and the reduction of tradeoffs between systems, sectors, and goals (Schipfer et al. 2024).

It is acknowledged that artificial intelligence (AI) had the most impact on how the circular economy (CE) developed. AI has a lot of potential to help the total infrastructure needed to put circular practices into effect in real-world settings and accelerate the shift to a regenerative economic model. Furthermore, AI has proven to play a significant part in furthering global sustainability initiatives, establishing itself as a useful instrument for tackling important issues in an intelligent circular economy. These issues include supply chain management, waste management, sustainable manufacturing, energy source optimization, and reverse logistics (Özsoy 2023).

Nowadays, AI tools are being used more and more to regulate various waste treatment systems, including resource recovery systems; monitor processes cost-effectively; and forecast process performance (Behera et al. 2024). Machine learning (ML), especially its predictive analytics skills, is regarded as a crucial subset of AI, which includes a wide range of technologies with the capacity for autonomous reasoning and decision-making. AI systems differ from conventional rule-based systems in that they replicate human cognitive processes, including learning from data, adjusting to new inputs, and making decisions on their own. Logistics optimization relies heavily on machine learning's well-established function in pattern recognition and forecasting, which provides actionable insights from complicated datasets, even though AI encompasses more sophisticated applications like natural language processing and picture recognition (Nunes 2025).

Therefore, this book chapter aims to present the use of AI in Critical Metal Bio-Circularity, provided with case studies from the World, which can be seen as a new paradigm for sustainable resource recovery. Chapter two explains the foundations of critical metal bio-circularity, providing details from sub-chapters such as principles of bioleaching, biosorption, and bio-circularity. Chapter three introduces the AI-powered framework for resource recovery, including information about ML models, deep learning applications, and reinforcement learning. Chapter four showcases case studies and applications, where the audience can learn about AI-driven bio-circularity of lithium from batteries, recovery of rare earth elements from electronic waste, and the role of AI in urban mining and industrial symbiosis. Chapter five discusses the challenges and ethical considerations, and the final chapter summarizes the future outlook of the topic.

2 Foundations of Critical Metal Bio-circularity

This section provides useful information for the foundations of critical metal bio-circularity relevant to the three subtopics.

3 The AI-Powered Framework for Resource Recovery

The following sections discuss the usage of AI-powered frameworks for resource recovery. Some of the technologies in this section have not previously been used with AI, but they provide a fundamental understanding of how to employ technology.

3.1 Data Collection and Pre-processing for AI Models

Data collection, wrangling, and preprocessing are essential processes in the AI/ML model-building lifecycle. These processes precede all model-building activities, culminating in feature engineering for model development (Rahman 2023).

3.2 ML Models for Process Optimization

This section can be divided into two categories.

3.2.1 Predicting Bioleaching Efficiency

ML is an emerging field of artificial intelligence that has demonstrated accuracy, dependability, and cost-effectiveness in a variety of applications. Furthermore, ML can be used to evaluate regression-based ML algorithms to determine which random forest regression performed best. As a result, ML can be utilized to create an effective model for bioleaching processes (Mokarian et al. 2022).

3.2.2 Optimizing Biosorption Parameters

Liu et al. (2025) investigated the effectiveness of ML models in predicting biochar's adsorption capacity for dye removal. Nine models, CatBoost, XGBoost, Gradient Boosted Decision Trees, Random Forest, Histogram-Based Gradient Boosting, Kernel Extreme Learning Machine, Kriging, Light Gradient Boosting Machine, and AdaBoost, were employed to determine their prediction accuracy.

3.3 Deep Learning Applications

3.3.1 Image Recognition for Waste Sorting and Characterization

Zhang et al. (2021a, b) conducted a study to increase the accuracy of waste sorting using deep learning and to establish a basis for intelligent waste categorization using computer vision and mobile phone terminals. The work proposes a deep learning-based categorization model for recyclable garbage photos. The suggested trash categorization model incorporates a self-monitoring module into the residual network model, allowing it to integrate relevant information from all channel graphs, compress spatial dimension features, and have a global receptive field. Rahmetullahi (2025) employed a deep learning model using the YOLOv7 object detection algorithm to apply real-time identification of various types of waste.

3.3.2 Natural Language Processing (NLP) for Literature Mining and Knowledge Discovery

The NLP method allows machines to understand textual data, providing an automated route to knowledge discovery and information extraction from electronic text (VanGessel et al. 2023).

3.4 Reinforcement Learning for Autonomous System Control

Reinforcement learning (RL) excels at decision-making by providing support and control in real-time production environments. RL is based on the notion of goal-directed and interactive learning, led by a numerical reward signal (Nievas et al. 2024).

4 Case Studies and Applications

This section describes countries where AI technology is being used as case studies. The literature review reveals that more case studies on this topic are found in the United States than in Europe. Across Europe, this topic is being pursued primarily within the framework of Horizon 2020 projects supported by European Union Research Funds, but the outcomes of these projects are far from being applicable.

4.1 AI-Driven Bio-circularity of Lithium from Batteries

The following United States examples show how AI is being applied to improve important recycling processes like sorting, disassembly, and manufacturing optimization.

AMP Robotics: This business sorts and separates recyclable materials, including parts from batteries that have reached the end of their useful lives, using robotics driven by artificial intelligence. Their methods improve the efficiency and accuracy of the sorting process by identifying and classifying various items using computer vision and ML (ampsortation.com 2025).

Redwood Materials: A major force in the battery recycling sector, Redwood Materials is a well-known illustration of a business using technology to run extensive operations. Although they do not merely concentrate on "AI-driven bio-circularity," they employ a comprehensive strategy that combines data-driven techniques and automated procedures to recover precious minerals from batteries, such as nickel, cobalt, and lithium. Their facilities, which collaborate with manufacturers to close the loop on EV battery production, provide an effective, scalable example of a CE (Redwood.com 2025).

Li-Cycle: This company recycles using the "Spoke & Hub" technique. The "hubs" use a hydrometallurgical process to purify the "black matter," which is created when the "spokes" shred batteries. Their strategy uses AI and machine learning for process optimization and lifecycle assessment (LCA) with the goal of increasing material recovery rates and minimizing environmental impact (Glencore.com 2025).

Green Li-ion: A novel hydrometallurgical method developed by Green Li-ion transforms black mass directly into precursor cathode active materials (pCAM) and other useful products. According to an ISO-compliant LCA, their approach had a far smaller impact on climate change than conventional approaches. Although the precise AI application isn't described, sophisticated data analysis and optimization areas in which AI is crucial are probably necessary for the success of their streamlined approach (Greenli-ion.com 2025).

4.2 Recovery of Rare Earth Elements (REE) from Electronic Waste

A promising pilot project is being undertaken by Western Digital, a firm based in the United States, in partnership with Microsoft and other partners. While their recovery method is based on a chemical process, the project's success is dependent on an "advanced sorting ecosystem" that might use AI and automation to efficiently separate e-waste, including shredded hard drives, for REE recovery. The research successfully recovered REEs and other important metals from a significant volume of e-waste, proving the possibility of an economically feasible, large-scale operation (Western Digital 2025).

4.3 The Role of AI in Urban Mining and Industrial Symbiosis

AI has a big impact on urban mining and industrial symbiosis by streamlining operations and discovering new opportunities. Here are some case studies and examples of its use (Meetcenter.it 2024).

The iBot4CRMs Project: It is supported by Horizon Europe and aims to create AI-powered robotic devices that can extract CRMs from urban garbage. Pilot initiatives in various European nations, including Turkey, Greece, Spain, and Portugal, use robots to dismantle electric motors, desktop computers, and other devices to recover resources such as neodymium magnets, copper, and precious metals. This effort seeks to boost Europe's self-sufficiency in CRMs while decreasing dependency on traditional mining (Robotics and Automation News 2025).

Ecosym (AI-Powered Platform): Ecosym is an AI-powered platform that functions as a marketplace, connecting businesses that generate garbage with those that can use it as a raw resource. Its algorithms assess the composition of waste streams and offer ideal reuse scenarios, matching waste sources with potential customers to form industrial symbiosis networks (Ecosym 2025).

5 Challenges and Ethical Considerations

Data availability and quality, model interpretability, and control are the main challenges of AI-based studies in the field of resource recovery and environmental studies (Olamadı et al. 2024). As for the ethical implications, it is not limited only to the AI-driven aspects relevant to resource recovery, environmental, or ecological studies. The European Parliament (EP) made a declaration about the ethics of AI in 2020. In this report, it is highlighted that there is a huge gap in the national strategies of countries in the fields of environmental concerns, human psychology, and financial matters.

6 Future Outlook and Conclusion

Since the early 2000s, the pursuit of answers to environmental issues has accelerated with the creation of the Industry 4.0 idea. Notable advancements have been made, especially in resource recovery, with the integration of AI with nature-based solutions, launching circular economy strategies, and other components that might be regarded as sustainability tools. Because critical metal bio-circularity is so closely related to industry, it is one of the fields where AI may be used directly. In certain fields, artificial intelligence is still in its early stages. As a result, these uses will need to be characterized by particular standards in the years to come, initially at the national level and subsequently on a global level. Their legal bounds must be similarly determined on national and then international platforms in accordance with these criteria.

Consequently, the summary of the key findings of this summary can be listed as:

a. AI presents a viable setting for critical metal biocirculation if the required data is available.
b. AI has been applied and marketed as the result in the United States than in Europe. In EVs, this is very noticeable.
c. The use of AI for crucial metal biocirculation is still in the beginning with regard to national and international norms and legal regulations.

References

Abashina T, Vainshtein M (2023) Current trends in metal biomining with a focus on genomics aspects and attention to arsenopyrite leaching—a review. Microorganisms 11(1):186. https://doi.org/10.3390/microorganisms11010186

AMP Robotics (2025) New facility encompasses expanded production floor. https://ampsortation.com/articles/new-facility-encompasses-expanded-production-floor

Behera S, Dhal PK, Mohapatra B (2024) Application of artificial intelligence tools in wastewater and waste gas treatment systems: recent advances and prospects. J Environ Manage 370:122386. https://doi.org/10.1016/j.jenvman.2024.122386

Climate Sustainability (2024) https://climate.sustainability-directory.com/area/sustainable-resource-recovery/

Cluster Collaboration (2025) IS tooklit. https://www.clustercollaboration.eu/is-tooklit

Congress.gov (2025) Critical mineral resources: national policy and critical minerals list. https://www.congress.gov/crs-product/R47982

Ecosym (2025). https://www.ecosymglobal.com/

European Parliament (2020) The ethics of artificial ıntelligence: ıssues and ınitiatives. www.europarl.europa.eu/RegData/etudes/STUD/2020/634452/EPRS_STU(2020)634452_EN.pdf

Glencore (2025). https://www.glencore.com/

Green Li-ion (2025) Case studies successful implementation of battery recycling. https://www.greenli-ion.com/post/case-studies-successful-implementation-of-battery-recycling#:~:text=Li%2DCycle%20employs%20a%20Spoke,Scaling%20plans%20increase%20capacity%2010x

Holden NM, Slingerland B, Murphy JJ (2023) A review of industrial symbiosis: past, present, and future. Resour, Environ Sustain 14:100180. https://doi.org/10.1007/s43615-022-00180-y

International Energy Agency (2024) Global critical minerals outlook 2024. https://iea.blob.core.windows.net/assets/ee01701d-1d5c-4ba8-9df6-abeeac9de99a/GlobalCriticalMineralsOutlook2024.pdf

International Energy Agency (2025) Recycling of critical minerals. https://www.iea.org/reports/recycling-of-critical-minerals

Jung S, Lee M, Kim Y (2025) Assessment of a critical mineral recycling network: a case study on nickel recovery from production waste in Korean eco-industrial parks. J Ind Ecol 29(4):1197–1207

Liu C et al (2025) Enhanced machine learning prediction of biochar adsorption for dyes: parameter optimization and experimental validation. Carbon Res 4:46. https://doi.org/10.1007/s44246-025-00213-9

Meetcenter (2024) Challenge n. 6: regenerative AI for urban mining. https://www.meetcenter.it/en/challenge-n-6-regenerative-ai-for-urban-mining/#:~:text=AI%20can%20play%20a%20pivotal,circular%20processes%20to%20regenerate%20them
Mokarian P et al (2022) The advanced design of bioleaching process for metal recovery: a machine learning approach. Sep Purif Technol 291:120919. https://doi.org/10.1016/j.seppur.2022.120919
Nievas N et al (2024) Reinforcement learning for autonomous process control in Industry 4.0: advantages and challenges. Appl Artif Intell (38). https://doi.org/10.1080/08839514.2024.2383101
Nunes LJR (2025) The role of artificial intelligence (AI) in the future of forestry sector logistics. Future Transp 5(2):63. https://doi.org/10.3390/futuretransp5020063
Olamadı BD et al (2024) Artificial intelligence in environmental monitoring: advancements, challenges, and future directions. Hyg Environ Health Adv 12:100114
Özsoy T (2023) The role of artificial intelligence in facilitating the transition to a circular economy. İstanbul Nişantası Univ, J Soc Sci 2(2)
Rahman A (2023) Data collection, wrangling, and pre-processing for AI assurance. https://doi.org/10.1016/B978-0-32-391919-7.00022-6
Rahmetullah A (2025) WasteInNet: deep learning model for real-time identification of various types of waste. Clean Waste Syst 10:10098
Redwood (2025) https://www.redwoodmaterials.com/
Robotics and Automation News (2025) Urban mining Europe turns to AI-driven robotics to tackle critical raw materials shortage. https://roboticsandautomationnews.com/2025/06/05/urban-mining-europe-turns-to-ai-driven-robotics-to-tackle-critical-raw-materials-shortage/91468/#:~:text=A%20notable%20initiative%20in%20this,extracting%20CRMs%20from%20urban%20waste
Scafa M, Luzzini D, Rossi R (2025) A critical review of industrial symbiosis models. In: Volume 7: transdisciplinary engineering methods for social ınnovation of Industry 4.0, p 1184. https://doi.org/10.3233/978-1-61499-898-3-1184
Schipfer F, Burli P, Fritsche U, Eickelkamp T (2024) The circular bioeconomy: a driver for system integration. Energ Sustain Soc 14:34. https://doi.org/10.1186/s13705-024-00461-4
Silverado Policy Accelerator (2024) Strategic defense critical minerals: a targeted list for national and economic security. https://silverado.org/reports-and-publications/strategic-defense-critical-minerals/
U.S. Department of Energy (2023) 2023 critical materials assessment. https://www.energy.gov/sites/default/files/2023-05/2023-critical-materials-assessment.pdf
University of Michigan, Center for Sustainable Systems (2024) Critical materials factsheet. https://css.umich.edu/publications/factsheets/material-resources/critical-materials-factsheet
VanGessel F et al (2023) Natural language processing for knowledge discovery and information extraction from energetics corpora. Propellants, Explos, Pyrotech 48. https://doi.org/10.1002/prep.202300109
Volesky B (2003) Biosorption of heavy metals. CRC Press
Western Digital (2025) At scale hard disk drive rare earth material capture program launched. https://www.westerndigital.com/company/newsroom/press-releases/2025/2025-04-17-at-scale-hard-disk-drive-rare-earth-material-capture-program-launched#:~:text=In%20a%20multi%2Dparty%20pilot,step%20toward%20closing%20that%20loop
Zante G, Goulart M, Pimenta P (2024) A toolbox for improved recycling of critical metals and materials in low-carbon technologies. RSC Sustain 2:320–347. https://doi.org/10.1039/D3SU00390F
Zhang L, Li J, Wang H (2021a) Global supply risk assessment of the metals used in clean energy technologies. J Clean Prod 331:129602. https://doi.org/10.1016/j.jclepro.2021.129602
Zhang Q et al (2021b) Recyclable waste image recognition based on deep learning. Resour Conserv Recycl 171:105636

The Future of AI in Critical Metal Recovery from E-waste: An Innovative Approach

Himanshu, Khan Jabroot, Deepika Saxena, and Atul Khalangre

Abstract The issue of sustainable recovery of electronic trash (e-waste) has turned global because more electronic products, such as lithium, cobalt, and rare earth components, are required, which are critical to the electronics, clean power, and military industries. Despite its relative success, the traditional recycling method has its disadvantages in terms of scalability, precision, and impact on the environment. This chapter investigates the radical potential of artificial intelligence (AI) in modifying the recovery and recycling of critical metals in e-waste streams. Such technologies as computer vision, machine learning algorithms, and digital twins allow us to implement intelligent sorting, autonomous disassembly, predictive process optimization, and improved material recovery efficiency with the help of artificial intelligence. AI can be combined with robotic systems, bioleaching models, and sensor-based monitoring to enable a drastic improvement in metal production levels and purity, and reduce operating costs and chemical consumption of the industry. Other such technological and ethical challenges, covered by the chapter, are system interoperability, algorithmic bias, and data scarcity. The world case examples present constructive applications and benefits, with a special focus on artificial intelligence-enhanced participation in the establishment of circular networks of supply and sustainability of the environment. With legislative support, cross-sector collaboration, and digital infrastructure, AI has the potential to accelerate the transition to a closed-loop system of vital metal recovery, reducing reliance on mining and leading to green economic

Himanshu · K. Jabroot · D. Saxena (✉)
Department of Horticulture, School of Agriculture, Lovely Professional University, Jalandhar, Delhi, Phagwara, Punjab, India
e-mail: Deepika.20662@lpu.co.in

Himanshu
e-mail: rohillah86@gmail.com

K. Jabroot
e-mail: khanjabroot029@gmail.com

A. Khalangre
Department of Food Technology, CIRI-Agroalimentare, University of Bologna, Bologna, Italy
e-mail: atul.khalangre2@unibo.it

A. P. Das and S. Priyadarsini (eds.), *Recovery and Recycling of Critical Metals from Electronic Waste for Environmental Sustainability*, Earth and Environmental Sciences Library, https://doi.org/10.1007/978-3-032-29064-9_17

development. This chapter provides a prospect into the effect of artificial intelligence further into the future in managing e-waste.

Keywords Smart recovery systems · Intelligent process control · Circular material economy · Data-driven sustainability

1 Introduction

In large part due to the availability of the so-called critical metals, which include gold, silver, palladium, copper, lithium, and the rare earth elements, all of which are crucial to modern electronics, renewable energy systems, and electric vehicle technologies, the exponential growth of electronic devices around the globe has been accompanied by a huge increase in electronic waste, or e-waste, generating complex resource-related, economic and environmental challenges. The traditional recycling procedures of e-waste, including pyrometallurgical, hydrometallurgical, and mechanical separation processes, are often inefficient and less selective in terms of recovery as well as environmental sustainability, resulting in loss of precious metals and health hazards as byproducts (Ferreira-Filipe et al. 2025). Artificial intelligence (AI) has turned out to be a breakthrough technology that can transform the essential metal recovery of e-waste, with the application of intelligent automation, predictive analytics, and optimization of recycling processes. Besides boosting material recovery, algorithms in AI-based techniques, such as machine learning, computer vision, robotics, and data-centered decision support, can reduce the costs of energy, greenhouse gas emissions, and risks associated with the handling of e-waste manually (Mohamed 2024). One example of an AI application is the realization of autonomous sorting robots through the method of transfer learning and multispectral imaging. This was demonstrated in an Indian study that incorporated mobile AI-based robots in recycling plants to detect and sort different e-waste materials, such as printed circuit boards, plastics, and metals. The overall benefits of both operational efficiency and safety improvement were announced by such an application, which not only improved the recovery rate of crucial metals but also significantly reduced the exposure that workers would have to hazardous chemicals (Ilankoon et al. 2024). They projected the application of an AI-based decision support system that integrates visual identification with predictive modeling in order to optimize metal concentration before extraction. This system made higher selectivity of the valuable metals, such as gold and palladium, possible; supervised learning algorithms were applied, and large datasets of e-waste material composition were used to predict the optimal recovery paths using the minimum energy and chemical reagents possible. AI can help to build the benefit of a circular economy by transforming e-waste, which has been previously discarded, into a reliable source of secondary resources, which is evident by quantifiable operational enhancements in recovery and lower carbon footprint, as demonstrated by the study. Despite such changes, there are several significant gaps in research that do not allow the adoption and scaling of AI-driven e-waste recycling solutions. To begin with, it is not possible to generalize prediction models to various e-waste types and electronic devices composition because there are no standardized and publicly available datasets to train AI algorithms (Vega-Muratalla et al. 2024). Second, AI systems and the existing recycling infrastructure remain inactive because most facilities continue to utilize traditional manual or semi-mechanical methods, and this presents a barrier regarding funding, technical expertise, and restructuring of operations. Third, long-term studies on the economic viability, environmental impact, and scalability of AI-based recovery systems in different geographies and regulatory systems have not been conducted, even though pilot studies have reported short-term efficiency benefits. Interdisciplinary collaboration between the AI experts, material scientists, environmental engineers, and policymakers is also necessary as quickly as possible to resolve the technological and regulatory challenges and ensure that the AI applications are not only technically correct but also socially and environmentally responsible. Ethical concerns related to data privacy, the transparency of AI decision-making, and the potential replacement of jobs should also be treated in order to permit the responsible application at scale (Ahmed et al. 2025). These gaps should be addressed to make the most use of AI in e-waste recycling. This will enable sustainable extraction of valuable metals and lessen the resource waste and environmental destruction. By enhancing the efficiency of recycling tasks and the overall sphere of resource management and industrial sustainability, AI can revolutionize the recycling process and reduce the reliance on virgin mining and the environmental impact of e-waste. Combined with robotics advances, sensor technologies, and big data analytics, the current refinement of AI algorithms will result in more and more advanced solutions that have the potential to streamline chemical processing pathways, anticipate the condition of recycling equipment, and respond to complex streams of e-waste in real-time. To summarize, the e-waste issue of AI application in critical recovery of metals can offer a revolutionary path to technical revolution, economic effectiveness, and environmental sustainability, despite the persistence of challenges and gaps in research. More studies, investment, and cross-disciplinary collaboration will be needed in order to deploy AI to its full potential. This will ensure that critical metal recovery and recycling are made greener, safer, and sustainable, besides being made more efficient. This will eventually contribute to the attainment of the long-term goal of a circular economy that conserves resources, limits waste, and has a great positive influence on the environment.

2 Critical Metal Scarcity: Drivers and Geopolitical Impacts

The necessity for essential metals has grown since the world is getting more and more dependent on high-tech industries, on renewable sources of energy, and on electric vehicles, highlighting the issues of scarcity and the corresponding geopolitical consequences. Lithium, cobalt, nickel, rare earth elements (REEs), platinum group metals, and indium are crucial to batteries, semiconductors, wind, solar, and a range of other electronic devices. These metals have unequal geographical distribution, increasing

2.3 *Trade Wars, Export Controls, and Supply Chain Volatility*

The trade policies and market volatility are increasingly pointing towards the geopolitical dimension of shortages of essential metals. Even as a measure to protect home businesses, countries with high reserves often have their exportation curtailed or practice strategic hoarding. The ban on exporting rare earths in China is one of the best illustrations of how national policy can impact global supply chains and increase costs, and force diversification of suppliers or technical innovation in other areas. The supply of cobalt in DRC is disrupted by artisanal mining, political, and labor problems, which have also contributed to the volatility of the market, showing the necessity of ethical and transparent supply sourcing. The trade conflicts, tariffs, and diplomatic tensions between the world economies influence global manufacturing, EV adoption, and renewable energy projects, which also raise vulnerabilities and contribute to the increase of them. To reduce the dependence on high-risk regions, both governments and companies are increasingly considering recycling, strategic storage, and alternative supplies. Indonesia prohibited the export of nickel ore in 2021 to enhance domestic processing and the production of EV batteries (Shabani et al. 2025). The policy indicated the interdependence between trade policy, resource control, and market stability because it increased the local industries at the expense of global networks of the supply of nickel (Deberdt and Le Billon 2023).

2.4 *E-waste Recycling as a Resilience Strategy*

Waste recycling has become an essential resilience strategy in order to cope with a lack of environmental sustainability and supply. As an example, a ton of discarded smartphones can hold up to 350 kilos of palladium, 30 kg of copper, and 300 g of gold, and electronic waste is becoming a secondary source of important metals. The metals can be extracted with high efficiency due to the development of high-quality recycling methods such as pyrometallurgical processing, hydrometallurgical extraction, and sorting with the help of AI. Besides reducing the dependence on primary mining, e-waste urban mining can contribute to eliminating resource shortages, reducing environmental harm, and advancing the concepts of the circular economy (Diallo et al. 2024). The discarded electronics in Japan have recovered over 90% of the rare earth metals with the aid of AI-enhanced e-waste recycling plants in the nation. These factories reduce the reliance on imported raw materials by integrating robots, optical sorting, and chemical extraction methods to promote friendly environmental practices. Similarly, EU member states have implemented producer responsibility laws, which minimize the complexity of goods produced by manufacturers, hence making urban mining more efficient. E-waste recovery is also a measure to control geopolitical threats. By diversifying the supply sources by secondary metal recycling streams, countries will be able to reduce their vulnerability to trade barriers, political instabilities, and fluctuations of prices in the main metal markets. High-quality recycling infrastructure, along with AI-based optimization, is gaining recognition with policymakers and industry as a key investment to industrial and national resilience (Alagarsamy et al. 2025).

3 AI as a Catalyst for Green Metallurgy

Green metallurgy characterizes the metallurgical technologies that use less polluting materials and reagents, utilise resources more effectively, and reduce environmental impacts (including greenhouse gas emissions, the generation of waste, and energy use). Artificial intelligence (AI) is being employed to hasten green metallurgy, predicting material behavior, increasing process parameters, reducing defects, and steering innovation in cleaner directions. In this part, how AI is assisting green metallurgical procedures using the current case studies. We also examine the way that AI analytics are maximizing resource efficiency.

3.1 Introduction to Green Metallurgical Techniques

Green metallurgy is the next step to the sustainability of metal processing, being focused on minimizing the waste generation, consuming energy, and greenhouse gas emissions, and maximizing the recovery rates and preserving the purity of the materials. Traditional metallurgical processes, such as hydrometallurgy and pyrometallurgy, have been a major contributor to industrial pollution and CO_2 emissions in the world. However, the application of artificial intelligence (AI) has begun to transform these conventional approaches into smart, efficient, and environmentally friendly solutions. Machine learning (ML), predictive analytics, and real-time data interpretation enable the optimization of temperature control, dosing of reagents, and energy input in most metallurgical processes with the help of Artificial Intelligence (AI). To drive the use of cleaner pathways, digital twins of metallurgical reactors, such as model real-time reactions and predictors of the effect of input factors on metal yields, purity, and emissions. To achieve the sustainability objective, green metallurgical methods such as solvent extraction, eco-friendly leaching, hydrogen-based reduction, and bio-assisted recovery are becoming more integrated with AI algorithms, which automate the changes. Therefore, it can be stated that artificial intelligence (AI) is the central driver in the modern metal industry, which leads to the emergence of innovation, guides the decisions on an information basis, and addresses the issue of environmental responsibility and manufacturing (Liu et al. 2024). They utilized the environmentally friendly lixiviants to demonstrate AI-based leaching optimization of cobalt and nickel recovery of spent lithium-ion batteries. The study reduced CO_2 emission and reagent use by 18% and 22%, respectively, of traditional acid-based leaching with a neural network model based on the kinetics of reactions and pH-temperature data. It also studied the use of artificial intelligence (AI) to control adaptive roasting in the extraction of rare earth elements (REE). They have applied a decision-tree algorithm to refine the roasting parameters to prevent overburning and reduce the wastage by nearly 15%. These examples prove that AI is needed to develop predictive control mechanisms to ensure optimal conditions of the process, reduce material losses, and increase recycling (Bashiri et al. 2024).

3.2 Role of AI in Cleaner Processing Routes

Cleaner processing pathways in metallurgy aim at the replacement of chemically dangerous and carbon-intensive chemicals with less dangerous or renewable compounds, along with process control systems that enhance accuracy. This becomes simpler when AI uses predictive algorithms to identify the most viable pathways of enhancing multivariate processes like temperature, reaction time, and reagent interactions. The capability of AI to analyze massive data of hydrometallurgical reactors, smelting units, and pilots of hydrogen reduction systems will help it to offer real-time suggestions on low-energy and environment-friendly methods, which will decrease human error and its adverse impact on the environment (Zeb et al. 2024). They provided an AI-based sustainability analysis of the hydrometallurgical recovery of cobalt and lithium based on used batteries. They employed machine learning-based multi-criteria decision analysis to evaluate 20 new recycling methods based on their evaluation factors, such as recovery efficiency, CO_2 emission, water consumption, and energy consumption (Premathilake et al. 2025). Compared to thermal recycling, the model determined that solvent-based and selective leaching methods were the most sustainable, thus lowering the total process energy consumption by 30%. The usage of AI-based process control in the Alumina Decarbonisation Pilot Project at Rio Tinto Refinery Yarwun to perform hydrogen-based calcination. Machine learning algorithms to monitor the activity of furnaces in real-time resulted in the reduction of fossil fuel utilization by 90% and carbon dioxide release by more than 98% without lowering the quality of calcification by adjusting the hydrogen-air ratio (Khalili et al. 2025). Moreover, an extensive artificial intelligence implementation of the mining, smelting, and processing facilities is illustrated by the use of the so-called Kun'an AI Model, which was presented in the non-ferrous metals industry. With the aid of AI-based ore-grade forecasting and predictive emission control, as well as optimization of smelting parameters, this system has increased energy efficiency by 12% and decreased slag wastes by 17%. It helps in the emission-aware process design and enables the predictive control of cleaner processing directions. To achieve the goals of ensuring the economic sustainability of metallurgical operations and at the same time achieving environmental goals, human metallurgists on the one hand, along with AI systems on the other, collaborate to constantly adjust to unpredictable conditions, like feed variations or impurities of reagents (Patil et al. 2025).

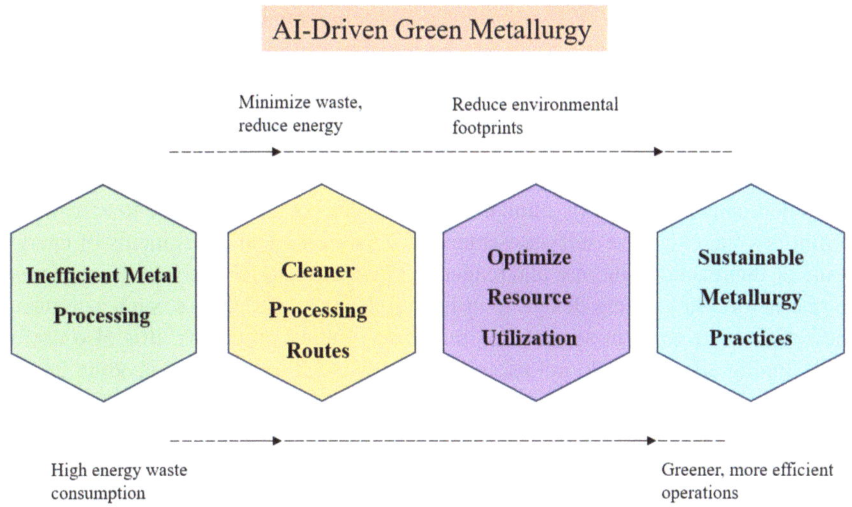

Fig. 1 The transformative impact of AI on cleaner metal processing routes

As presented in Fig. 1, AI-driven green metallurgy involves several key transitions, notably focusing on cleaner processing routes to minimize waste and reduce energy consumption.

3.3 *Optimizing Resource Efficiency Through AI Analytics*

Resource efficiency is the central tenet of green metallurgy when the idea is to maximize metal recovery and reduce its inputs, such as energy, water, and chemicals. This optimization is required through intelligent automation, data-driven forecasting, and pattern recognition of the AI analytics. Machine learning algorithms can predict material flow behaviour, identify inefficiencies in complex systems, and optimize the production-to-resource-use ratio. AI-assisted life cycle assessment (AI-LCA) systems also enable real-time evaluation of environmental impacts, allowing firms to shift to more environmentally friendly metallurgical operations without compromising profitability. In China, at the Bayan Obo mines, AI-enhanced tailings reprocessing of the rare earth elements was demonstrated to be effective. The system minimized chemical waste (25%) and maximized recovery (35%) with the optimization of the dose and temperature of leaching agents using deep reinforcement learning (Kalupahana et al. 2025). AI-driven predictive analytics was used to estimate key mineral demand and optimize the supply chain decisions, thus reducing over-extraction and aligning the production with the sustainability goals (Arun Kumar 2026). They used AI to track the processes of copper smelting, which earned the team 20% of fuel economy improvement and decreased the amount of slags by 14% because of the continuous optimization of oxygen flows and temperature distributions. Its applicability to interdisciplinary issues can be underscored in the applications of AI in these cases. It transforms the metallurgical processes that are lifeless into dynamic learning systems that can be adapted and can self-improve. Using AI coupled with the concept of a digital twin, a full picture of the process, including the input of ore and the final products of the processing process, can be observed, and inefficient areas that cannot be seen by human operators. The AI ensures close to zero waste conditions via extensive self-assessment of sensor data and adjustments of energy sources, the reaction kinetics, and reagents. There are also AI analytics that are used to restructure the process design to utilize circles in materials use, such as reusing industrial waste to form second raw materials in alloy processes. It is also useful in designing alloys that do not use rare elements such as cobalt or neodymium by predicting alternative alloys that have desirable mechanical or magnetic properties (Chaudhari et al. 2025).

The growing body of literature indicates that the application of AI in different metallurgical operations has led to measurable improvements in the recovery of materials, energy conservation, and reduction of emissions. Table 2 provides a summary of significant current research and industrial applications, indicating the way AI-driven innovations are transforming the concept of sustainability in green metallurgy.

4 Cognitive Robotics in Component Recognition and Extraction

E-waste management and metal recovery are evolving as a result of the introduction of Cognitive Robotics based on a fusion of robotic automation, machine learning, and artificial intelligence. The challenge of breaking down and separating elements with precious metals such as gold, palladium, neodymium, and cobalt has become especially troublesome since the e-waste flow of the world reaches more than 62 million tons per year. Although this may be a fact, it would be time-consuming, hazardous, and economically unsustainable to disassemble large volumes of hands using the traditional method. Cognitive robotics offers an innovative alternative: intelligent machines with the ability to realize, learn, and adapt to the evolving form of discarded electronic devices. Based on computer vision and reinforcement learning innovations, as well as autonomous manipulation, these robots are able to identify valuable components, selectively disassemble them, and separate metallic fractions with previously unachievable accuracy. Figure 2 illustrates an automation–complexity framework highlighting the progression from manual component recognition to advanced AI- and robotics-based systems for efficient sorting, dismantling, and recovery of critical metals from electronic waste.

Table 2 Applications of artificial intelligence in advancing green metallurgy

Processing sector	AI Integration & method	Application	Efficiency outcome	References
Lithium & cobalt (battery recycling)	Machine learning (multi-criteria decision model)	Sustainability assessment of hydrometallurgical recycling routes	30% lower energy use, 25% emission reduction, improved recyclability	Öztürk et al. (2023)
Alumina refining (bauxite to alumina)	AI-based predictive control with hydrogen calcination	AI-controlled hydrogen–air mixing during calcination	98% CO_2 reduction and 90% reduction in fossil fuel dependency	Ajwani et al. (2025)
Rare earth elements (tailings reprocessing)	Deep reinforcement learning model for reagent dosage	Optimization of leaching kinetics and acid concentration	35% higher REE recovery and 25% lower chemical waste	Kursunoglu (2025)
Copper smelting industry	AI digital twin and sensor-driven optimization	Real-time monitoring of furnace temperature and oxygen flow	20% energy efficiency improvement and 14% slag reduction	Pech-Rodríguez et al. (2024)
Nickel & cobalt (secondary extraction)	Convolutional neural networks for reaction pathway prediction	Prediction of solvent extraction equilibria and phase separation	18% faster process optimization and reduced solvent losses	Nurfajrin et al. (2025)
Steel manufacturing	Deep learning image recognition for microstructure optimization	Alloy design using scrap-based inputs and microstructure prediction	22% defect reduction and 16% resource savings via alloy substitution	Raabe et al. (2024)
Gold recovery from E-waste	AI-guided adaptive hydrometallurgy	Optimized cyanide-free leaching with thiourea alternatives	27% increase in gold yield and 40% toxicity reduction	Gulliani et al. (2023)
Non-ferrous metals (China's "Kun'an" AI model)	Large-scale industrial AI model	Predictive smelting optimization, emission control, and ore grading	17% waste reduction and 12% increase in operational efficiency	Gerold et al. (2024)
Multi-metal recovery (urban mining plants)	AI-driven process integration and digital twins	Combined recovery of Cu, Ag, Pd, and Au from printed circuit boards	33% increase in throughput, 28% energy savings	Mohamed et al. (2025)

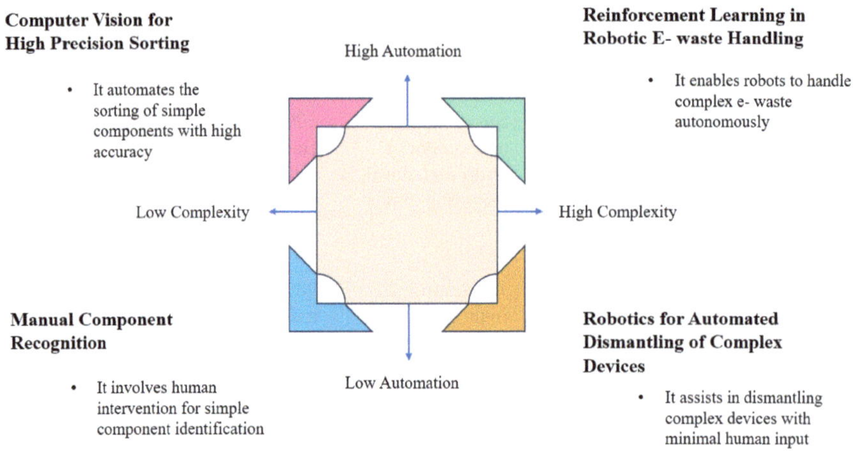

Fig. 2 Conceptual matrix of automation and task complexity in critical metal recovery

4.1 Computer Vision for High-Precision Sorting

Computer vision is involved in robotic e-waste recognition methods. Robot systems are now able to recognize and classify objects such as integrated circuits, printed circuit boards (PCBs), and capacitors based on their color, shape, and spectral properties, and features of deep convolutional neural networks (CNNs) and 3D object mapping. Through integrating AI vision and hyperspectral imaging (HSI) and X-ray fluorescence (XRF), it is possible to identify materials using their spectral fingerprint and the difference between rare earth magnets, copper traces, and plastic insulation in milliseconds. The sorting system that was created by RoboSort Technologies was a computer vision-based system that was able to identify recyclable metals in mixed e-waste streams with 96% accuracy. It was based on the technique of training CNNs on over 200,000 annotated consumer electronics samples, as well as a dual-camera image setup using multispectral images. The vision model of the robot would have been able to provide higher recovery efficiency and purity, and even distinguish between oxidized and unoxidized copper (Nišić et al. 2024). In comparison with traditional CNN-based models was an improvement in PCB component detection and an 18% lower amount of incorrect classification. More and more AI-based segmentation methods are applied to isolate interrelated components besides the process of visual classification. Such models can dynamically adapt to deformation, dust, and light, which is normal in disassembly facilities. The cognitive robotic systems are thus based on the perceptual level of computer vision, the development of computer vision, which started with the recognition of pictures but has evolved to a dynamic and adaptive understanding of pictures in real-time (Rojas et al. 2025).

4.2 Robotics for Automated Dismantling of Complex Devices

The physical representation of the intelligent recovery is robotic disassembly, which is more than familiar. Modern intelligent robots utilize multi-axis manipulators, force-feedback systems, and tactile sensors to break down devices without damaging the internal components that are considered vital to them. With AI, these systems are able to map internal architecture in advance and anticipate the placement of screws, solder joints, and other useful components, such as CPU processors or batteries, and can begin extraction. The Fraunhofer Institute of Manufacturing Engineering and Automation (Germany) has an example of such development in the AutoDisMANTLE 2025 project. Its robot arm (equipped with haptic feedback and visual servoing), with a success rate of more than 85, breaks down computers and cellphones automatically, pulling out screws, wires, and modules. The unusual aspect of this system is its cognitive planning engine that can simulate human flexibility by realigning grip and trajectory during operation when one of its components does not cooperate. Another example of a cutting-edge technique is the Daisy 2.0 robotic line (2024) that amalgamates computer vision and cognitive motion planning to take apart over 1.2 million iPhones annually. The technology is nearly flawless in the extraction of cobalt and rare earth magnets contained in lithium-ion batteries. As compared to former deterministic robots, Daisy 2.0 is compatible with the variation in designs through probabilistic learning that enables adaptation to the model variations. These technologies are the transition to adaptive intelligence by radically reducing the volume of manual effort required, increasing recovery output, and improving the safety of the workers. The costliness of installation, as well as the need to have high-quality data to train adaptive algorithms on diverse devices, however, remains a limitation to the mainstream implementation (Khanam et al. 2024).

4.3 Reinforcement Learning in Robotic E-waste Handling

Reinforcement learning (RL) is the next advance in robot cognition to handle e-waste. By means of incentive signals for efficiency, recovery purity, and time reduction, a robot can learn the most optimal disassembly or sorting process through trial-and-error as a result of reinforcers, unlike in supervised learning, which depends on labeled data. This is a skill needed to manage different and unpredictable waste streams. They applied deep reinforcement learning (DRL) to control robotic picking of circuit boards and capacitors out of shredded electronic waste. The robot reduced the destruction of the parts by 35% and enhanced the accuracy of extraction by 27% over time (Kim et al. 2025). They applied RL-based adaptive control to robotic dismantling arms that varied the torque and angle dynamically during disassembly, saving 15% of energy. These systems also utilize digital twins, virtual representations of a particular environment, to simulate thousands of interactions before being put in the real world. This enhances the flexibility of robots as well as reducing training and

Table 3 Recent global implementations of cognitive robotics in E-waste recovery and critical metal extraction

Project	Target material and processing sector	Core technology used	Key functionality	Accuracy	References
RoboSort Vision	Mixed e-waste sorting	CNN + Multispectral Imaging	Metal identification & segregation	96% classification accuracy	Charpentier et al. (2023)
ETH–Empa PCB Classifier	PCB recognition	Graph Neural Networks	Component mapping & labeling	18% reduction in misclassification	Chen et al. (2023)
Daisy 2.0	iPhone dismantling	Cognitive motion planning + ML	Rare earth & cobalt recovery	1.2 M devices/ year	Lu et al. (2023)
RECLAIM Project	Robotic dismantling arms	RL-based torque control	Energy-efficient disassembly	15% energy reduction	Das and Koskinopoulou (2025)
AURORA Line	Industrial e-waste	Computer vision + 3D mapping	Automated sorting	92% metal identification accuracy	Oke and Potgieter (2024)

material wastage. The RL-powered robots enhance their ability to make decisions by updating the policy networks periodically, and it is consistent with the concept of green metallurgy, which integrates sustainability with precision (Schröder and Iwasaki 2024).

The move towards sustainable recovery of resources in the world is being propelled by perceptual, decision, and self-learning robotics. The latest developments of cognitive robots capable of disassembling and extracting metal out of e-waste indicate the way this technology links artificial intelligence with green metallurgy (Table 3).

5 Deep Learning Models for Metallurgical Yield Forecasting

The use of deep learning (DL) models in metallurgical processes has changed the lives of prediction and optimization in metal production and recovery in industries completely. These models project the results of numerous metallurgical systems that no one has ever heard of with high accuracy and efficiency, with the use of advanced algorithms and large datasets.

Figure 3 illustrates an integrated analytical framework in which predictive modelling, neural network algorithms, and real-time process estimation converge to enable accurate and dynamic forecasting of metallurgical yield.

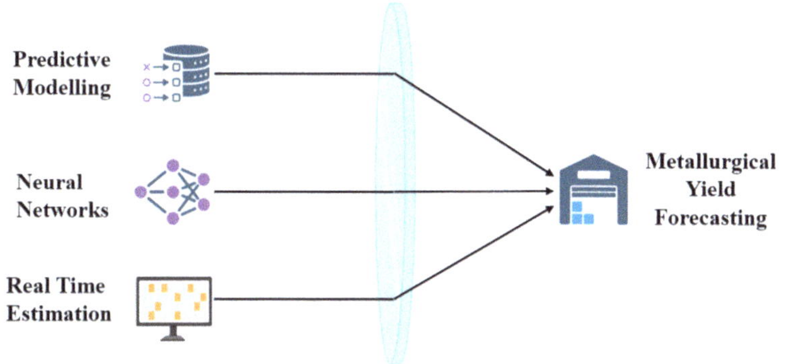

Fig. 3 AI-enabled system for real-time metallurgical yield forecasting

5.1 Predictive Modeling in Pyro- and Hydrometallurgical Systems

Since predictive modeling enables accurate projections of metal recovery, enhances efficiency in the process, and lowers the energy, it is vital in both hydrometallurgical and pyrometallurgical systems. Deep learning algorithms such as artificial neural networks (ANNs), convolutional neural networks (CNNs), and ensemble learning systems can be employed to use the enormous quantities of data provided by these processes, often provided by sensors, operational records, and material composition experiments. Mathematically predictive models can be effective in capturing complex interactions among operating conditions, feed composition, and environmental conditions to improve extraction efficiency. They demonstrate that machine learning algorithms can perform copper recovery predictions from flotation processes. In order to connect operational effectiveness with sustainability, predictive modeling may be used (León et al. 2025). Deep learning technologies used to process lithium-ion battery cathodes in the pyrometallurgical recycling have led to much higher metal recovery and reduced impact on the environment. These examples demonstrate that deep learning is essential in inferring the outcomes of processes, particularly in complex metallurgical procedures, where traditional statistical models may fail to notice dynamic variations and nonlinear relationships (Antony Jose et al. 2024).

5.2 Neural Networks for Process Parameter Optimization

Neural networks are quite efficient at optimizing process parameters, which means that metallurgists can find the most favorable operating conditions that allow them to increase yield, decrease energy expenditures, and maintain consistent product quality using neural networks. Complex neural network structures can deal with multivariate inputs and can learn nonlinear relationships amongst process variables, should they be optimized with algorithms such as Levenberg–Marquardt or differential evolution. The same research on artificial neural networks used to predict MAG welding conditions in the desired bead shape was one of the examples cited where the ANN-based systems could optimize complex manufacturing processes. They employed a Levenberg–Marquardt ANN to react sodium fluorosilicate with molten aluminum so that the formation of Al-Si cast alloys could be maximized. To maximize silicon recovery, as well as minimize resource usage, the research established optimal operating temperatures, stirring rates, and minimum response times. These demonstrations indicate that neural networks can adapt to the fluctuation in the quality of raw materials or process perturbation besides forecasting favorable conditions, which is essential in massive metallurgical operations. Neural networks also address economic as well as environmental objectives as they continuously adjust parameters to assist in the realization of process stability, enhancement of total yield, and operational efficiency (Miladinović et al. 2024).

5.3 Real-Time Metal Recovery Estimation

One of the revolutionary applications of deep learning to the metallurgy field is the real-time estimation of metal recovery, which allows adaptive optimization, dynamic control, and immediate decision-making. To provide feedback nearly immediately on metal recovery performance, the deep learning models, including CNNs and reinforcement learning-based systems, can be used to analyze the live data streams of sensors and logs of processes. They used a CNN-based system with an accuracy of between 77.5% and 95 to estimate fill levels of scrap containers in a copper tube mill. These capabilities enabled operators to boost productivity, reduce downtime, and optimize material handling. Systems with deep learning capabilities have been applied to predict waste formation, control slag composition, and enhance the efficiency of the pyrometallurgical copper production process in general. To reduce the losses of materials and ensure the stable recovery rates, such systems unite real-time sensor data with the past process records and prediction models in order to allow adaptive changes in the operational parameters. Deep learning solutions in real time can considerably improve the efficiency of operations and sustainable metallurgy by making continuous monitoring and control of the processes possible (Koinig et al. 2024).

To emphasize the existing multitude and industrial relevance of the modern deep learning applications in the metallurgical yield forecasting, a summary of the studies on prediction accuracy, operational efficiency, and cutting-edge AI techniques is given in Table 4.

Table 4 Innovative deep learning applications in metallurgical yield forecasting

Objective	Features used	AI technique	Industrial impact	References
Predict copper recovery in flotation	Sensor data, ore composition, reagent dosages	Stacking ensemble learning	Reduced process inefficiencies, improved recovery	He et al. (2024)
Forecast gold concentrate grade	Multi-step process variables, flow rates	Deep neural networks	Enabled 30-min ahead prediction, improving scheduling	Zhang et al. (2024)
Optimize Al–Si cast alloy production	Reaction time, temperature, and stirring speed	ANN with Levenberg–Marquardt algorithm	Increased silicon yield, minimized resource use	Mahran et al. (2025)
Real-time scrap container monitoring	Images, fill-level sensors	CNN + computer vision	Reduced downtime and manual checking	Alexopoulos et al. (2023)
Predict steel yield and tensile strength	38 microalloy features	Neural network regression	More precise alloy selection, reduced test cycles	Shannak et al. (2024)
Leaching efficiency for rare earth elements	Temperature, acid concentration, pH	Explainable AI	Enabled interpretable predictions for process control	Nguyen et al. (2025)
Forecast hydrometallurgical copper output	Historical production, flow rates	Recurrent neural networks (RNN)	Better resource planning and recovery estimates	Kenzhaliyev et al. (2025)

6 Digital Twins for Process Simulation and Optimization

6.1 Principles of Digital Twins in Recycling Plants

Digital twins (DTs) are gradually gaining recognition as the groundbreaking tools of controlling and optimizing complex industrial operations, particularly recycling plants. Digital twins are simply dynamic, real-time digital versions of physical systems that constantly update themselves with data related to sensors, Internet of Things objects, process monitoring systems, and previous operation records. DTs may represent every stage of the movement of materials in recycling plants, including collection and sorting, shredding, separation, and chemical recovery. With this virtual model, operators can actively identify potential bottlenecks or equipment failures, predict how the system will behave in different conditions, and model different conditions that can happen. Besides providing an informative and visual understanding, the twin allows predictive maintenance plans to take place, thereby reducing unplanned downtimes and enhancing the life of equipment. DTs ensure a safer and cheaper way of managing the system and also facilitate the creation of a virtual laboratory in which process changes can be tested without disrupting actual operations (Li et al. 2024).

6.2 AI-Driven Virtual Prototyping for Process Refinement

With artificial intelligence (AI) added to them, the possibilities of digital twins are considerably expanded, making it possible to refer to so-called AI-powered virtual prototyping. The approach utilizes advanced analytics and machine learning in the DT with the aim of predicting outcomes, setting trends, and proposing better operational solutions. AI is able to handle all the large volumes of real-time data generated at recycling sites and can also identify small correlations that otherwise would be overlooked by human operators. This saves the trial-and-error testing of the actual systems since one can iterate and improve on the process parameters in a virtual environment. In the industrialized offshore construction, as an example, a study by They evidences that AI-connected DTs enhance the workflow sequences, predict potential conflicts, and regenerate the operational parameters before the implementation process. The same strategies can be implemented in the processes of metallurgy and recycling to simulate variations in the input materials, equipment, or even the dosage of chemical reagents to establish the most efficient and environmentally friendly operations. Besides accelerating the innovation process, AI-based virtual prototyping can also be used to test the scenario in extreme or unpredictable conditions, which provides a risk-free environment to continuously refine the process. The capacity of these systems to adapt dynamically with new data as new information is received provides the ability of recycling facilities to be flexible in response to changes in their material streams or market demands (Kiprono et al. 2023).

6.3 Benefits for Operational Transparency and Efficiency

When digital twins are integrated into recycling processes, there are considerable benefits of transparency, efficiency, and decision-making in the operations. By relying on DTs, operators, engineers, and management can track the key performance indicators and detect the inefficiencies, as well as make evidence-based decisions, which will allow them to see the entire picture of plant performance in real-time. Since the stakeholders can trace the flow of materials within the plant, the amount of energy consumed, and the usage of equipment, the transparency of the operations is further enhanced. Predictive analytics in the twin enables proactive maintenance, which reduces unscheduled downtime, utilizes machines to their full capacity, and extends the lifetime of equipment. Moreover, DTs contribute to the attainment of the sustainability targets through seeking methods to enhance the restoration of the essential resources, reduce the amount of waste generation, and use less energy (Sathya et al. 2024). They argued that integration of simulation models and production data can significantly contribute to the optimization of the planning, allow a rapid restructuring of production, and ensure continuous optimization of processes. The digital twins can be a teaching and training tool to the operators of the plant as well, providing them with a safe environment to explore the scenarios of the working and understand the dynamics of the system without threatening the output. Taken together, DTs establish an iterative system where simulations are informed using real-time information and operational changes are informed using simulation knowledge so that resiliency, better performance, and continuous learning are encouraged (Xing et al. 2025).

7 Ethical AI and Bias in Automated Decision Systems

7.1 Potential Risks of Algorithmic Bias in E-waste Tech

Artificial intelligence is taking a bigger and bigger part in e-waste management and recycling technology, and this is a grave ethical issue since algorithmic bias may occur. Examples of automated decision-making systems are robotic sorting, predictive maintenance, and process optimization; most of these processes rely on previous and real-time information. In the case of unbalanced, unrepresentative, or inadequate datasets, AI systems can give biased results. These outcomes may be poor sorting, inequitable resource distribution, or even prioritizing certain waste streams above others. In situations involving rare or hybrid materials, such as the use of a trained robotic sorting system that has been grounded on the high-density plastics, the system may not perform as well, leading to an inadequate recovery of the metals and an increased environmental impact. Also, biased prediction models of hazardous waste treatment or recovery of materials might be used inadvertently to benefit a specific plant activity or a specific geographical location, which can further increase existing disparities in environmental protection or resource accessibility. Besides the inefficiency in operations, algorithmic bias poses the problem of accountability as it can prove difficult to audit, justify, or fix judgments by opaque AI models. This is why it is very important that designers and operators have effective monitoring systems and biasing detection procedures (Phogat et al. 2025).

7.2 *Responsible Data Governance and Access*

To minimize prejudice and ensure the ethical use of AI in environmental technologies, data governance should be responsible. To ensure that datasets accurately represent the diversity of e-waste streams and operating situations, data governance frameworks should establish clear rules and requirements for collecting, validating, and preparing data. The openness in the data application is also vital since it helps the stakeholders to understand the way AI models analyse and react to the data. The safe storage of sensitive information, such as profiles of dangerous materials or proprietary industrial processes, needs access control systems that ensure that access control is done without compromising the sufficient amount of accessible models to facilitate model evaluation and training. Data governance must be responsible so as to reduce bias and make AI use in environmental technology ethical. Data governance systems should also have clear guidelines on data collection, validation and preparation so as to have datasets that are representative of the variety of e-waste streams and operating environments. As it makes the stakeholders learn how the AI models interpret and respond to the data, the openness in the use of data is also significant. The access control systems should be such that they provide adequate accessibility to model assessment and training and also the secure management of sensitive information, like profiles of hazardous substances or distinct industry operations (Herrington and Gordon 2024).

7.3 *AI Ethics in Environmental Technology*

AI ethics in the environmental technology include the values of justice, accountability, transparency, and sustainability, which can be regarded as a guide to the ethical construction and implementation of automated systems. In e-waste management, ethical AI contributes to making sure that technology progress does not increase environmental or social injustice. The deployment of automated dismantling robots, e.g. needs to consider workforce integration or reskilling and should consider the risk of job displacement. Transparency in the process of making AI decisions is also necessary to allow operators and regulators to assess forecasts, verify findings, and take the necessary step. To ensure that AI-based optimizations favor material recovery, consume lesser energy, and have no detrimental effects, environmental sustainability remains one of the most valued ethical issues. The importance of having multiple stakeholders involved in the process of shaping AI regulatory models, including but not limited to the involvement of technologists, legislators, and community representatives, has become the subject of the recent literary discussion. These participatory methods will ensure that ethical, social and environmental concerns are embedded throughout the life of the AI systems since data collection, model deployment and post implementation monitoring. Organizations can more efficiently dispose e-waste with

AI and not violate social norms and environmental responsibilities by incorporating high ethical standards (Goyal and Gupta 2024).

8 Integrating IoT and AI for Closed-Loop Traceability

8.1 Smart Labeling and IoT Tracking of E-waste Flows

Artificial intelligence (AI) and the Internet of Things (IoT) have a revolutionary potential of ensuring closed-loop traceability in e-waste management. Since the manufacturing process up to the end-of-life process, there can be tracking of individual parts and devices with smart labeling systems such as RFID tags, QR codes, and Near-Field Communication (NFC) devices. Such labels provide real-time data regarding the whereabouts, state, and condition of e-waste products in combination with Internet of Things-based sensors. This enables proper management of material flow within recycling plants or the supply chain. This vast amount of information could be checked with the help of AI algorithms to detect anomalies, predict bottlenecks, and enhance the processes of collecting and sorting information. To illustrate, the usage of smart sensors to monitor the lithium-ion batteries in real-time can possibly avoid hazardous scenarios, ensure proper segregation, and ensure the recovery of such valuable metals as nickel, cobalt, and lithium. Smart labeling and IoT tracking can become the core of the effective traceability system as the physical objects are linked with streams of digital data. This enhances efficiency in its operations and provides clear and verifiable material histories to the stakeholders and regulators (Kumar et al. 2024).

8.2 Blockchain Integration for Material Source Authentication

The IoT allows real-time monitoring, but the blockchain technology can improve these systems by providing safe, unmodifiable, and visible records of material origin. Using AI and IoT in combination with blockchain, one can trace the recycling process and check the sources of materials used, ensuring compliance with ethical sourcing and environmental policies. Examples of records kept in blockchain include keeping track of the source of every product, the steps it follows to be processed, and the final recovery outcomes when e-waste elements are collected in separate locations. This is to prevent fraud and facilitate accountability. AI may also enhance the performance of blockchain by analyzing those data streams to identify tendencies of inefficiency, potential contamination, or the illicit distribution of materials. According to current studies, blockchain-based systems of traceability have the benefit of certifying recycled materials to be reused in high-value sectors such as electronics or battery manufacture, and also significantly boost confidence among manufacturers, recyclers, and regulators. With the combination of blockchain, IoT, and AI, the stakeholders can develop a closed-loop system that is fully transparent, auditable, and reliable, and real-time monitoring, verification, and optimization of all material flows (Zante et al. 2024).

8.3 AI for Lifecycle and Supply Chain Intelligence

Artificial intelligence (AI) offers lifecycle and supply chain visibility to the IoT and blockchain systems by analyzing the vast amounts of data that they generate, as well as monitoring and verification. The machine learning algorithms can identify the optimal recycling paths, verify the amount of material that can be recycled from the devices that are discarded, and predict the time during which a product will cease its useful purpose. Examples of how AI can be used include selecting high-value parts to recover a selective number, forecasting the volume of e-waste that will be generated depending on the usage patterns of customers at the facility, and allocating processing resources efficiently in recycling plants. Also, by planning material flows, dynamically changing the course of collection, and sharing responsibilities across multiple locations, AI-based analytics could make the supply chain more efficient. The insights that AI models come up with also help in strategic decision-making, such as investment in advanced recycling technology, designing products in a circular way, or developing long-term producer responsibility guidelines. Together, AI, IoT, and blockchain create a comprehensive system of closed-loop traceability that will allow managing e-waste sustainably, maximizing the recovery of the most important metals, and holding all participants in the supply chain accountable. This combined strategy will make e-waste management a data-driven, efficient, and environmentally friendly system by enabling actionable intelligence at each *point*, in collection and transport, processing, and reintegration into manufacturing (Quinto et al. 2025).

9 Conclusion

Due to the introduction of artificial intelligence (AI), digital twins, cognitive robots, and Internet of Things (IoT) technologies, critical metal recovery of electronic trash is experiencing a major change. The developments are offering new frontiers of sustainable management of resources. Digital twins provide virtual simulation benefits, operating strategies, reduced downtime, and process visibility, whereas AI-based predictive models and neural networks enhance yield prediction in metallurgy, process optimization, and estimating recovery through real-time recovery estimation. Cognitive robots are capable of automating component-detection and disassembly with the help of computer vision and reinforcement learning, making the complex equipment easier and more accurate to handle. Meanwhile, the blockchain integration and IoT-enabled smart labeling are used to ensure closed-loop traceability that ensures provenance, compliance, and responsibility throughout the e-waste supply chain. Application of case studies demonstrates that these innovations could be applied in practice, where it is stressed that sustainability measures, operational performance, and metal recovery could be gained. These technologies can both address the immediate solution to critical problems of insufficient access to needed metals and contribute to the objectives of the circular economy, since they help to eliminate waste and optimally utilize resources to ensure reuse. The convergence of AI, automation, and traceability is a promise of a future in which intelligent, transparent, and sustainable systems are offered as the means of providing economic and environmental value to the process of managing e-waste, signifying a radical shift in the management of e-waste.

References

Ahmed I, Seku K, Nidamanuri SB, Koyyala KK (2025) Handling metal waste to salvage with automation. In: Sadasvivuni KK, Bacanin N, Kim J, Vashisht BN (eds) Harnessing automation and machine learning for resource recovery and value creation. Elsevier, pp 187–212. https://doi.org/10.1016/B978-0-443-27374-2.00007-8

Ajwani V, Ajwani N, Rai A (2025) Artificial intelligence approach to predict ladle aluminium content and to protect its refractory lining. Can Metall Q 64(2):786–792. https://doi.org/10.1080/00084433.2024.2367180

Alagarsamy S, Vishnuram P, Shanmugam Y, Thentral T, Gono R, Gono M (2025) Advancements and challenges in lithium-ion battery lifecycle management towards a sustainable circular economy for electric vehicles. IEEE Open J Power Electron 6:1491–1533. https://doi.org/10.1109/OJPEL.2025.3604475

Alexopoulos K, Catti P, Kanellopoulos G, Nikolakis N, Blatsiotis A, Christodoulopoulos K, Ziata E (2023) Deep learning for estimating the fill-level of industrial waste containers of metal scrap: a case study of a copper tube plant. Appl Sci 13(4):2575. https://doi.org/10.3390/app13042575

Antony Jose S, Cook CAD, Palacios J, Seo H, Torres Ramirez CE, Wu J, Menezes PL (2024) Recent advancements in artificial intelligence in battery recycling. Batteries 10(12):440. https://doi.org/10.3390/batteries10120440

Arun Kumar A (2026) Artificial intelligence for climate neutrality: enhancing environmental sustainability. In: Singh R, Filho WL, Kumar V (eds) Data-driven sustainability: harnessing technology for a greener future and environmental resilience, Springer Nature Switzerland, Cham, pp 175–189. https://doi.org/10.1007/978-3-032-13412-7_11

Bashiri A, Habibi M, Sufali A, Shekarsokhan S, Maleki R, Razmjou A (2024) Artificial intelligence models for efficiency estimation of adsorbents in rare-earth element recovery based on feature engineering. Ind Eng Chem Res 63(42):17930–17948. https://doi.org/10.1021/acs.iecr.4c01935

Basuhi R, Bhuwalka K, Moore EA, Diersen I, Malik RH, Young E, Olivetti EA (2024) Clean energy demand must secure sustainable nickel supply. Joule 8(11):2960–2973. https://doi.org/10.1016/j.joule.2024.10.008

Biancardi A, D'Adamo I, Santibanez-Gonzalez ED, Bascur JFV (2025) A flexible and circular management of copper in Chile: new perspectives toward sustainable development. Global J Flex Syst Manag 26(4):813–837. https://doi.org/10.1007/s40171-025-00464-w
Charpentier NM, Maurice AA, Xia D, Li WJ, Chua CS, Brambilla A, Gabriel JCP (2023) Urban mining of unexploited spent critical metals from E-waste made possible using advanced sorting. Resour Conserv Recycl 196:107033. https://doi.org/10.1016/j.resconrec.2023.107033
Chaudhari SS, Shahakar M, Gholap P, Dhamdhere PB, Agrawal SS, Mahajan RG, Yashwant S (2025) Enhancing energy efficiency in metallurgical operations using AI: a state-of-the-art review. JMMF 73(6):1595–1616. https://doi.org/10.18311/jmmf/2025/48703
Chen X, Wu Y, He X, Ming W (2023) A comprehensive review of deep learning-based PCB defect detection. IEEE Access 11:139017–139038. https://doi.org/10.1109/ACCESS.2023.3339561
Chidunchi I, Kulikov M, Safarov, R, Kopishev E (2024) Extraction of platinum group metals from catalytic converters. Heliyon 10(3). https://doi.org/10.1016/j.heliyon.2024.e25283
Das AR, Koskinopoulou M (2025) Towards sustainable manufacturing: a review on innovations in robotic assembly and disassembly. IEEE Access 13:100149–100166. https://doi.org/10.1109/ACCESS.2025.3576441
Deberdt R, DiCarlo J (2024) Pericentricity on the Congolese copperbelt: how the DRC shapes Chinese cobalt supply chains and the low-carbon transition. Globalizations 21(8):1496–1517. https://doi.org/10.1080/14747731.2024.2374432
Deberdt R, Le Billon P (2023) Green transition mineral supply risks: comparing artisanal and deep-sea cobalt mining in a time of climate crisis. Extr Ind Soc 14:101232. https://doi.org/10.1016/j.exis.2023.101232
Depraiter L, Goutte S, Porcher T (2025) Geopolitical risk and the global supply of rare earth permanent magnets: insights from China's export trends. Energy Econ 146:108496. https://doi.org/10.1016/j.eneco.2025.108496
Diallo S, Tran LH, Larivière D, Blais JF (2024) Mass balance and economic study of a treatment chain for rare earths, base metals and precious metals recovery from used smartphones. Miner Eng 215:108824. https://doi.org/10.1016/j.mineng.2024.108824
Ferreira-Filipe DA, Duarte AC, Hursthouse AS, Rocha-Santos T, Silva ALP (2025) Biobased strategies for E-waste metal recovery: a critical overview of recent advances. Environments 12(1):26. https://doi.org/10.3390/environments12010026
Fichani K, Teseletso LS, Kaavera J, Dintwe TK, Shemang E, Matshediso BI (2025) The impact of junior miners on the global supply of high-purity manganese sulfate monohydrate for the electric vehicle battery market. J South Afr Inst Min Metall 125(5):233–242. https://doi.org/10.17159/2411-9717/3196/2025
Gerold E, Luznik L, Samberger S, Antrekowitsch H (2024) Sustainable extraction and recycling of non-ferrous metals: a review from a European perspective. Philos Trans A Math Phys Eng Sci 382(2284):20240173. https://doi.org/10.1098/rsta.2024.0173
Goyal S, Gupta S (2024) A comprehensive review of current techniques, issues, and technological advancements in sustainable E-waste management. e-Prime: Adv Electr Eng Electron Energy 9:100702. https://doi.org/10.1016/j.prime.2024.100702
Gulliani S, Volpe M, Messineo A, Volpe R (2023) Recovery of metals and valuable chemicals from waste electric and electronic materials: a critical review of existing technologies. RSC Sustain 1(5):1085–1108. https://doi.org/10.1039/D3SU00034F
He G, Liu M, Zhao H, Huang K (2024) Ensemble prediction modeling of flotation recovery based on machine learning. Int J Min Sci Technol 34(12):1727–1740. https://doi.org/10.1016/j.ijmst.2024.11.012
Herrington R, Gordon S (2024) Delivering critical raw materials: ecological, ethical, and societal issues. In: Peppoloni S, Capua DG (eds) Geoethics for the future. Elsevier, pp 235–247. https://doi.org/10.1016/B978-0-443-15654-0.00018-9
Ilankoon IMSK, Dilshan RADP, Dushyantha N (2024) Co-processing of e-waste with natural resources and their products to diversify critical metal supply chains. Miner Eng 211:108706. https://doi.org/10.1016/j.mineng.2024.108706

Kalupahana R, Dushyantha N, Ratnayake AS (2025) Circular hydrometallurgy to overcome the limitations of conventional extractive metallurgy. J Sustain Metall 11(4):3322–3342. https://doi.org/10.1007/s40831-025-01247-4

Kenzhaliyev B, Azatbekuly N, Aibagarov S, Amangeldy B, Koizhanova A, Magomedov D (2025) Predicting industrial copper hydrometallurgy output with deep learning approach using data augmentation. Minerals 15(7):702. https://doi.org/10.3390/min15070702

Khalili Y, Yasemi S, Abdi M, Ghasemi Ertian M, Mohammadi M, Bagheri M (2025) A review of integrated carbon capture and hydrogen storage: AI-driven optimization for efficiency and scalability. Sustain 17(13):5754. https://doi.org/10.3390/su17135754

Khanam R, Hussain M, Hill R, Allen P (2024) A comprehensive review of convolutional neural networks for defect detection in industrial applications. IEEE Access 12:94250–94295. https://doi.org/10.1109/ACCESS.2024.3425166

Kim KS, Shin YJ, Lee WJ, Kang SK (2025) Toward Advanced Degradable Soft Robotics Incorporating Transient Soft Electronics: Materials, Fabrication, and Applications. Adv Electron Mater 12(1):e00400. https://doi.org/10.1002/aelm.202500400

Kiprono NR, Smolinski T, Rogowski M, Chmielewski AG (2023) The state of critical and strategic metals recovery and the role of nuclear techniques in the separation technologies development. Separations 10(2):112. https://doi.org/10.3390/separations10020112

Koinig G, Kuhn N, Fink T, Lorber B, Radmann Y, Martinelli W, Tischberger-Aldrian A (2024) Deep learning approaches for classification of copper-containing metal scrap in recycling processes. Waste Manag 190:520–530. https://doi.org/10.1016/j.wasman.2024.10.022

Kumar A, Kumar Y, Kumari A, Kumar S (2024) IoT-enabled systems for E-waste monitoring and recycling. In: Gupta A, Kumar R, Kumar V (eds) Integrated waste management: a sustainable approach from waste to wealth, Springer Nature Singapore, Singapore, pp 375–394. https://doi.org/10.1007/978-981-97-0823-9_18

Kursunoglu S (2025) A review on the recovery of critical metals from mine and mineral processing tailings: recent advances. J Sustain Metall 11:2023–2050. https://doi.org/10.1007/s40831-025-01126-y

León F, Rojas L, Bazán V, Martínez Y, Peña A, Garcia J (2025) A systematic review of copper heap leaching: key operational variables, green reagents, and sustainable engineering strategies. Processes 13(5):1513. https://doi.org/10.3390/pr13051513

Li C, Zhu Z, Mao X (2024) Copper industrial product monitoring system based on digital twin technology. In: Proceedings of the international conference on algorithms, software engineering, and network security, pp 531–536. https://doi.org/10.1145/3677182.3677276

Liu A, Hu G, Wu Y, Guo F (2024) Life cycle environmental impacts of pyrometallurgical and hydrometallurgical recovery processes for spent lithium-ion batteries: present and future perspectives. Clean Technol Environ Policy 26(2):381–400. https://doi.org/10.1007/s10098-023-02640-x

Lu Y, Pei W, Peng K (2023) State of the art of automatic disassembly of WEEE and perspective towards intelligent recycling in the era of Industry 4.0. Int J Adv Manuf Technol 128(7):2825–2843. https://doi.org/10.1007/s00170-023-12043-3

Mahran GM, Omran ANM, Gomaa HE, Mahamed MAEH (2025) Optimizing the processing parameters of producing Al–Si alloys using sodium fluosilicate via artificial neural network. Mater Adv 6:4705–4713. https://doi.org/10.1039/D5MA00142K

Miladinović S, Gajević S, Savić S, Miletić I, Stojanović B, Vencl A (2024) Tribological behaviour of hypereutectic Al-Si composites: a multi-response optimisation approach with ANN and Taguchi grey method. Lubricants 12(2):61. https://doi.org/10.3390/lubricants12020061

Mohamed D, Fayad A, Mohamed AMO, Al Nahyan MT (2025) The role of e-waste in sustainable mineral resource management. Waste 3(3):27. https://doi.org/10.3390/waste3030027

Mohamed AMO (2024) Nexuses of critical minerals recovery from e-waste. Acad Environ Sci Sustain 1(2). https://doi.org/10.20935/AcadEnvSci7363

Mubarok MWSU, Kartini E (2024) Nickel diplomacy: strengthening Indonesia's role in global battery and electric vehicles supply chain. In AIP Conf Proc 3213(1):020003. https://doi.org/10.1063/5.0240397

Nguyen QL, Lai HN, Nguyen HT, Pham LL, Herdean A, Kuzhiumparambil U, Vo PH (2025) Explainable artificial intelligence for predicting rare earth elements leaching from secondary resources. J Hazard Mater 496:139479. https://doi.org/10.1016/j.jhazmat.2025.139479

Nišić D, Lukić B, Gordić Z, Pantelić U, Vukićević A (2024) E-waste management in Serbia, focusing on the possibility of applying automated separation using robots. Appl Sci 14(13):5685. https://doi.org/10.3390/app14135685

Nurfajrin ZD, Fajar AT, Maghfirah A, Goto M (2025) Accelerated screening of deep eutectic solvents for nickel/cobalt extraction: integrating quantum chemistry, machine learning, and experimental validation. Ind Eng Chem Res 64(12):6655–6663. https://doi.org/10.1021/acs.iecr.4c03736

Oke EA, Potgieter H (2024) Discarded e-waste/printed circuit boards: a review of their recent methods of disassembly, sorting and environmental implications. J Mater Cycles Waste Manag 26(3):1277–1293. https://doi.org/10.1007/s10163-024-01917-7

Öztürk M, Evin E, Özkan A, Banar M (2023) Comparison of waste lithium-ion batteries recycling methods by different decision making techniques. Environ Res Technol 6(3):226–241. https://doi.org/10.35208/ert.1243162

Patil Y, Tatiya M, Dharmadhikari DD, Shahakar M, Patil SK, Mahajan RG, Kurhade AS (2025) The role of AI in reducing environmental impact in the mining sector. JMMF 73(5):1365. https://doi.org/10.18311/jmmf/2025/48521

Pech-Rodríguez WJ, Sahin NE, Suarez-Velázquez GG, Rocha-Rangel E (2024) Industry 4.0: A new window for the future waste management systems. In: Kumar V, Bhat SA, Kumar S, Verma P (eds) Environmental engineering and waste management: recent trends and perspectives, Springer Nature Switzerland, Cham, pp 685–714. https://doi.org/10.1007/978-3-031-58441-1_23

Phogat P, Kumar S, Wan M (2025) A scientometrics study of advancing sustainable metal recovery from e-waste: processes, challenges, and future directions. RSC Sustain 3:2434–2454. https://doi.org/10.1039/D5SU00049A

Premathilake DS, Guimarães LF, Espinosa DCR, Tenório JAS, Vaccari M, Junior ABB (2025) Advancing hydrometallurgical recycling of spent lithium-ion batteries: an AI-based readiness and sustainability assessment. RSC Sustain 3:4975–4991. https://doi.org/10.1039/D5SU00417A

Quinto S, Law N, Fletcher C, Le J, Antony Jose S, Menezes PL (2025) Exploring the E-waste crisis: strategies for sustainable recycling and circular economy integration. Recycling 10(2):72. https://doi.org/10.3390/recycling10020072

Raabe D, Jovičević-Klug M, Ponge D, Gramlich A, Kwiatkowski da Silva A, Grundy AN, Ma Y (2024) Circular steel for fast decarbonization: thermodynamics, kinetics, and microstructure behind upcycling scrap into high-performance sheet steel. Annu Rev Mater Res 54(1):247–297. https://doi.org/10.1146/annurev-matsci-080222-123648

Robin RJ (2024) Viability and eco-consequences of synthetic and natural graphite for lithium-ion battery anodes in the USA. IEEE Eng Manag Rev 52(3):131–147. https://doi.org/10.1109/EMR.2024.3372697

Rojas L, Peña Á, Garcia J (2025) AI-driven predictive maintenance in mining: a systematic literature review on fault detection, digital twins, and intelligent asset management. Appl Sci 15(6):3337. https://doi.org/10.3390/app15063337

Sathya D, Saravanan G, Thangamani R (2024) Reinforcement learning for adaptive mechatronics systems. Comput Intell Tech Mechatron, 135–184. https://doi.org/10.1002/9781394175437.ch5

Schröder M, Iwasaki F (2024) From nickel to Electric cars? Indonesia's resource cum automotive industry policy. J Asia Pac Econ 29(4):2065–2086. https://doi.org/10.1080/13547860.2023.2231192

Shabani A, Aminaei M, Toroghinejad MR (2025) Data-driven prediction of ultimate tensile strength in low alloy steel: a machine learning approach. Comput Mater Sci 258:114045. https://doi.org/10.1016/j.commatsci.2025.114045
Shannak SD, Cochrane L, Bobarykina D (2024) Strategic analysis of metal dependency in the transition to low-carbon energy: a critical examination of nickel, cobalt, lithium, graphite, and copper scarcity using IEA future scenarios. Energy Res Soc Sci 118:103773. https://doi.org/10.1016/j.erss.2024.103773
Simaei E, Rahimifard S (2024) AI-based decision support system for enhancing end-of-life value recovery from e-wastes. Int J Sustain Eng 17(1):80–96. https://doi.org/10.1080/19397038.2024.2306293
Vega-Muratalla VO, Ramírez-Márquez C, Lira-Barragán LF, Ponce-Ortega JM (2024) Review of lithium as a strategic resource for electric vehicle battery production: availability, extraction, and future prospects. Resour 13(11):148. https://doi.org/10.3390/resources13110148
Walke S, Naniwadekar M, Thakar CM, Mandake MB, Tapre RW, Jadhav S (2024) Optimization of chemical engineering processes in the mining and metal industry. JMMF 72(4):377. https://doi.org/10.18311/jmmf/2024/43879
Xing L, Chau KY, Pan SH, Sadiq M (2025) Enhancing circular supply chain performance in platform-based economies: the role of AI, blockchain integration, and the IoT in promoting transparency. Corp Soc Resp Env Ma 33(2):2295–2312. https://doi.org/10.1002/csr.70274
Zante G, Elgar CE, Hartley JM, Mukherjee R, Kettle J, Horsfall LE, Abbott AP (2024) A toolbox for improved recycling of critical metals and materials in low-carbon technologies. RSC Sustain 2(2):320–347. https://doi.org/10.1039/D3SU00390F
Zeb A, Linnosmaa J, Seppi M, Saarela O (2024) Developing deep learning surrogate models for digital twins in mineral processing—a case study on data-driven multivariate multistep forecasting. Min Eng 216:108867. https://doi.org/10.1016/j.mineng.2024.108867
Zhang S, He X, Ding Y, Shi Z, Wu B (2024) Supply and demand of platinum group metals and strategies for sustainable management. Renew Sustain Energy Rev 204:114821. https://doi.org/10.1016/j.rser.2024.114821
Zhou K, Mao B, Zhang Y, Chen Y, Xiang Y, Yu Z, Qu J (2024) A cable-actuated soft manipulator for dexterous grasping based on deep reinforcement learning. Adv Intell Syst 6(10):2400112. https://doi.org/10.1002/aisy.202400112

Future Prospective of Critical Metal Bio Recovery: Advancing Green Technologies for a Circular and Sustainable Future

Abhinashi Singh Sodhi, Sonu Bhatia, and Navneet Batra

Abstract Worldwide surge in electronic waste (e-waste) generation presents serious environmental and resource security challenges due to toxic constituents and the presence of economically valuable critical metals such as copper, gold, silver and palladium. Conventional methods for metal recovery, while effective, often involve energy intensive and environmentally damaging processes. In recent years, various biorecovery methods such as biosorption, bioleaching, bioaccumulation, and biomineralization. These approaches have been extensively studied with emphasis on microbial mechanisms and physicochemical factors that influence their efficiency. Advancements including biosorbents integrated chemical pretreatment and microbial consortium, have significantly enhanced recovery potential. This chapter explores the current advances, mechanisms, and applications of biorecovery approaches for critical metals. Furthermore, it highlights challenges such as scalability, process optimization, metal selectivity, and biocompatibility of leachates remain critical barriers. Future research directions towards integrating biorecovery technologies into circular economy framework are paving way to achieve sustainable e-waste management and critical metal supply security.

Keywords Critical metals · Biorecovery · Green technology · Biotechnology · Sustainable development

1 Introduction

Global electronic waste (e-waste) has been rising at an alarming pace, driven by rapid technological advancement, short product life cycle, and increasing consumer demand for electronics. Every year, millions of tonnes of electrical and electronic equipment ranging from mobile phones, laptops to refrigerators, air conditioners and

A. S. Sodhi (✉) · S. Bhatia · N. Batra
Department of Biotechnology, Goswami Ganesh Dutta Sanatan Dharma College, Sector-32-C, Chandigarh 160030, India
e-mail: abhinashisingh90@gmail.com

A. P. Das and S. Priyadarsini (eds.), *Recovery and Recycling of Critical Metals from Electronic Waste for Environmental Sustainability*, Earth and Environmental Sciences Library, https://doi.org/10.1007/978-3-032-29064-9_18

medical devices are discarded worldwide, making e-waste one of the fastest growing solid waste streams. Countries such as United States of America, China, India, Russia contributes a major share of e-waste annually. India alone generates 25% of the total annual global e-waste (Awasthi and Li 2017; Quinto et al. 2025).

Global E-Waste Monitor (2024) has estimated the generation of e-waste to rise upto 82 million tonnes by 2030 and 110 million tonnes by 2050 (Global e-waste Monitor, United Nations 2024). This surge not only poses serious environmental and health concerns because of hazardous constituents, but also results in the wastage of valuable resources, including precious metals and rare earth elements such as copper (Cu), silver (Ag), aluminium (Al), gold (Au), nickel (Ni), iron (Fe), palladium (Pd), rhodium (Rh), and platinum (Pt). It has been well documented that only a fifth of the world's total e-waste is ever reused. Moreover, improper disposal of e-waste in landfills, open dumps, or through incineration leads to the release of toxic substances into the air and their infiltration into soil and groundwater, thereby causing widespread environmental contamination (Fig. 1).

Conventional approaches mainly focus on separating and dismantling the electronic waste followed by pyrometallurgy, hydrometallurgy and electrometallurgy. These methods while efficient in handling high amount of e-waste, requires large volume of hazardous chemicals, heat and are energy intensive thus contributing towards significant operational costs as well as environmental stress and carbon footprint (Ahirwar and Tripathi 2021; Ferreira-Filipe et al. 2025).

To address these challenges, the interest of scientific community and industry is shifting on developing and implementing bio-based methods for recovery of metals from e-waste that are more environment friendly and operate under mild conditions. This shift aligns with the global transition from a linear manufacturing model to a circular economy, which is often associated with a reduced carbon footprint. Microorganism based methodologies such as biohydrometallurgy, bioleaching, biosorption,

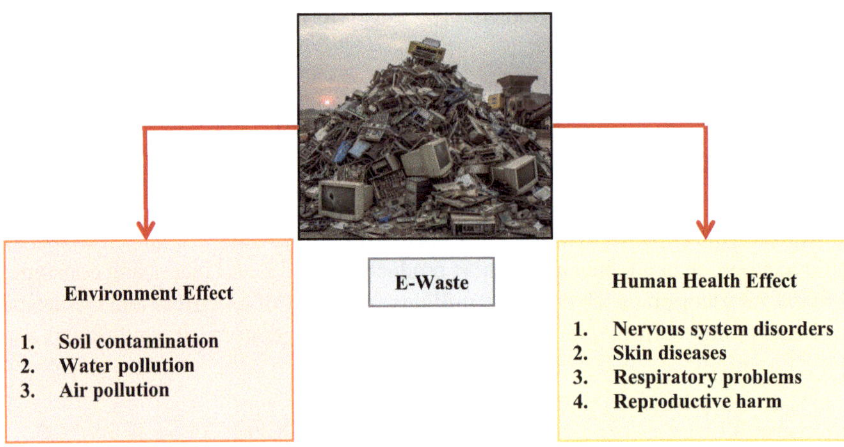

Fig. 1 Environmental and human health consequences associated with electronic waste

bioaccumulation, and biomineraliztion are being explored extensively, with a strong emphasis on harnessing the potential of underlying microbial mechanisms.

Furthermore, the integration of advanced tools, including biosorbents, siderophores and microbial consortium systems that combine biological and conventional methods, has demonstrated significant improvements in the efficiency of critical metal recovery. Supercritical fluid based processes often termed as green technologies are also paving the way for next generation of metal recovery processes.

Although the biorecovery of critical metals from e-waste offers a promising pathway towards a more profitable and sustainable economy while ensuring environmental safety, several challenges remain that need to be addressed. This chapter addresses e-waste management, with a particular emphasis on the recovery of critical metals using bio-based methods. Further, composition of electronic waste, concentration, and economic value of the metals it contains will be discussed. Moreover, the principles and mechanisms of biorecovery, including biohydrometallurgy, bioleaching, biosorption, bioaccumulation, and biomineraliztion, along with the microbial processes that drive these methods will be discussed. Key factors influencing biorecovery efficiency such as pH, temperature, and type of microbial strain, are analyzed to highlight their role in process optimization. Current insights are provided through case studies of laboratory and pilot-scale applications, illustrating recovery yields, process improvements, and scalability considerations. Finally, the chapter situates bio-based metal recovery within the broader context of circular economy principles and sustainable frameworks.

2 Composition and Impact of Critical Metals in Electronic Waste

E-waste comprises a variety of discarded or non-functional electronic equipments including cell phones, LEDs, home appliances, industrial and medical devices (Fig. 2). These waste materials contains a variety of critical metals including aluminium, copper, gold, silver, iron, platinum and palladium (Priyadarsini and Das 2024, 2026). These metals are incorporated into the electronic components for their properties such as high electrical conductivity, corrosion resistance, and malleability, ensuring reliable connections and performance. They play a crucial role in circuit boards, connectors, and switches, where their unique characteristics help prevent tarnishing and signal degradation, even when present in small quantities.

The total value of metals embedded in e-waste generated in 2022 was estimated at around US $91 billion, including $19 billion (Cu), $15 billion (Au), and $16 billion (Fe) (Harshan and Rajan 2025).

A list of various types of e-waste and their prominent critical metal composition is given in Table 1.

These metal ions released from e-waste can exert severe and persistent impacts not only on the ecosystem but also on public health. For instance, heavy metals such

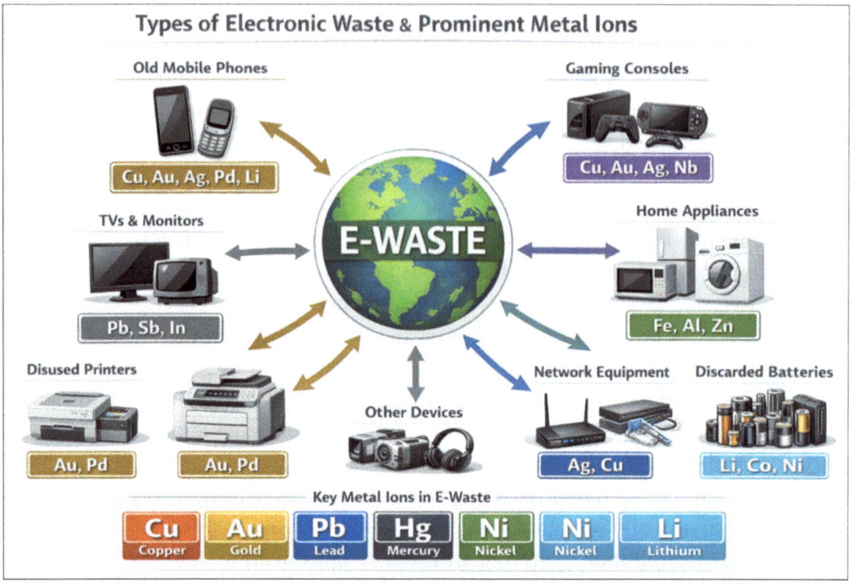

Fig. 2 Various types of e-waste and prominent metal ions

as arsenic (As), cadmium (Cd), chromium (Cr), lead (Pb), nickel (Ni), commonly released from batteries, and circuit boards, can cause neurological damage, developmental disorders, kidney problems, and carcinogenic effects, while also contaminating soil and water ecosystems. Arsenic, originating from semiconductor devices, poses developmental, reproductive, and carcinogenic risks and leads to groundwater contamination and toxicity to aquatic life. Lead, widely present in circuit boards and soldering materials, is particularly harmful to children, causing neurological and cognitive impairment, and readily contaminates soil and water through leaching. Cadmium, used in rechargeable batteries and resistors, is associated with kidney damage, bone demineralization, and cancer, and accumulates in soil, food crops, and aquatic system. Copper, prevalent in wiring and circuit boards, can cause gastrointestinal disturbances alongwith liver and kidney damage at high exposure levels, while elevated environmental concentrations are toxic to aquatic organisms and contribute to soil and water contamination. Lithium, primarily originates from discarded lithium ion batteries that are used in electric vehicles, laptops, mobiles and other electronic products. Release of lithium in the environment lead to skin and respiratory irritation, neurological effects, endocrine system related disorders and fire hazards, while also leading to soil and water contamination, ecosystem toxicity, and long-term environmental degradation. Fine particulates of gold mainly originating from discarded television monitors, printed circuit boards and semiconductor chips, where it is used for its excellent conductivity and corrosion resistance. These can cause skin irritation, eye irritation, and gastrointestinal discomfort upon ingestion (Mrozik et al. 2021; Brindhadevi et al. 2023; Fatema et al. 2025).

Table 1 Concentration and distribution of critical metals in various e-waste fractions

E-waste	Ag	Al	Au	Cu	Fe	Li	Ni	Pb	Pd	Sn	References
Printed circuit boards (PCBs)	0.83%	–	0.29%	19.3%	–		3.46%	1.53%	–	5.53%	Gorewoda et al. (2020)
Lithium-ion batteries	–	4.27%	–	2.55%	0.38%	3.03%	0.5%	–	–	–	Roy et al. (2021), Biswal and Balasubramanian (2023)
Television board	280 ppm	10%	20 ppm	10%	28%	–	0.3%	1.0%	10 ppm		Abdelbasir et al. (2020)
Refrigerators	42–450 ppm	1.3–2.0%	–	2.0–4.1%	–	–	–	0.021–2.5%	–	1.6–2.0%	Li et al. (2019), Tian et al. (2022), Manikandan et al. (2023)
Mobile phone scrap	1380 ppm	1.0%	350 ppm	13%	05%	–	0.3%	0.3%	210 ppm		Abdelbasir et al. (2020)
PC board scrap	1000 ppm	–	250 ppm	20%	–	–	1%	1.5%	110 ppm		Hageluken (2006), Ilyas and Lee (2014a, b)
Calculator scrap	260 ppm	–	50 ppm	3.0%	–	–	0.5%	0.1%	5.0 ppm	–	Hageluken (2006), Ilyas and Lee (2014a, b)
Electric wires and cables	–	–	–	58.3%	–	–	–	–	–	–	Wędrychowicz et al. (2023)

Ag Silver; *Al* Aluminium; *Au* Gold; *Cu* Copper; *Fe* Iron; *Li* Lithium; *Ni* Nickel; *Pb* Lead; *Pd* Palladium; *Sn* Tin

Toxicity and bioaccumulative nature of these metal ions highlight an urgent requirement for sustainable bio-recovery methods and bioremediation strategies to selectively recover valuable metals while simultaneously detoxifying contaminated environments.

3 Principles and Mechanisms of Biorecovery

The recovery of critical metals from e-waste can be effectively achieved through bio-based processes that harness the metabolic activities of variety of microorganisms (bacteria or fungi) (Fig. 3). These processes offers a promising alternative to physical and chemical based extraction methods, which often involve hazardous reagents and generate toxic by-products. By operating under milder conditions and relying on naturally occurring microorganisms, bio-based methods results in lower energy consumption, significantly reducing environmental pollution, and enables the selective recovery of critical metals.

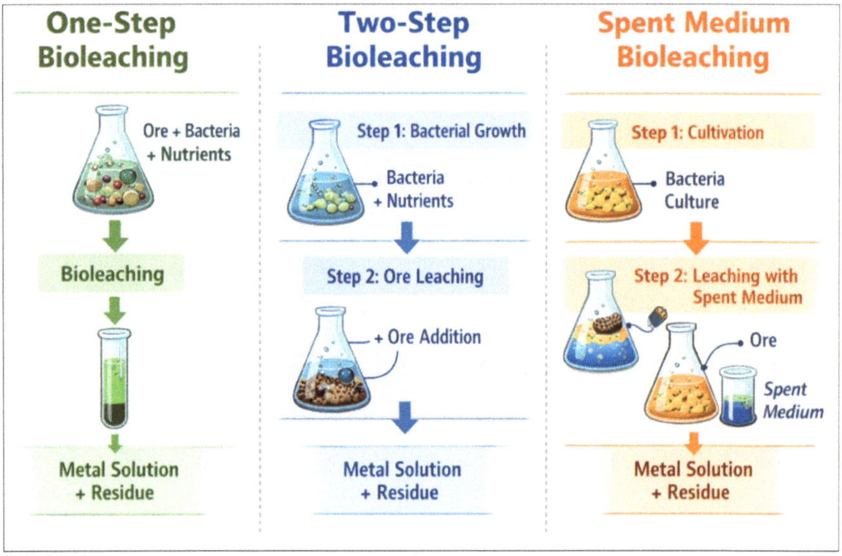

Fig. 3 Schematic representation of three types of bioleaching strategies (1) One-step bioleaching, (2) Two-step bioleaching, (3) Spent Medium Bioleaching

3.1 Bio-hydrometallurgy

Conventional approaches, including pyrometallurgy and hydrometallurgy techniques, have been employed for the recovery of metals from electronic waste. Pyrometallurgy involves high-temperature (upto 900 °C) processes such as roasting, smelting, and refining, where heat and reducing agents are used to convert ores into metals. It is fast and suitable for large-scale operations but highly energy-intensive and often generate polluting compounds such as SO_2 and CO_2. In contrast, hydrometallurgy relies on the use of chemicals or reagents at relatively low temperatures, where metals are dissolved from ores or wastes through leaching, purified by techniques like solvent extraction or precipitation, and finally recovered by methods such as electrowinning. However, these methods are highly energy-demanding and generate substantial amount of secondary pollutants (such as brominated and chlorinated di-benzofurnas and dioxins), thereby increasing the overall operational costs of the process (Khatri et al. 2018; Faramarzi et al. 2020). Biohydrometallurgy is a broad discipline and includes a variety of techniques such as bioleaching, biosorption, and biomineralization. These techniques are discussed in subsequent sections. It is an eco-friendly and cost-effective technique to recover metals even at a low concentration from e-waste. It offers advantages such as low operational cost, less energy consumption, reduced chemical reagents usage, and near zero secondary waste effluent generation (Magoda and Mekuto 2022). The process employs microorganisms for dissolution of metals from waste materials by producing organic acids. A variety of bacterial strains including *Acidithiobacillus ferrooxidans*, *Chromobacterium violaceum*, *Magnetospirillum* sp., *Sulfobacillus thermosulfidooxidans* help in the solubilization and efficient recovery of base as well as precious metals such as Cu, Al, Au, Fe, Ni, Zn. Moreover, cyanogenic bacteria are increasingly being investigated for their potential to improve the dissolution and recovery of heavy metals from electronic waste (Ghosh et al. 2024).

3.2 Bioleaching

Bioleaching is a process of using microorganism to extract metals from ores and other materials. Both bacteria and fungi play major roles in bioleaching systems. Acidophilic bacteria such as *Acidithiobacillus ferrooxidans* and *Leptospirillum ferrooxidans* oxidize metal sulfides, releasing soluble metal ions, while fungi such as *Aspergillus niger* and *Penicillium simplicissimum* contribute through the production of organic acids that solubilize metals. This low energy and eco-friendly approach offers a sustainable alternative to conventional pyrometallurgical and hydrometallurgical techniques for metal recovery (Johnson 2014; Yaashikaa et al. 2022). Bioleaching can be performed in three ways, One step bioleaching, Two step bioleaching and Spent medium bioleaching (Fig. 3). In One step bioleaching, the selected microbial strain is inoculated into the bioleaching medium alongwith the e-waste and placed under optimum conditions for overall microbial action. Whereas, in Two step bioleaching the selected microbial strain is cultivated in the suitable growth medium without e-waste till the time maximum microbial growth rate is achieved followed by addition of e-waste at a specific particle size and pulp density. In spent medium method, the microbial strain is cultivated in a culture medium without e-waste, until it achieves maximum growth rate and cell density. Subsequently, the e-waste is introduced into the cell-free culture supernatant, which is obtained by centrifugation or filtration of microbial broth. The metabolites retained in this supernatant promote the solubilization and recovery of metals from the e-waste (Işildar et al. 2016; Dave et al. 2018; Wang et al. 2021; Adetunji et al. 2023).

Numerous studies have been conducted to explore the potential of bioleaching in remediation of electronic waste. For instance, bioleaching of computer PCBs using *Aspergillus niger* resulted in significant recovery of 97% Cu and 74% Ni at pH of 5.15 and a pulp density of 10 g/l. Fungi mediate biodegradation of the sample by secreting organic acids resulting in the surface morphology modification and erosion of the particles of the e-waste during the bioleaching process. Increased acid production accelerates metal solubilization, leading to enhanced metal recovery rates (Arshadi et al. 2020). *Pleurotus florida* and *Pseudomonas* spp. exhibited highest recovery efficiencies between 10.5% and 18.0% of copper and iron respectively, attributed to bio-catalysis for laccase enzymes (Kaur et al. 2022). *Aspergillus fumigatus* A2DS, from mining wastewater successfully extracted 57% of copper and 32% of nickel under one-step leaching at 30 °C for seven days. Analysis confirmed that the fungus effectively dissolved copper compounds present in printed circuit boards (PCBs) (Patel and Lakshmi 2021). Similarly, *Acidithiobacillus ferrooxidans* was employed for extraction of copper from waste printed circuit boards resulting in copper recovery rate of 32.44% after 7 days at pH 2.0 (Annamalai and Gurumurthy 2019). Two step bioleaching process was found to be suitable for extracting 100% of Cu by *Aspergillus niger* from PCBs (Sharada et al. 2022). Bioleaching of critical metals from waste PCBs using *Penicillium expansum* resulted recovery rate of 70% for elements like La, Tb, Pr, Nd, and Gd within 24 h. The extraction was driven by fungal activities such as organic acid production, pH regulation, and enhanced phosphate availability (Baez et al. 2024).

These studies collectively demonstrate that bioleaching is an effective and adaptable approach for recovering valuable metals from a variety of electronic waste. Continued research focusing on process optimization, microbial consortia, and scale-up strategies could further enhance its efficiency and enable industrial-level application of this environmentally friendly metal recovery technology.

3.3 Biosorption and Bioaccumulation

Biosorption is a process that involves the binding of metal ions or pollutants onto the surface of biosorbent commonly, microbes, agricultural waste and biopolymers

(Shrestha et al. 2021). This interaction occurs through various cellular components including cell wall or extracellular polysaccharides which contain functional groups capable of binding metal ions. For instance, chitosan powder can act as a smart 'magnet' for extracting certain metals from spent lithium ion batteries, specifically cobalt (Co) and nickel (Ni). Results showed 80% absorption of Co and Ni under optimal conditions of pH 4.0, 10 g/l chitosan and 40 times dilution within 5 min. For reuse, the metals could be easily stripped from chitosan using mild sulfuric acid, achieving full recovery for cobalt and nickel and over 82% for manganese. These findings clearly demonstrate chitosan as a low-cost and sustainable material for selectively recovering valuable metals from battery waste, offering both environmental and resource sustainability benefits (Roshanfar et al. 2025).

On the other hand, the process of active uptake and storage of metals within the microbial cell is known as bioaccumulation. Although, biosorption and bioaccumulation are sometimes considered similar processes, there is a distinct difference between the two. Biosorption mainly occurs on the cell surface through interactions with functional groups such as carboxyl, hydroxyl, or amino groups, and it does not require metabolic energy. In contrast, bioaccumulation is an energy dependent, metabolically driven process involving translocation of metal ions across cell membrane followed by their internal compartmentalization, often mediated by metal chelating proteins or vesicles (Filote et al. 2021). *Bacillus megaterium* demonstrated a high biosorption capacity, removing approximately 65% of copper from printed circuit boards (PCBs) under optimized conditions, including a contact time of 60 min, pH 5.0, and an agitation speed of 150 rpm (Nivetha and Karthick 2018). Similarly, *Frankia* bacterial strains, DDNSF-03 and DDNSF-04 showed removal of toxic metals mercury, cadmium and lead from PCBs. SEM analysis revealed morphological changes in the bacterial cells thus indicating metal accumulation. AAS spectrum analysis demonstrated a significant reducing metal concentration, initially from Hg 4.3, Cd 8.3, Pd 4.6 (ppm) in the e-waste to final treated e-waste of Hg 0.09, Cd 5.09, Pb 0.49 (ppm), with DDNSF-03 and Hg 2.15, Cd 5.6, Pb 2.82 (ppm) with DDNSF-04 (Narayanasamy et al. 2021).

Overall, both biosorption and bioaccumulation represent promising biological strategies for metal recovery, each with distinct mechanisms and advantages. While biosorption offers a rapid, energy-efficient, and reusable method ideal for large-scale treatment of metal-laden effluents, bioaccumulation enables selective intracellular sequestration, potentially useful for trace metal recovery or detoxification.

3.4 Biomineralization

Biomineralization is a promising microbial method for carrying out remediation of metal contaminated environment. In this, process microorganisms convert toxic metal ions to insoluble mineral precipitates, immobilizing them and preventing environmental harm (Lai et al. 2023; Mallick et al. 2025). Microorganisms can promote mineralization by catalyzing redox reactions including pH shift that lead to the precipitation of contaminants such as Pb, Cd, Cr, Fe, and U. In some microorganisms, the release of inorganic phosphate enhances the formation of insoluble metal–phosphate complexes, whereas other bacterial groups facilitate the precipitation of metal hydroxides or carbonates through metabolite-driven alkalinization and hydrocarbon (alkane) production. Furthermore, the ability of microorganisms to selectively precipitate and concentrate metal ions highlights their potential as an eco-friendly approach for the recovery of critical metals from secondary resources and industrial waste streams.

4 Factors Influencing Biorecovery of Metals

4.1 Pre-treatment of E-waste

The direct biological processing of e-waste is challenging due to its complex chemistry and structure, thus necessitating pre-treatment of the waste material. Therefore, preparatory mechanical and physical processes are applied for disintegrating the complex structure and increasing the surface area for microbial interaction and concentrating the target metals, thereby optimizing the subsequent bioleaching efficiency. E-waste is subjected to a series of steps including dismantling, size reduction, milling, sieving and screening, followed by separation techniques such as magnetic, electrostatic, and density based separation (Kaya 2016; Arya and Kumar 2020; Castro and Bassin 2022; Kumar et al. 2022; Premathilake et al. 2023; Erkmen et al. 2025). These preliminary steps are crucial in segregating and upgrading the metal fractions, thereby facilitating an initial concentration of metals prior to subsequent hydrometallurgical or electrometallurgical processing. This strategy enhances the diffusion of desirable elements while minimizing excessive use of chemicals, prolonged operations and reduced environmental risks (Khaliq et al. 2014; Erkmen et al. 2025).

This approach enhances the transfer of desired elements while minimizing chemical usage, operation duration, and environmental impact.

The pre-treatment process consists of the:

(1) Sorting
(2) Stabilization/discharge
(3) Dismantling or disassembly
(4) Grinding or crushing, and
(5) Separation (Hua et al. 2020; Ali et al. 2022).

4.2 Particle Size

Mechanical size reduction of electronic waste through shredding or grinding is a critical pre-treatment strategy that significantly enhances metal recovery by increasing the exposed surface area and liberating metal-bearing phases from polymeric and ceramic matrices (Chicardi et al. 2025). Several bio-based recovery studies have demonstrated a strong dependence of metal extraction efficiency on particle size. For instance, shredded dust originating through mechanical dismantling of e-waste is proved to be an enriched source of precious metals including copper, silver, iron, nickle and gold. Pre-treatment using ferric chloride as chemical lixiviant resulted in improved bioloeaching efficiency of *Pseudomona balearica* SAE1 with recovery rate of 95% for copper and 36% for silver suggesting the potential role of economical role of chemical lixiviants in metal recovery (Thakur and Kumar 2021). *Penicillium expansum* achieved substantial recovery of rare earth elements from waste PCBs within 24 h, including La (40%), Tb (50%), and Pr, Nd, and Gd (70%), when the waste was milled to particle size below 0.5 mm (Baez et al. 2024). Likewise, recovery efficiencies of Au and Ag increased markedly to 95.4% and 96.2%, respectively, when waste light-emitting diode (LED) materials were ground to $\leq$ 600 μm prior to bioleaching (Zhang et al. 2024). These findings are consistent with earlier observations by Nagy et al. (2017), who reported a Ga concentration of 254 ppm predominantly in the 106–1000 μm size fraction following mechanical grinding of LED waste.

Collectively, these studies underscore the pivotal role of particle size reduction in enhancing bio-based recovery of critical and precious metals from electronic waste streams. Size reduction facilitates selective metal fractionation by increasing the area-to-volume ratio and enhancing metal bioavailability, thereby promoting more efficient microbial interaction and leaching kinetics.

4.3 Microbes

Microbes play a significant role in bioremediation of different types of environmental waste by employing different mechanisms to solublize the metals and exhibit varying degree of tolerance towards the harsh conditions of e-waste environments. Primarily, two types of microbial classes are used in e-waste bioremediation, chemolithotrophes (*Acidithiobacillus*) and heterotrophes (*Pseudomonas*). Moreover, fungal species of *Aspergillus* and *Penicillium* are commonly employed for metal solubilization due to their ability to produce organic acids. Further, their addition as single type of microbe or in a consortium also influences their rate of bioremediation (Dutta et al. 2023; Harshan and Rajan 2025).

A list of microorganisms, metals recovered and types of e-waste they act upon is mentioned in Table 2.

Table 2 Microorganisms employed in the biorecovery of metals from different types of electronic waste

Microbe	Metal recovered	Type of E-waste	References
Acidthiobacillus thiooxidans	Al, Cu, Mn, Zn	PCBs	Hong and Valix (2014)
Thermosulfdooxidans	Al, Cu, Ni, Zn	PCBs	Rodrigues et al. (2015)
Aspergillus niger	Au, Cu, Ni	PCBs and computer motherboards	Madrigal-Arias et al. (2015)
Aspergillus nomius JAMK1	Cu, Ni, Pb	Spent batteries	Chatterjee et al. (2020)
A. ferrooxidans *Leptospirillum ferrooxidans* *A. thiooxidans*	Cu	TV circuit boards	Bas et al. (2013)
Aspergillus niger DDNS1	Ag, Cu, Fe	PCBs	Narayanasamy et al. (2018)
Chromobacterium violaceum	Au	Grey dust-like ESM	Tay et al. (2013)
Acidithiobacillus ferrooxidans *Acidiphilium acidophilum*	Cu, Zn, Ni, Pb	Computer PCBs	Priya and Hait (2020)
Pseudomonas balearica SAE1	Au, Ag	Computer PCBs	Kumar et al. (2018)
Magnetotactic bacteria	Pb, Ni, Cd, Zn,	Spent Li-ion and Ni–Cd batteries	Sannigrahi et al. (2024)

4.4 pH

Bio-based metal recovery is greatly influenced by the system's pH, which is considered as a key operational parameter governing metal solubilization (Du et al. 1995). Moreover, pH of the system has a profound effect on growth and activity of acidophilic bacteria (Kumar and Nagendran 2007). However, the optimum pH requirement can vary depending on the microbial species involved, the type of substrate, and the specific metal being leached (Adetunji et al. 2023). For instance, the cultivation of *Chromobacterium violaceum* increased Au and Cu mobilization from 7.78 to 10.8%, and 4.9 to 11.4%, when grown at a pH of 11.0 (gold), and 10.0 (copper) (Chi et al. 2011). Similarly, gold recovery from PCBs by *Pseudomonas balearica* SAE1 enhanced significantly from 44% to 68.5% by shifting the pH from 7.0 to 9.0 (Kumar et al. 2018). Therefore, maintaining an appropriate and stable pH according to the microbial consortium and waste composition is essential for achieving efficient and sustainable bioleaching performance.

4.5 Temperature

Temperature is a critical parameter affecting the recovery rate of metals from e-waste by influencing reaction rates, chemical stability, and microbial activity. While higher temperatures generally increase reaction rates, potentially enhancing leaching efficiency for some metals, excessive heat can lead to degradation of reagents or the loss of metals through vaporization or undesirable alloy formation (Priya and Hait 2017). For instance, leaching rate of copper, iron and lead increase with temperature. However, above 30 °C, rate of metal recovery decrease that can be associated to citrate destruction (Torres et al. 2018). In another study, thermal decomposition of PCBs was carried out for recovery of copper, nickel, zinc and lead. Roasting of circuit boards 300 °C resulted in 93% recovery of Cu and 100% Ni, Zn and Pb respectively (Panda et al. 2020). Further, maintaining an optimum temperature for microbial activity is critical for efficient metal recovery (Nnaemeka et al. 2025).

Therefore, optimal temperature is crucial and highly dependent on the specific e-waste composition and the chosen recovery method.

4.6 Aeration

Aeration plays a crucial role in bioleaching of metals, as it directly influences the growth, metabolism, and activity of microorganisms involved in metal solubilization from e-waste. Most of the acidophilic bacteria and fungi grow well under aerobic environment resulting in optimum growth and bioleaching efficiency.

A. ferrooxidans derives energy for growth by oxidizing ferrous ions (electron donor) i.e., oxidation of Fe^{2+} to Fe^{3+} with dissolved oxygen (O_2) acts as a terminal electron acceptor (Liang et al. 2016; Liu et al. 2020; Biswal and Balasubramanian 2023).

Acidithiobacillus ferrooxidans isolated from the acid mine drainage demonstrated that Co extraction from spent lithium ion batteries reached 74% when leaching medium was aerated, whereas Co leaching was reduced to 52% under static, non-aerated conditions (Putra et al. 2022). For instance, aeration rate of 0.07–1 L/min for a stirred bioleaching tank inoculated with *Sulfobacillus thermosulfidooxidans* resulted in high amount of leaching of Al, Cu, Zn, Ni and reduced the lag time of the bacterial culture (Ilyas and Lee 2014a, b). Air flow rate between 50 and 250 L/h resulted in efficient uranium bioloeaching 100 L/h being the optimum value to achieve 100% of U bioleaching (Zare Tavakoli et al. 2017). Similarly, an aeration rate of 20 L/min resulted in the recovery of 95% Cu from waste PCBs (Chen et al. 2015). Bioleaching of 96% Cu at a flow rate of 0.3 ml/min was obtained *A. ferrooxidans* strain ATCC 23,270 (Murali et al. 2022). The above mentioned studies actively demonstrates the positive effect of optimized aeration rate for efficient bioleaching of metals. However, aeration rate increased beyond a limit can negatively affect the whole bioleaching process by excessive turbulence, shear stress thus adversely affecting the microbial growth (Tezyapar Kara et al. 2023).

4.7 Pulp Density

Pulp Density refers to the ratio of solid material (mass of e-waste or ore) to the total volume of the liquid medium (leachate or culture medium) used in bioleaching or microbial metal recovery processes. A higher pulp density often lead towards toxicity of the e-waste due to limited oxygen mass transfer rate thus reducing microbial growth and metabolic activity. Significant bioleaching efficiency was demonstrated by *Chromobacterium violaceum*, *Roseovarius tolerans* DSM 11457 and *Bacillus megaterium* at pulp densities of 0.5%, 1% and 4% respectively (Motaghed et al. 2014; Natarajan and Ting 2015; Gu et al. 2018; Kudpeng et al. 2020; Vardanyan et al. 2022; Adetunji et al. 2023). Pulp density of 2% resulted in bioleaching efficiency of Co and Li upto 72% and 89% respectively. Meanwhile, it was observed that increase in pulp density from 1 to 4% decreased the leaching of Co and Li from 52 to 10% and 80 to 37% respectively (Niu et al. 2014). Metal leaching efficiency of *Acidithiobacillus thiooxidans* from spent lithium ion batteries at 50 g/l of pulp density was 92%, 88%, and 78% for Li, Ni, and Co (Heydarian et al. 2025). Pulp density determines the balance between metal availability and microbial tolerance. Hence, optimizing pulp density is crucial for maximizing biorecovery of metals from e-waste.

5 Advances in Biorecovery Technologies from E-waste

5.1 Novel Biosorbents

Biosorption has emerged as an effective and sustainable approach for the recovery of critical metals from e-waste. This process employs naturally derived biological materials such as microbial biomass, algae, fungi, or agricultural residues that possess functional groups capable of binding and concentrating metal ions from complex waste matrices. This low-cost, sustainable, and scalable approach not only reduces the ecological footprint of metal recovery but also supports circular economy principles by transforming waste into a resource. Major advantage of these materials is their high absorption capacity and low cost resulting in their application in the recovery of precious metals such as gold and platinum (Pinto et al. 2023). Redox-active biochar was applied to enhance bioleaching from PCBs. Biochar facilitated the Fe(II)/Fe(III) redox cycle, promoting effective copper leaching. Key microbial species, *Alicyclobacillus* spp. and *Sulfobacillus* spp., cooperated in this process, with biochar regulating their ratio to boost efficiency (Wang et al. 2016). Similarly, itaconic acid-grafted poly(vinyl alcohol)-encapsulated wood pulp (IA-g-PVA-en-WP) in combination with DL-malic acid and hydrogen peroxide was employed as leaching mixture to extract metals from PCBs. The technique resulted in the recovery of metals upto 93.03% for Cd, 95.18% for Pb, and 97.08% for Ni (Varshney et al. 2017).

5.2 *Microbial Consortium*

The use of microbial consortium in place of a single microbe proved to be more advantageous in enhancing metal solubilization efficiency, improving process stability, and enabling the simultaneous degradation of complex waste components during bioleaching of electronic waste. These attributes are possible due to the synergistic interactions between different microbial species within the consortium and their enhanced resilience to environmental variability, characteristics that single microbes alone cannot exhibit effectively (Latorre et al. 2016; Tapia et al. 2022). Bioleaching efficiency of copper (69%) and zinc (91%) from PCBs was achieved by employing acidophilic iron-oxidizing consortium consisting of *Tissierella*, *Acidiphilium* and *Leptospirillum* through iron oxidation (Tapia et al. 2022). The use of *Frankia* consortium resulted in significant recovery of gold (90%) and silver (75%) from fine particles of PCB due to phosphatase enzyme and organic acid production (Narayanasamy et al. 2022). Similarly, lithium recovery from waste lithium–iron phosphate (LFP) batteries using acidophilic bacterial consortium comprising of *Acidithiobacillus* and *Acidiphilium* resulted in bioleaching yield of 98% (Gu et al. 2025). García-Balboa et al. (2022) reported that a mixed acidophilic consortium comprising *Acidiphilium multivorum* and *Leptospirillum ferriphilum* achieved exceptionally high recovery efficiencies from e-scrap materials, with more than 99% recovery of Cu, Co, and Al, approximately 98% of Zn, and moderate recovery of Cd (53%). In another study, three distinct microbial consortia (SC1, SC2, and SC3), predominantly belonging to the *Firmicutes* phylum demonstrated strong bioleaching performance when applied to spent lithium-ion batteries, achieving Li and Cu recovery efficiencies of 86% and 87%, respectively, along with Co recovery of 42% (Rautela et al. 2026). Similarly, a bacterial consortium enriched from natural acid mine drainage showed high efficacy in the bioleaching of PCBs, recovering approximately 95% of Cu (Xiang et al. 2010).

5.3 *Supercriticial Fluid Technology*

Supercritical CO_2 was employed to extract silver from printed circuit boards pre-treated with supercritical water. Recovery efficiency of 72.07% was achieved using supercritical CO_2 at 50 °C for 90 min at a pressure of 300 bars (Fallah Haghighi et al. 2025).

In another study, combination of supercritical CO_2 and aqueous acid as co-solvent was used for the extraction of copper from PCBs. Results showed 82% of copper recovery under 4 h. In this approach PCBs were subjected to pre-treatment with $scCO_2$ and aqueous sulfuric acid at 120 °C and 148 atm for 30 min, followed by leaching ambient temperature in a mixture of 2 M sulfuric acid and 0.2 M hydrogen peroxide (Hsu et al. 2021). Supercritical water pre-treatment was used to process PCBs in order to concentrate and extract silver. Recovery of 98% was achieved at 230 bar pressure, 400 °C temperature, and 45 min retention time followed by supercritical carbon dioxide extraction at 217 bar pressure, 51 °C, and 40 min (Fayaz et al. 2022).

5.4 Siderophores

Siderophores are metal chelating compounds produced by a variety of microorganisms. Biologically, these compounds enables microbes to acquire metal ions essential for their growth, survival and virulence (Schalk 2025). Due to their strong metal chelating ability, siderophores have also been investigated for environmental bioremediation where they can be used as effective agents to extract metals from contaminated environments. Studies have reported high recovery efficiencies for lithium and molybednum in a cost-effective and rapid manner, demonstrating potential of this approach as an eco-friendly method (Ambaye et al. 2020; Gulliani et al. 2023). A study by Zheng et al. (2024) demonstrated recovery of gallium and indium from pretreated spent surface-mounted device (SMD) LEDs using siderophore desferrioxamine E (DFOE). DFOE, a naturally occurring hydroxamate-type siderophore, achieved the highest Ga leaching efficiency (12%) at pH 4 and the best In recovery (24%) at pH 7.3 within 24 h. In another study, siderophores from *Pseudomonas* sp. strain ASA235 demonstrated recovery of 14.8% of lanthanum, alongside trace amounts of other REEs, from NiMH battery anodes. Similarly, siderophore from *Pseudomonas aeruginosa*, achieved over 99% recovery of lithium from the cathode material of spent lithium-ion batteries. All these findings significantly highlight the potential of microbial siderophores as sustainable and eco-friendly agents for metal recovery from electronic waste, with future prospects for process scaling and optimization (Vishwakarma and Hait 2024; Hegazy et al. 2025).

Table 3 enlists the comparative analysis of various e-waste handling methods.

Table 3 Comparison of E-waste handling methods

Method	Description	Advantages	Disadvantages
Physical methods	Involves mechanical or physical processes such as dismantling, shredding, magnetic separation and gravity based sorting	Simple and cost-effective for preliminary separation. Reduces bulk volume of waste. Allows recovery of base metals and reusable components. Environmentally safer (no chemicals used).	Inefficient for fine metal recovery. Requires labor-intensive dismantling. Generates dust and particulate pollution. Cannot remove hazardous materials like heavy metals.
Chemical methods	Use of acids, bases, or other chemical reagents for leaching metals from e-waste (e.g., hydrometallurgy, pyrometallurgy, solvent extraction)	High recovery efficiency of precious metals (Au, Ag, Cu, Pd). Faster processing time. Suitable for large-scale industrial applications. Can handle complex e-waste mixtures.	Produces toxic effluents and secondary pollution. Requires proper waste treatment systems. High energy and chemical consumption. Hazardous to workers and environment if not managed properly.
Biological methods	Utilizes microorganisms (bacteria, fungi) to leach and recover metals from e-waste through bioleaching or biosorption Common microbes: *Acidithiobacillus ferrooxidans*, *Aspergillus niger*, *Pseudomonas* sp.	Eco-friendly and sustainable. Low energy and chemical input. Can recover metals from low-grade waste. Safer for workers and environment.	Slow process; longer retention time. Sensitive to environmental conditions (pH, temperature, oxygen). Limited scalability for industrial use. May not efficiently recover all metals.

6 Pilot-Scale Applications and Commercial Viability of Biobased Recovery of Metals

A pilot scale study was conducted by Falco et al. (2014) to determine the efficiency of microbe based recovery of metals from electronic waste. Air-lift bioreactor with a sulphur packed bed was used to perform recovery of Zn and Mn from spent alkaline and Zn-C batteries. *Acidithiobacillus thiooxidans* was used to produce an acid-rich solution capable of dissolving metals from used alkaline and zinc–carbon batteries. The process involved two main steps: a leaching reactor, where the bacteria-generated acidic medium was mixed with powdered battery waste to extract metals, and a recovery reactor, where those metals were recovered through electrolysis. After 12 days, the bacteria produced a medium with an acidity of 350 mM [H^+], which successfully extracted 100% of the zinc (Zn) and 67% of manganese (Mn) from battery powder. During electrolysis, metallic zinc was deposited at the cathode, and manganese oxide formed at the anode, all achieved in a single step at room temperature. Overall, the study demonstrated a sustainable and cost-effective method for recovering valuable metals using biotechnology. The process not only reduces environmental impact but could also offset recycling costs by selling the recovered zinc and manganese products.

6.1 Bioscope Technologies

Bioscope Technologies is a UK-based company providing a sustainable way particularly based on microbes of recovering valuable metals from electronic waste and replacing traditional treatments that use high-temperature and chemical recovery. The company recover precious metals by following an innovative approach. E-waste is converted into a fine, sand-like powder, which is then analyzed through comprehensive Inductively Coupled Plasma (ICP) testing. This detailed analysis helps in identifying the material composition accurately, allowing the metals to be easily categorized into base metals (such as Cu, Ni, Pb, and Sn), valuable metals (including Au, Ag, and Pd), and rare earth elements (like neodymium, yttrium, and holmium). Using its proprietary large-scale bioleaching technology, Bioscope then recovers valuable metals through a low-impact, biologically driven process that minimizes chemical use and supports sustainable metal recovery. Currently the company has copper and gold extraction underway and is developing processes for palladium and platinum (www.bioscope.tech).

6.2 Biometallica

BioMetallica, a Singapore and Hong Kong based company is a pioneer in developing microbe based recovery methodologies of precious metals, especially platinum group metals (PGMs) like Pt, Pd and Rh from e-waste and spent catalysts. Their approach combines engineered bacteria (in bioreactors) with downstream metal reduction/separation techniques. They have developed a bacterial strain that produces biogenic cyanide at more than double the level of non-engineered strains, enabling more efficient dissolution of PGMs from waste streams.

Compared to conventional recycling methods, BioMetallica operations are performed at room temperature, generate zero wastewater or related secondary waste, use about six times less energy and cost roughly one-third as much. The whole process involves 4-steps: pre-treatment of e-waste/catalyst feedstock → bio-leaching (microbial dissolution of metals) → reduction to the metallic state → separation & purification into nanoparticles of Rh, Pd, Pt with a bioleaching efficiency of 99%, 95%, 92% respectively. Recycling one ton of waste feedstock prevents the release of 44t CO_2e of greenhouse gas emission. This technology is significant in the context of the growing need for sustainable metal recovery, as ore grades decline and e-waste volumes surge, methods that recover critical metals in an eco-friendly way are increasingly important. In summary, BioMetallica is positioned to help turn waste into wealth by unlocking precious metals from complex waste streams using biotechnology, lowering environmental footprint while delivering economic value (www.biometallica.com).

6.3 Bactech Environmental

BacTech Environmental is a Canadian company dedicated towards promoting environmental sustainability within the mining and resource industries. Its core innovation lies in the use of bacterial oxidation commonly known as bioleaching, a process originally developed by BacTech Mining Corporation to extract precious and base metals from complex ores, concentrates, and mine tailings. In conventional mining operations, sulphide-rich tailings react with oxygen and water in the atmosphere to produce metal laden acidic runoff, commonly known as acid rock drainage that pollutes surrounding ecosystems. BacTech's bioleaching technology addresses this issue by using naturally occurring bacteria to accelerate the oxidation of sulphides under controlled conditions. This not only prevents further acid rock drainage (ARD) formation but also stabilizes toxic compounds, rendering the remaining tailings environmentally benign.

The process creates optimal growth conditions for the bacteria referred to as a "Garden of Eden" environment, allowing them to complete in just six days what would typically take around twenty years through natural weathering. As a result, BacTech's system offers both environmental protection and economic benefit by enabling the recovery of valuable metals such as gold, silver, cobalt, and nickel from waste materials. While the technology is currently being applied to mine waste, its potential can extend further particularly in the bio-based recovery of metals from e-waste, offering a sustainable pathway for circular resource utilization and waste reduction (www.bactechgreen.com).

6.4 N2S, UK

N2S, United Kingdom is another firm working on developing solutions to tackle the problem of electronic waste (e-waste). Their proprietary bioleaching processes employs naturally occurring bacteria to enable the recovery of precious metals Ag, Au, Cu, Pt and Pd from printed circuit boards in a more sustainable way (www.n2s.co.uk).

7 Challenges

The present day scenario of bio-based recovery of critical metals clearly demonstrates its significant potential and growing commercial viability as a sustainable alternative to conventional metallurgical processes. However, a variety of challenges still need to be addressed before these methods can realize their full potential for e-waste treatment (Fig. 4). Firstly, to process large amount of e-waste, scaling up, process optimization and purification will pose biggest challenge for rapid and efficient metal extraction. Due to the variability of the e-waste, finding the optimal operating conditions that can maximize both microbial activity and metal solubilization is complex. Secondly, the heterogeneity of e-waste poses a significant challenge in selecting suitable microbial strains that can thrive under harsh and acidic environment for bio-based metal recovery. E-waste comprises a complex mixture of metals, plastics and ceramics which can interfere with microbial activity and reduce the efficiency of bioleaching processes. Moreover, the e-waste consist of large amount of heavy metals including Cd, Cr, Hg, and Pb. These metals can hinder the overall biorecovery process by inhibiting the growth of microbes. Thirdly, presence of non-metallic materials including thermoset resins and glass fibers constitute upto 70% of e-waste mass and can physically encapsulate metals thus hindering the microbial access. Lastly, the successful implementation is often hindered by a lack of stringent e-waste rules. Countries across the world have different policies for handling, collection, and transboundary movement of e-waste that enables many developed countries exporting their e-waste to poor nations and bypassing proper recycling norms and environmental safeguards. Another major policy gap lies in insufficient incentives for formal recyclers and limited public awareness about safe disposal practices. The absence of harmonized recycling standards and certification systems makes it difficult to ensure environmentally sound recovery processes. Additionally, rapid technological advancements and short product life cycles outpace regulatory updates, leaving outdated laws unable to address emerging waste streams such as solar panels, EV batteries, and IoT devices (Kumar and Yaashikaa 2020; Arya and Kumar 2020; Ghulam and Abushammala 2023; Harshan and Rajan 2025).

8 Future Perspectives

Metal recovery has traditionally relied on techniques such as pyrometallurgy and hydrometallurgy, which are still widely used at an industrial scale. These methodologies are dependent on high operating temperatures and harsh chemical reagents,

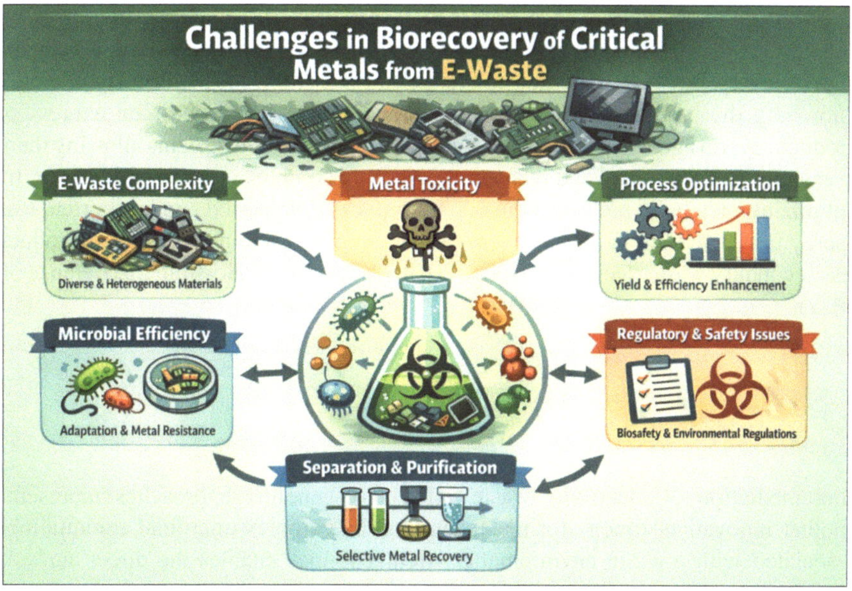

Fig. 4 Major challenges in biorecovery of critical metals from E-waste

leading to the generation of secondary liquid wastes, hazardous gaseous emissions, and a substantial carbon footprint. Although bio-based recovery methods have emerged as environmentally friendly alternatives, they have notable limitations, such as slow reaction kinetics, low selectivity for specific metals, and limited tolerance to metal toxicity within complex waste matrices. To overcome these limitations, advanced methods based upon synthetic biology, peptide based metal binding and metagenomics have gained increasing attention as they enhance microbial capacities and improve overall recovery efficiency. Additionally, these approaches operate under mild conditions, avoid high temperature processing, and produces minimum secondary waste, making them promising candidates for sustainable and efficient metal recovery systems. Apart from technological advancements, research and industry–academia collaborations play a pivotal role in shaping the translation of laboratory scale innovations into scalable and economically viable, solutions for e-waste metal recovery.

8.1 Synthetic Biology

Synthetic biology is an multidisciplinary field involving the constructing of new biological systems or redesigning the native organisms with enhanced properties to perform specific functions. By employing the principles of genetic engineering and molecular biology, this approach enables the development of microbes as precision tools with vast applicability. In context to metal biorecovery, engineered microbes can be endowed with enhanced traits such as improved enzymatic catalytic efficiency, high recovery rate and increased tolerance to metal toxicity, thereby significantly improving the process efficiency and robustness. These microbes can express or produce several agents including organic acids, metal binding proteins allowing them to selectively accumulate and recover metals from complex mixtures. Moreover, by optimizing metabolic pathways the overall process time period can be shorten, less energy intensive with higher recovery rate in comparison to native strains. Produce less secondary waste (De Lorenzo et al. 2018; Capeness and Horsfall 2020; Han et al. 2022).

8.2 Metagenomics

Bioremediation of electronic waste using metagenomic approaches represents another innovative strategy for understanding the complex microbial communities associated with e-waste environments. Metagenomics enables the direct analysis of collective microbial genomes from metal-rich niches, including mine tailings, acid mine drainage, landfill sites, and e-waste recycling facilities. Being culture-independent, this approach facilitates the discovery of novel and previously uncharacterized microorganisms, as well as the identification of functional genes involved in metal transformation and recovery that would otherwise remain inaccessible using conventional culture-dependent techniques. Moreover, functional and sequence-based metagenomic analyses, novel genes and metabolic pathways involved in metal binding, transport, reduction, and bioleaching can be identified. Metagenomics also supports the design of efficient microbial consortia by revealing synergistic interactions among community members that enhance multi-metal recovery and tolerance to metal toxicity. Furthermore, metagenome-derived genes can be integrated into synthetic biology frameworks to engineer tailor-made microorganisms or enzymes with improved selectivity, kinetics, and stability for targeted metal recovery.

Overall, metagenomics bridges the gap between environmental microbial diversity and sustainable e-waste recycling technologies, offering a scalable and sustainable alternative to conventional physicochemical methods for critical earth metal recovery (Martinez et al. 2015; De Sousa et al. 2018; Pattanaik et al. 2020).

In a study, shotgun metagenomic sequencing and next-generation sequencing (NGS) were employed to characterize the microbial communities present in e-waste bioreactors and to understand their functional roles in metal recovery. Taxonomic profiling revealed that the bioreactors were dominated by metal-resistant bacterial genera such as *Pseudomonas*, *Cupriavidus*, *Desulfovibrio* and *Geobacter* which are known to thrive in metal-rich environments. Functional annotation of the metagenomic data uncovered genes associated with metal reduction, transport, and biofilm formation, including resistance-nodulation-division (RND) transporters, metallothioneins, and oxidoreductases, which are crucial for heavy metal sequestration and tolerance (Ghosh and Singh 2021).

8.3 Peptide Based Biorecovery

Peptide-based metal recovery is another rapidly evolving approach for the biorecovery of metals. This approach possess several beneficial properties that makes it a significant method for sustainable bio-recovery of wide range of metals from e-waste:

1. High selectivity for critical and rare earth metals.
2. Molecular level recognition of metals.
3. More resistant to harsh process environment.
4. Compatible with biological systems (microbial surfaces and membrane proteins) and other techniques including ultrasound, genetic engineering and immobilization.
5. Non-toxic.

All these properties are crucial while recovering critical metals from complex e-waste mixtures even at low concentrations in an environment friendly manner and maintaining material stability.

Numerous studies have highlighted the effectiveness of peptide based approaches for recovering metals from electronic waste. For example, the application of small peptides in combination with ultrasound has resulted in significantly enhanced Cu recovery from waste PCBs. Maximum Cu leaching efficiencies of $95.4 \pm 0.1\%$, $92.5 \pm 0.1\%$, and $87.4 \pm 0.1\%$ were achieved using glutathione, glycylglycine, and glycyl-L-glutamine, respectively. These variations in leaching efficiency can be attributed to differences in the functional groups present in the peptides, which influence their metal-binding capacities. Moreover, ultrasound-assisted leaching copper recovery was 2.1–3.9-fold higher than those obtained without ultrasound treatment (Zhang et al. 2023). Similarly, 95.14% of gold recovery was achieved from e-waste and electroplating wastewater using protein (hemoglobin) crystals as metal binding peptide (Jin et al. 2025). In another approach, the cell surface of *Escherichia coli* was genetically engineered to express lithium binding peptide resulting in the high amount of lithium adsorption (3240.187 μmol g DCW^{-1}) from battery polluted water (Selvamani et al. 2023; Martinez-Rodriguez et al. 2025).

The above mentioned studies actively demonstrates that small peptides represent an effective and environmentally benign strategy for metal recovery from e-waste, as the resulting metal-chelated peptides are non-toxic and can significantly reduce the cost and complexity of downstream processing.

8.4 Research and Industry-Academia Collaborations

Collaborative efforts between research institutions and industry are crucial for translating laboratory-scale innovations into practical industrial applications. Such collaborations enable the translation of emerging scientific advances particularly in synthetic biology, metagenomics, and biohydrometallurgy into commercially viable and environmentally sustainable technologies.

Universities and research laboratories are fundamental centers for developing novel concepts and lab scale technologies. These institutions contribute by identifying new microbial strains, enzymes, and metabolic pathways involved in bioremediation and metal biorecovery from e-waste. However, academic research is often constrained by limited access to real industrial waste streams, pilot-scale facilities, and long-term operational data. This is where collaboration with industry becomes essential. Industry partners including e-waste recyclers, metallurgical companies, battery manufacturers, and clean-tech startups bring process expertise, infrastructure, and market perspective. Through such partnerships, industry can contribute in the following ways:

1. Access to realistic e-waste feedstocks and datasets to address material heterogeneity, contamination, and industrial process constraints.
2. Assessment of techno-economic viability and sustainability, including life-cycle impacts, energy efficiency, cost, and regulatory compliance.
3. Facilitation of technology scale-up and commercialization through integration of biological processes with existing recycling workflows.
4. By integrating academic innovation with industrial pragmatism, these collaborations contribute to resilient supply chains for critical metals, reducing dependence on primary mining and geopolitically sensitive resources.

Further, these type of collaborations lead towards development of skilled workforce by providing opportunities for students and researchers to work on industrial projects and developing industry relevant skills. This helps build a skilled workforce prepared to tackle future challenges in the circular economy.

Industry–academia collaborative projects play a crucial role in securing sustained investments for research and innovation, as they are generally more attractive to government funding agencies and private investors. Such partnerships demonstrate a clear translational pathway from fundamental research to pilot-scale validation and eventual commercialization. Recognizing the strategic importance of critical metals several countries have established dedicated policy frameworks and funding programs to support research and development in sustainable metal recovery, including bio-based approaches. Governments are increasingly integrating critical raw material security, circular economy principles, and green recovery technologies into national research agendas. Furthermore, national funding agencies often prioritize projects that align with policy objectives such as resource efficiency, waste valorization, reduced environmental footprint, and supply chain resilience. This policy-driven funding environment has significantly accelerated research on biological recovery of critical metals by supporting large-scale collaborative programs, pilot demonstrations, and technology readiness level (TRL) advancement.

Key National Policies and Strategies for Critical Minerals and Materials includes:

Critical Minerals and Materials Program, U.S. Department of Energy, United States of America (https://www.energy.gov/cmm/critical-minerals-and-materials-program).

National Critical Minerals Mission (NCMM), Ministry of Mines, Government of India (https://mines.gov.in/webportal/content/national-critical-mineral-mission).

Canadian Critical Mineral Strategies, Government of Canada (https://www.canada.ca/en/campaign/critical-minerals-in-canada/canadian-critical-minerals-strategy.html).

Critical Raw Materials Act (CRMA), European Union (https://www.consilium.europa.eu/en/infographics/critical-raw-materials/).

Critical Minerals Strategy (2023–2030), Department of Industry, Science and Resources, Government of Australia (https://www.industry.gov.au/publications/critical-minerals-strategy-2023-2030).

More recently, international programs and cross-border initiatives have gained prominence, alongside growing investment from the private sector driven by sustainability and corporate responsibility goals. Together, public and private funding mechanisms have accelerated innovation and facilitated the transition of metal recovery technologies from research to practical application (Phogat et al. 2025).

9 Conclusion

As global demand for critical and rare earth metals continues to accelerate, the limitations and geopolitical vulnerabilities of primary mining have become increasingly evident. In this context, bio-based metal recovery methodologies became a cornerstone of green technologies and sustainable resource management strategy.

Advances in microbial mechanisms including biohydrometallurgy and bioleaching, are positioning bio-based metal recovery as a scalable and practical solution for circular economy applications in the e-waste sector.

Certain challenges related to process optimization and e-waste regulations require interdisciplinary research focused solutions. It can be done by integrating innovative methods with existing technologies whether physical, chemical or biological. Further, the integration of biorecovery into national resource management frameworks can significantly mitigate supply chain risks arising from trade restrictions and geopolitical tensions. Economically, it can also create new opportunities for bio-based industries, innovation in waste valorization, and the generation of green jobs, thereby contributing to sustainable economic growth. Aligning biorecovery strategies with national sustainability agendas and global frameworks such as the United Nations Sustainable Development Goals particularly SDGs 9 (Industry, Innovation, and Infrastructure), 12 (Responsible Consumption and Production), and 13 (Climate Action) can enhance industrial resilience while supporting climate and environmental commitments. Critical metal biorecovery represents a transformative pathway toward sustainable material management and green technological advancement. By closing material loops, reducing dependence on finite primary resources, and securing a stable supply of essential metals, biorecovery not only mitigates environmental and geopolitical risks but also safeguards future technological innovation and industrial growth.

Conflict of Interest The authors declare that their is no conflict of interest related to the present book chapter.

References

Abdelbasir SM, McCourt KM, Lee CM, Vanegas DC (2020) Waste-derived nanoparticles: synthesis approaches, environmental applications, and sustainability considerations. Front Chem 8:782

Adetunji AI, Oberholster PJ, Erasmus M (2023) Bioleaching of metals from e-waste using microorganisms: a review. Minerals 13(6):828

Ahirwar R, Tripathi AK (2021) E-waste management: a review of recycling process, environmental and occupational health hazards, and potential solutions. Environ Nanotechnol Monit Manag 15:100409

Ali H, Khan HA, Pecht M (2022) Preprocessing of spent lithium-ion batteries for recycling: need, methods, and trends. Renew Sustain Energy Rev 168:112809

Ambaye TG, Vaccari M, Castro FD, Prasad S, Rtimi S (2020) Emerging technologies for the recovery of rare earth elements (REEs) from the end-of-life electronic wastes: a review on progress, challenges, and perspectives. Environ Sci Pollut Res 27(29):36052–36074

Annamalai M, Gurumurthy K (2019) Enhanced bioleaching of copper from circuit boards of computer waste by *Acidithiobacillus ferrooxidans*. Environ Chem Lett 17(4):1873–1879

Arshadi M, Esmaeili A, Yaghmaei S (2020) Investigating critical parameters for bioremoval of heavy metals from computer printed circuit boards using the fungus *Aspergillus niger*. Hydrometallurgy 197:105464

Arya S, Kumar S (2020) Bioleaching: urban mining option to curb the menace of E-waste challenge. Bioengineered 11(1):640–660

Awasthi AK, Li J (2017) Management of electrical and electronic waste: a comparative evaluation of China and India. Renew Sustain Energy Rev 76:434–447

Baez AG, Muñoz LP, Timmermans MJ, Garelick H, Purchase D (2024) Molding the future: optimization of bioleaching of rare earth elements from electronic waste by *Penicillium expansum* and insights into its mechanism. Biores Technol 402:130750

Bas AD, Deveci H, Yazici EY (2013) Bioleaching of copper from low grade scrap TV circuit boards using mesophilic bacteria. Hydrometallurgy 138:65–70

Biswal BK, Balasubramanian R (2023) Recovery of valuable metals from spent lithium-ion batteries using microbial agents for bioleaching: a review. Front Microbiol 14:1197081

Brindhadevi K, Barceló D, Chi NTL, Rene ER (2023) E-waste management, treatment options and the impact of heavy metal extraction from e-waste on human health: scenario in Vietnam and other countries. Environ Res 217:114926

Capeness MJ, Horsfall LE (2020) Synthetic biology approaches towards the recycling of metals from the environment. Biochem Soc Trans 48(4):1367–1378

Castro FD, Bassin JP (2022) Electronic waste: environmental risks and opportunities. In: Hazardous waste management. Elsevier, pp 421–458

Chatterjee A, Das R, Abraham J (2020) Bioleaching of heavy metals from spent batteries using *Aspergillus nomius* JAMK1. Int J Environ Sci Technol 17(1):49–66

Chen S, Yang Y, Liu C, Dong F, Liu B (2015) Column bioleaching copper and its kinetics of waste printed circuit boards (WPCBs) by *Acidithiobacillus ferrooxidans*. Chemosphere 141:162–168
Chi TD, Lee JC, Pandey BD, Yoo K, Jeong J (2011) Bioleaching of gold and copper from waste mobile phone PCBs by using a cyanogenic bacterium. Miner Eng 24(11):1219–1222
Chicardi E, Lopez-Paneque A, García-Orta VHG, Sepúlveda-Ferrer RE, Gallardo JM (2025) Enrichment methods for metal recovery from waste from electrical and electronic equipment: a brief review. Metals 15(2):140
Dave SR, Sodha AB, Tipre DR (2018) Microbial technology for metal recovery from e-waste printed circuit boards. J Bacteriol Mycol 6(4):241–247
De Lorenzo V, Prather KL, Chen GQ, O'Day E, von Kameke C, Oyarzún DA et al (2018). The power of synthetic biology for bioproduction, remediation and pollution control: the UN's Sustainable Development Goals will inevitably require the application of molecular biology and biotechnology on a global scale. EMBO Rep 19(4), EMBR201745658
De Sousa CS, Hassan SS, Pinto AC, Silva WM, De Almeida SS, Soares SDC, Azevedo V et al (2018) Microbial omics: applications in biotechnology. In: Omics technologies and bio-engineering. Academic Press, pp 3–20
Du YG, Tyagi RD, Sreekrishnan TR (1995) Operational strategy for metal bioleaching based on pH measurements. J Environ Eng 121(7):527–535
Dutta D, Rautela R, Gujjala LKS, Kundu D, Sharma P, Tembhare M, Kumar S (2023) A review on recovery processes of metals from E-waste: a green perspective. Sci Total Environ 859:160391
Erkmen AN, Ulber R, Jüstel T, Altendorfner M (2025) Towards sustainable recycling of critical metals from e-waste: bioleaching and phytomining. Resour Conserv Recycl 215:108057
Falco L, Martinez A, Di Nanno MP, Thomas H, Curutchet G (2014) Study of a pilot plant for the recovery of metals from spent alkaline and zinc-carbon batteries with biological sulphuric acid and polythionate production. Lat Am Appl Res 44(2):123–129
Fallah Haghighi H, Khorshidi J, Zarei T, Bakhshan Y (2025) Silver recovery from E-waste: optimization of hybrid supercritical water oxidation and supercritical CO_2 extraction process. Int J Environ Sci Technol 22(7):6049–6070
Faramarzi MA, Mogharabi-Manzari M, Brandl H (2020) Bioleaching of metals from wastes and low-grade sources by HCN-forming microorganisms. Hydrometallurgy 191:105228
Fatema K, Hassan MN, Hasan S, Roy H (2025) E-waste recycling in an optimized way for copper recovery by leaching and a case study on E-waste generation and management in Dhaka city. Heliyon 11(1)
Fayaz SM, Abdoli MA, Baghdadi M, Karbassi AR (2022) Extraction of silver from computer printed circuit boards wastes by supercritical fluids: pretreatment study. Int J Environ Sci Technol 19(6):4883–4890
Ferreira-Filipe DA, Duarte AC, Hursthouse AS, Rocha-Santos T, Silva ALP (2025) Biobased strategies for E-waste metal recovery: a critical overview of recent advances. Environments 12(1):26
Filote C, Roşca M, Hlihor RM, Cozma P, Simion IM, Apostol M, Gavrilescu M (2021) Sustainable application of biosorption and bioaccumulation of persistent pollutants in wastewater treatment: current practice. Processes 9(10):1696
García-Balboa C, Martínez-Alesón García P, López-Rodas V, Costas E, Baselga-Cervera B (2022) Microbial biominers: sequential bioleaching and biouptake of metals from electronic scraps. Microbiol Open 11(1):e1265
Ghosh VK, Singh K (2021) Metagenomic analysis of microbial communities in E-waste bioreactors 7(6):1–4
Ghosh S, Sarkar B, Thongmee S (2024) Biohydrometallurgical process for recovery of metals from effluents of electronic industries. In: Emerging innovative trends in the application of biological processes for industrial wastewater treatment. Elsevier, pp 149–163
Ghulam ST, Abushammala H (2023) Challenges and opportunities in the management of electronic waste and its impact on human health and environment. Sustainability 15(3):1837

Gorewoda T, Eschen M, Charasińska J, Knapik M, Kozłowicz S, Anyszkiewicz J, Kurylak W (2020) Determination of metals' content in components mounted on printed circuit boards from end-of-life mobile phones. Recycling 5(3):20

Gu T, Rastegar SO, Mousavi SM, Li M, Zhou M (2018) Advances in bioleaching for recovery of metals and bioremediation of fuel ash and sewage sludge. Biores Technol 261:428–440

Gu J, Nie Y, Li Z, Yang K, An L, Jiang H, Zhao X (2025) Efficient bioleaching of Li from waste lithium–iron phosphate batteries by acidophilic bacterial consortium: Enrichment condition, Li recovery, and brief carbon footprint analysis. Chem Eng J 509:161450

Gulliani S, Volpe M, Messineo A, Volpe R (2023) Recovery of metals and valuable chemicals from waste electric and electronic materials: a critical review of existing technologies. RSC Sustain 1(5):1085–1108

Hageluken C (2006) Improving metal returns and eco-efficiency in electronics recycling-a holistic approach for interface optimisation between pre-processing and integrated metals smelting and refining. In: Proceedings of the 2006 IEEE international symposium on electronics and the environment, 2006. IEEE, pp 218–223

Han P, Teo WZ, Yew WS (2022) Biologically engineered microbes for bioremediation of electronic waste: wayposts, challenges and future directions. Eng Biol 6(1):23–34

Harshan K, Rajan AP (2025) Unveiling the potential of microbial biominers in bioleaching for heavy metal recovery from E-waste—a comprehensive review. J Hazard Mater Adv, 100691

Hegazy AS, Soliman HM, Mowafy AM, Mohamedin AH (2025) Bioleaching of lanthanum from nickel metal hydride dry battery using siderophores produced by Pseudomonas sp. World J Microbiol Biotechnol 41(2):39

Heydarian A, Vakilchap F, Mousavi SN, Mousavi SM (2025) Bacterial acidic agents-assisted multi-elemental (Ni Co, and Li) leaching of used lithium-ion batteries at high pulp densities. Sci Rep 15(1):16517

Hong Y, Valix M (2014) Bioleaching of electronic waste using acidophilic sulfur oxidising bacteria. J Clean Prod 65:465–472

Hsu E, Durning CJ, West AC, Park AHA (2021) Enhanced extraction of copper from electronic waste via induced morphological changes using supercritical CO_2. Resour Conserv Recycl 168:105296

Hua Y, Zhou S, Huang Y, Liu X, Ling H, Zhou X, Yang S (2020) Sustainable value chain of retired lithium-ion batteries for electric vehicles. J Power Sources 478:228753

Ilyas S, Lee JC (2014a) Bioleaching of metals from electronic scrap in a stirred tank reactor. Hydrometallurgy 149:50–62

Ilyas S, Lee JC (2014b) Biometallurgical recovery of metals from waste electrical and electronic equipment: a review. ChemBioEng Rev 1(4):148–169

Işıldar A, van de Vossenberg J, Rene ER, van Hullebusch ED, Lens PN (2016) Two-step bioleaching of copper and gold from discarded printed circuit boards (PCB). Waste Manage 57:149–157

Jin XQ, Li JL, Li CY, Chen LL, Wang XT, Shi WP, Yin DC (2025) Efficient and selective gold recovery from practical e-waste and electroplating wastewater using protein crystals. Resour Conserv Recycl 222:108450

Johnson DB (2014) Biomining—biotechnologies for extracting and recovering metals from ores and waste materials. Curr Opin Biotechnol 30:24–31

Kaur P, Sharma S, Albarakaty FM, Kalia A, Hassan MM, Abd-Elsalam KA (2022) Biosorption and bioleaching of heavy metals from electronic waste varied with microbial genera. Sustainability 14(2):935

Kaya M (2016) Recovery of metals and nonmetals from electronic waste by physical and chemical recycling processes. Waste Manage 57:64–90

Khaliq A, Rhamdhani MA, Brooks G, Masood S (2014) Metal extraction processes for electronic waste and existing industrial routes: a review and Australian perspective. Resources 3(1):152–179

Khatri BR, Sodha AB, Shah MB, Tipre DR, Dave SR (2018) Chemical and microbial leaching of base metals from obsolete cell-phone printed circuit boards. Sustain Environ Res 28(6):333–339

Kudpeng K, Bohu T, Morris C, Thiravetyan P, Kaksonen AH (2020) Bioleaching of gold from sulfidic gold ore concentrate and electronic waste by *Roseovarius tolerans* and *Roseovarius mucosus*. Microorganisms 8(11):1783

Kumar RN, Nagendran R (2007) Influence of initial pH on bioleaching of heavy metals from contaminated soil employing indigenous *Acidithiobacillus thiooxidans*. Chemosphere 66(9):1775–1781

Kumar A, Holuszko ME, Song S (2022) Recycling technologies–physical separation. In: Electronic waste: recycling and reprocessing for a sustainable future, pp 95–134

Kumar A, Saini HS, Kumar S (2018) Bioleaching of gold and silver from waste printed circuit boards by *Pseudomonas balearica* SAE1 isolated from an e-waste recycling facility. Curr Microbiol 75(2):194–201

Kumar PS, Yaashikaa PR (2020) Recent trends and challenges in bioleaching technologies. In: Biovalorisation of wastes to renewable chemicals and biofuels, pp 373–388

Lai H, Ding X, Cui M, Zheng J, Chen Z, Pei J, Zhang J (2023) Mechanisms and influencing factors of biomineralization based heavy metal remediation: a review. Biogeotechnics 1(3):100039

Latorre M, Cortés MP, Travisany D, Di Genova A, Budinich M, Reyes-Jara A, Maass A (2016) The bioleaching potential of a bacterial consortium. Biores Technol 218:659–666

Li X, Ren Q, You X, Yang Y, Shan M, Wang M (2019) Material flow analysis of discarded refrigerators from households in urban and rural areas of China. Resour Conserv Recycl 149:577–585

Liang G, Li P, Liu W, Wang B (2016) Enhanced bioleaching efficiency of copper from waste printed circuit boards (PCBs) by dissolved oxygen-shifted strategy in *Acidithiobacillus ferrooxidans*. J Mater Cycles Waste Manage 18(4):742–751

Liu X, Liu H, Wu W, Zhang X, Gu T, Zhu M, Tan W (2020) Oxidative stress induced by metal ions in bioleaching of LiCoO2 by an acidophilic microbial consortium. Front Microbiol 10:3058

Madrigal-Arias JE, Argumedo-Delira R, Alarcón A, Mendoza-López MR, García-Barradas O, Cruz-Sánchez JS, Ferrera-Cerrato R, Jiménez-Fernández M (2015) Bioleaching of gold, copper and nickel from waste cellular phone PCBs and computer goldfinger motherboards by two Aspergillus nigerstrains. Braz J Microbiol 46:707–713

Magoda K, Mekuto L (2022) Biohydrometallurgical recovery of metals from waste electronic equipment: current status and proposed process. Recycling 7(5):67

Mallick S, Pradhan T, Das S (2025) Bacterial biomineralization of heavy metals and its influencing factors for metal bioremediation. J Environ Manage 373:123977

Manikandan S, Inbakandan D, Nachiyar CV, Namasivayam SKR (2023) Towards sustainable metal recovery from e-waste: a mini review. Sustain Chem Environ 2:100001

Martinez P, Vera M, Bobadilla-Fazzini RA (2015) Omics on bioleaching: current and future impacts. Appl Microbiol Biotechnol 99(20):8337–8350

Martinez-Rodriguez GA, Rojas-Contreras JA, Vázquez-Ortega PG, Reyes-Jáquez D, Medrano-Roldán H, Urtiz-Estrada N, Zazueta-Álvarez DE et al (2025) Biotechnological strategies for the recovery of lithium and other metals from a secondary source: the role of microorganisms and metal-binding peptides. Recycling 11(1):4

Motaghed M, Mousavi SM, Rastegar SO, Shojaosadati SA (2014) Platinum and rhenium extraction from a spent refinery catalyst using *Bacillus megaterium* as a cyanogenic bacterium: statistical modeling and process optimization. Biores Technol 171:401–409

Mrozik W, Rajaeifar MA, Heidrich O, Christensen P (2021) Environmental impacts, pollution sources and pathways of spent lithium-ion batteries. Energy Environ Sci 14(12):6099–6121

Murali A, Plummer MJ, Shine AE, Free ML, Sarswat PK (2022) Optimized bioengineered copper recovery from electronic wastes to increase recycling and reduce environmental impact. J Hazard Mater Adv 5:100031

Nagy S, Bokányi L, Gombkötő I, Magyar T (2017) Recycling of gallium from end-of-life light emitting diodes. Arch Metall Mater 62(2B):1161–1166

Narayanasamy M, Dhanasekaran D, Vinothini G, Thajuddin N (2018) Extraction and recovery of precious metals from electronic waste printed circuit boards by bioleaching acidophilic fungi. Int J Environ Sci Technol 15(1):119–132
Narayanasamy M, Dhanasekaran D, Thajuddin N (2021) Bioremediation of noxious metals from e-waste printed circuit boards by Frankia. Microbiol Res 245:126707
Narayanasamy M, Dhanasekaran D, Thajuddin N (2022) Frankia consortium extracts high-value metals from e-waste. Environ Technol Innov 28:102564
Natarajan G, Ting YP (2015) Gold biorecovery from e-waste: an improved strategy through spent medium leaching with pH modification. Chemosphere 136:232–238
Niu Z, Zou Y, Xin B, Chen S, Liu C, Li Y (2014) Process controls for improving bioleaching performance of both Li and Co from spent lithium ion batteries at high pulp density and its thermodynamics and kinetics exploration. Chemosphere 109:92–98
Nivetha SR, Karthick R (2018) Studies on the biosorption on removal of heavy metal copper in printed circuit boards. Int Res J Eng Technol 5(3):753–757
Nnaemeka IC, Nonso UC, Ikechukwu OM, Anezichukwu AF, Ikechukwu NA, Chukwudi EB et al (2025) Examining the efficiency of microbe-assisted metal extraction: a review of bio-hydrometallurgical leaching techniques. Hybrid Adv, 100407
Panda R, Jadhao PR, Pant KK, Naik SN, Bhaskar T (2020) Eco-friendly recovery of metals from waste mobile printed circuit boards using low temperature roasting. J Hazard Mater 395:122642
Patel F, Lakshmi B (2021) Bioleaching of copper and nickel from mobile phone printed circuit board using *Aspergillus fumigatus* A2DS. Braz J Microbiol 52(3):1475–1487
Pattanaik A, Samal DK, Sukla LB, Pradhan D (2020) Advancements and use of omic technologies in the field of bioleaching: a review. Biointerface Res Appl Chem 11:10185–10204
Phogat P, Kumar S, Wan M (2025) A scientometrics study of advancing sustainable metal recovery from e-waste: processes, challenges, and future directions. RSC Sustain
Pinto J, Colónia J, Abdolvaseei A, Vale C, Henriques B, Pereira E (2023) Algal sorbents and prospects for their application in the sustainable recovery of rare earth elements from E-waste. Environ Sci Pollut Res 30(30):74521–74543
Premathilake DS, Botelho Junior AB, Tenorio JAS, Espinosa DCR, Vaccari M (2023) Designing of a decentralized pretreatment line for EOL-LIBs based on recent literature of lib recycling for black mass. Metals 13(2):374
Priya A, Hait S (2017) Comparative assessment of metallurgical recovery of metals from electronic waste with special emphasis on bioleaching. Environ Sci Pollut Res 24(8):6989–7008
Priya A, Hait S (2020) Biometallurgical recovery of metals from waste printed circuit boards using pure and mixed strains of *Acidithiobacillus ferrooxidans* and *Acidiphilium acidophilum*. Process Saf Environ Prot 143:262–272 v
Priyadarsini S, Das AP (2024) Analytical and structural characterization of waste lithium-ion batteries for their effective recycling strategy. Environ Chem Ecotoxicol. https://doi.org/10.1016/j.enceco.2024.12.004
Priyadarsini S, Das AP (2026) Lithium in waste water: impacts on pollution and human health. Decarbonization of Wastewater Pollutants as a Sustainable Solution, 353–364. https://doi.org/10.1016/B978-0-443-33062-9.00010-2
Putra RA, Al Fajri I, Hariyadi A (2022) Metal bioleaching of used lithium-ion battery using *Acidophilic ferrooxidans* isolated from acid mine drainage. Key Eng Mater 937:193–200
Quinto S, Law N, Fletcher C, Le J, Antony Jose S, Menezes PL (2025) Exploring the e-waste crisis: strategies for sustainable recycling and circular economy integration. Recycling 10(2):72
Rautela R, Yadav BR, Kumar S (2026) Process optimization of bioleaching for sustainable recovery of metals from waste lithium-ion batteries. Bioresour Technol, 133839
Rodrigues ML, Leão VA, Gomes O, Lambert F, Bastin D, Gaydardzhiev S (2015) Copper extraction from coarsely ground printed circuit boards using moderate thermophilic bacteria in a rotating-drum reactor. Waste Manage 41:148–158

Roshanfar M, Sartaj M, Kazemeini S (2025) Sustainable recovery of critical metals from spent lithium-ion batteries using chitosan as biosorbent in citrate-sulfate media: a comprehensive isotherm, kinetic, and thermodynamic analysis. J Environ Manage 375:124400

Roy JJ, Cao B, Madhavi S (2021) A review on the recycling of spent lithium-ion batteries (LIBs) by the bioleaching approach. Chemosphere 282:130944

Sannigrahi S, Sarkar G, Jayasri MA, Suthindhiran K (2024) Bio-extraction of valuable metals from spent lithium-ion and nickel-cadmium batteries using magnetotactic bacteria and its role in electronic waste management. Sustain Environ 10(1):2345431

Schalk IJ (2025) Bacterial siderophores: diversity, uptake pathways and applications. Nat Rev Microbiol 23(1):24–40

Selvamani V, Jeong J, kannan Maruthamuthu M, Arulsamy K, Na JG, Hong SH (2023) Construction of the lithium binding peptide displayed recombinant Escherichia coli for the specific lithium removal from various metal polluted wastewater. J Environ Chem Eng 11(1):109029

Sharada HM, Abdel-Halim SA, Hafez MA, Elbarbary TA, Abdel-Fatah Y, Ibrahim IA (2022) Bioleaching of copper from electronic waste using *aspergillus niger*. Biointerface Res Appl Chem 12(6):8406–8425

Shrestha R, Ban S, Devkota S, Sharma S, Joshi R, Tiwari AP, Joshi MK (2021) Technological trends in heavy metals removal from industrial wastewater: a review. J Environ Chem Eng 9(4):105688

Tapia J, Dueñas A, Cheje N, Soclle G, Patiño N, Ancalla W, Lazarte A (2022) Bioleaching of heavy metals from printed circuit boards with an acidophilic iron-oxidizing microbial consortium in stirred tank reactors. Bioengineering 9(2):79

Tay SB, Natarajan G, Rahim MNBA, Tan HT, Chung MCM, Ting YP, Yew WS (2013) Enhancing gold recovery from electronic waste via lixiviant metabolic engineering in *Chromobacterium violaceum*. Sci Rep 3(1):2236

Tezyapar Kara I, Kremser K, Wagland ST, Coulon F (2023) Bioleaching metal-bearing wastes and by-products for resource recovery: a review. Environ Chem Lett 21(6):3329–3350

Thakur P, Kumar S (2021) Pretreatment of low-grade shredded dust e-waste to enhance silver recovery through biocyanidation by *Pseudomonas balearica* SAE1. 3 Biotech 11(11):454

Tian T, Liu G, Yasemi H, Liu Y (2022) Managing e-waste from a closed-loop lifecycle perspective: China's challenges and fund policy redesign. Environ Sci Pollut Res 29(31):47713–47724

Torres R, Segura-Bailón B, Lapidus GT (2018) Effect of temperature on copper, iron and lead leaching from e-waste using citrate solutions. Waste Manage 71:420–425

United Nations University (2024) Global e-waste monitor 2024 (report). United Nations University. 2024. Available online: https://Ewastemonitor.Info/Wp-Content/Uploads/2024/03/GEM_2024_1803_web_page_per_page_web.Pdf (accessed on 21 January 2025).

Vardanyan A, Vardanyan N, Aâtach M, Malavasi P, Gaydardzhiev S (2022) Bio-assisted leaching of non-ferrous metals from waste printed circuit boards—importance of process parameters. Metals 12(12):2092

Varshney S, Jain P, Srivastava S (2017) Application of ameliorated wood pulp to recover Cd (II), Pb (II), and Ni (II) from e-waste. J Mater Cycles Waste Manage 19(4):1446–1456

Vishwakarma A, Hait S (2024) Selective lithium extraction from spent lithium-ion batteries using siderophores produced by *Pseudomonas aeruginosa*: efficacy, kinetics, and artificial neural network modeling. Process Biochem 146:115–127

Wang S, Zheng Y, Yan W, Chen L, Mahadevan GD, Zhao F (2016) Enhanced bioleaching efficiency of metals from E-wastes driven by biochar. J Hazard Mater 320:393–400

Wang J, Faraji F, Ramsay J, Ghahreman A (2021) A review of biocyanidation as a sustainable route for gold recovery from primary and secondary low-grade resources. J Clean Prod 296:126457

Wędrychowicz M, Kurowiak J, Skrzekut T, Noga P (2023) Recycling of electrical cables—current challenges and future prospects. Mater 16(20):6632

Xiang Y, Wu P, Zhu N, Zhang T, Liu W, Wu J, Li P (2010) Bioleaching of copper from waste printed circuit boards by bacterial consortium enriched from acid mine drainage. J Hazard Mater 184(1–3):812–818

Yaashikaa PR, Priyanka B, Kumar PS, Karishma S, Jeevanantham S, Indraganti S (2022) A review on recent advancements in recovery of valuable and toxic metals from e-waste using bioleaching approach. Chemosphere 287:132230

Zare Tavakoli H, Abdollahy M, Ahmadi SJ, Khodadadi Darban A (2017) The effect of particle size, irrigation rate and aeration rate on column bioleaching of uranium ore. Russ J Non-Ferrous Met 58(3):188–199

Zhang J, Zhang X, Wang R, Wen K, Xu H, Xiang Z, Zhou X (2023) Copper recovery from waste printed circuit boards with small peptides enhanced by ultrasound. Sep Purif Technol 315:123680

Zhang A, Li S, Zhuang X, Song X, Gu W, Bai J, Wang J (2024) Separation and recycling of Au and Ag from waste light-emitting diodes. Environ Technol 45(15):3004–3015

Zheng K, Benedetti MF, Jain R, Pollmann K, van Hullebusch ED (2024) Recovery of gallium (and indium) from spent LEDs: strong acids leaching versus selective leaching by siderophore desferrioxamine E. Sep Purif Technol 338:126566

MIX
Papier aus verantwortungsvollen Quellen
Paper from responsible sources
FSC® C105338

If you have any concerns about our products,
you can contact us on
ProductSafety@springernature.com

In case Publisher is established outside the EU,
the EU authorized representative is:
Springer Nature Customer Service Center GmbH
Europaplatz 3, 69115 Heidelberg, Germany

Printed by Libri Plureos GmbH
in Hamburg, Germany